Mathematical Models and Methods for Real World Systems

PURE AND APPLIED MATHEMATICS

A Program of Monographs, Textbooks, and Lecture Notes

MONOGRAPHS AND TEXTBOOKS IN PURE AND APPLIED MATHEMATICS

Recent Titles

J. R. Weeks, The Shape of Space, Second Edition (2002)

M. M. Rao and Z. D. Ren, Applications of Orlicz Spaces (2002)

V. Lakshmikantham and D. Trigiante, Theory of Difference Equations: Numerical Methods and Applications, Second Edition (2002)

T. Albu, Cogalois Theory (2003)

A. Bezdek, Discrete Geometry (2003)

M. J. Corless and A. E. Frazho, Linear Systems and Control: An Operator Perspective (2003)

I. Graham and G. Kohr, Geometric Function Theory in One and Higher Dimensions (2003)

G. V. Demidenko and S. V. Uspenskii, Partial Differential Equations and Systems Not Solvable with Respect to the Highest-Order Derivative (2003)

A. Kelarev, Graph Algebras and Automata (2003)

A. H. Siddiqi, Applied Functional Analysis: Numerical Methods, Wavelet Methods, and Image Processing (2004)

F. W. Steutel and K. van Harn, Infinite Divisibility of Probability Distributions on the Real Line (2004)

G. S. Ladde and M. Sambandham, Stochastic versus Deterministic Systems of Differential Equations (2004)

B. J. Gardner and R. Wiegandt, Radical Theory of Rings (2004)

J. Haluska, The Mathematical Theory of Tone Systems (2004)

C. Menini and F. Van Oystaeyen, Abstract Algebra: A Comprehensive Treatment (2004)

E. Hansen and G. W. Walster, Global Optimization Using Interval Analysis, Second Edition, Revised and Expanded (2004)

M. M. Rao, Measure Theory and Integration, Second Edition, Revised and Expanded (2004)

W. J. Wickless, A First Graduate Course in Abstract Algebra (2004)

R. P. Agarwal, M. Bohner, and W-T Li, Nonoscillation and Oscillation Theory for Functional Differential Equations (2004)

J. Galambos and I. Simonelli, Products of Random Variables: Applications to Problems of Physics and to Arithmetical Functions (2004)

Walter Ferrer and Alvaro Rittatore, Actions and Invariants of Algebraic Groups (2005)

Christof Eck, Jiri Jarusek, and Miroslav Krbec, Unilateral Contact Problems: Variational Methods and Existence Theorems (2005)

M. M. Rao, Conditional Measures and Applications, Second Edition (2005)

K. M. Furati, Zuhair Nashed, and Abul Hasan Siddiqi, Mathematical Models and Methods for Real World Systems (2005)

Mathematical Models and Methods for Real World Systems

K. M. Furati
King Fahd University of Petroleum & Minerals
Dhahran, Saudi Arabia

Zuhair Nashed
University of Central Florida
Orlando, Florida, USA

Abul Hasan Siddiqi
King Fahd University of Petroleum & Minerals
Dhahran, Saudi Arabia

Chapman & Hall/CRC
Taylor & Francis Group
Boca Raton London New York Singapore

Published in 2006 by
Chapman & Hall/CRC
Taylor & Francis Group
6000 Broken Sound Parkway NW, Suite 300
Boca Raton, FL 33487-2742

Printed in the United States of America on acid-free paper
10 9 8 7 6 5 4 3 2 1

International Standard Book Number-10: 0-8493-3743-7 (Hardcover)
International Standard Book Number-13: 978-0-8493-3743-7 (Hardcover)

Library of Congress Cataloging-in-Publication Data

Catalog record is available from the Library of Congress

T&F informa

Taylor & Francis Group
is the Academic Division of T&F Informa plc.

Visit the Taylor & Francis Web site at
http://www.taylorandfrancis.com

and the CRC Press Web site at
http://www.crcpress.com

CONTENTS

PREFACE

The International Congress of Industrial and Applied Mathematics is organized at 4-year intervals under the auspices of the International Council of Industrial and Applied Mathematics (ICIAM). The ICIAM comprises 16 national societies: ANIAM (Australian and New Zealand Industrial and Applied Mathematics), CAIMS (Canada Applied and Industrial Mathematics Society), CSIA (Chinese Society for Industrial and Applied Mathematics), ECMI (European Consortium for Mathematics in Industry), ESMTB (Eupropean Society for Mathematics and Theoretical Biology), GAMM (Gescllschaft fur Angewandte Mathematik und Mechanike), IMA (Institute for Mathematics and Applications), ISIAM (Indian Society for Industrial and Applied Mathematics) JSIAM (Japan Society for Industrial and Applied Mathematics), Nortim (Nordiska Foreningen for Tillampad och Industriell Mathematik), SBMAC (Sociedade Brasiliera de Matematika Aplicade Computacional), SEMA (Sociedal Espanola de Matematica Applicada), SIMAI (Societa' Italiana di' Matematica, Applicata e Industiale), SMAI (Societa de Mathematiques Appliquees et Industrielles), SIAM (Society for Industrial and Applied Mathematics), and VSAM (Vietnamese Society for Applications of Mathematics). The objective of the national societies of ICIAM is similar. EMS (European Mathematical Society), LMS (London Mathematical Society), and SMS (Swiss Mathematical Society) are its associate members. The First Congress of Industrial and Applied Mathematics was held in Paris (1987), the second in Washington (1991), the third in Hamburg (1995), and the fourth in Edinburgh (1999). The sixth is scheduled to be held in Zurich (2007). It is the premier organization in the world for promoting teaching and research of applications of mathematics in diverse fields. Mini-symposiums are very important activities of such congresses. The member societies and distinguished workers of different areas are requested to submit proposals which are accepted after an appropriate reviewing process.

In recent years, all knowledgeable and responsible mathematicians are arguing vehemently for establishing linkage between mathematics and the physical world (besides many, we refer to professor Phillipe A. Griffiths' address "Trends for Science and Mathematics in 21st Century" (the inaugural function of an event of the WMY2000 in Cairo), and Professor Tony F.

Chan's article "The Mathematics Doctorate: A Time for Change" (Notices AMS, Sept. 2003)). Now, it is the general belief that mathematics cannot prosper in isolation. This book is an attempt to strengthen the linkages between mathematical sciences and other disciplines such as superconductors (an emerging area of science, technology, and industry), data analysis of environmental studies, and chaos. It also contains some valuable results concerning variational methods, fractal analysis, heat propagation, and multiresolution analysis having potentiality of applications.

The first two chapters are written by two distinguished industrial and applied mathematicians, Professor Dr. Helmut Neunzert, a distinguished industrial mathematician and the founding director of the prestigious Institute of Industrial Mathematics in Germany, and Dr. Noel G. Barton, Director of the Sydney Congress.

This book comprises chapters by those who were invited to the minisymposium in three parts on Mathematics of Real-World Problems. It is divided into four parts: Mathematics for Technology, Wavelet Methods for Real-World Problems, Classical and Fractal Methods for Physical Problems, and Trends in Variational Methods.

Part I contains chapters by H. Neunzert, N.G. Barton, and K.M. Furati and A.H. Siddiqi. Part II is based on the contributions of O. Christensen, S.H. Behiry et al., K.M. Furati et al., J. Kumar et al., O. Oğuz et al., A. Tokgozlu and Z. Aslan, and F.M. Khène. Part III is devoted to the chapters by M.A. Messaoudi and A.S. Al Shehri, F.D. Zaman and A.M. Al-Marzoug, R. Bhardwaj and R. Tuli, R. Bhardwaj and P. Kaur, and F. Nekka and J. Li. Part IV comprises chapters of M.S. Gockenbach and A.A. Khan, M.Yu. Rasulova et al., M.A. El-Gebeily and M.B.M. Elgindi, and M. Brokate and P. Manchanda. This book will be welcomed by all those having interest in acquiring knowledge of contemporary applicable analysis and its application to real-world problems.

The class of specialists who may have keen interest in the subject matter of this book is quite large as it includes mathematicians, meteorologists, engineers, and physicists.

Khaled M. Furati and A.H. Siddiqi would like to thank the King Fahd University of Petroleum & Minerals for providing financial assistance to attend the 5th ICIAM at Sydney. The help of Dr. P. Manchanda and Dr. Q. H. Ansari is acknowledged.

K. M. Furati, M. Z. Nashed,
and A. H. Siddiqi

CONTRIBUTING AUTHORS

1. M. K. Ahmad, Department of Mathematics, Aligarh Muslim University, Aligarh 202002, India

2. Z. Aslan, Department of Mathematics and Computing, Beykent University, Faculty of Science and Letters, İstanbul, Turkey;
   and
   Faculty of Engineering and Design, İstanbul Commerce University, Istanbul 34672, Turkey

3. U. Avazov, The Institute of Nuclear Physics, Ulughbek, Tashkent 702132, Uzbekistan

4. N. G. Barton, Sunoba Renewable Energy Systems, P.O. Box 1295, North Ryde BC, NSW 1670, Australia

5. S. H. Behiry, Department of Mathematics and Physics, Faculty of Engineering, Mansoura University, Mansoura, Egypt

6. R. Bhardwaj, Department of Mathematics, School of Basic and Applied Sciences, Guru Gobind Singh Indraprastha University, Kashmere Gate, Delhi 110006, India

7. M. Brokate, Institute of Applied Mathematics, Technical University of Munich, Munich, Germany

8. Z. Can, Department of Physics, Yildiz Technical University, Faculty of Science and Letters, İstanbul, Turkey

9. J. R. Cannon, Department of Mathematics, University of Central Florida, Orlando, FL 32816

10. O. Christensen, Department of Mathematics, Technical University of Denmark, Building 303, 2800 Lyngby, Denmark

11. M. B. M. Elgindi, Department of Mathematics, University of Wisconsin–Eau Claire, Eau Claire, WI 54702-4004

12. K. M. Furati, Mathematical Sciences Department, King Fahd University of Petroleum & Minerals, Dhahran 31261, Saudi Arabia

13. M. A. El-Gebeily, Mathematical Sciences Department, King Fahd University of Petroleum & Minerals, Dhahran 31261, Saudi Arabia

14. M. S. Gockenbach, Department of Mathematical Sciences, 319 Fisher Hall, Michigan Technological University, 1400 Townsend Drive, Houghton, MI 49931-1295

15. H. Hashish, Department of Mathematics and Physics, Faculty of Engineering, Mansoura University, Mansoura, Egypt

16. P. Kaur, Department of Mathematics, School of Basic and Applied Sciences, Guru Gobind Singh Indraprastha University, Kashmere Gate, Delhi 110006, India

17. A. A. Khan, Department of Mathematical Sciences, 319 Fisher Hall, Michigan Technological University, 1400 Townsend Drive, Houghton, MI 49931-1295

18. F. M. Khène, Research Institute, King Fahd University of Petroleum & Minerals, Dhahran 31261, Saudi Arabia

19. J. Kumar, Department of Mathematics, Gurunanak Dev University, Amritsar 143005, India

20. J. Li, 1 - Faculté de Pharmacie, 2 - Centre de Recherches Mathématiques, Université de Montréal, C.P. 6128, Succ. Centre-ville, Montréal, Québec, Canada H3C 3J7

21. P. Manchanda, Department of Mathematics, Gurunanak Dev University, Amritsar 143005, India

22. A. M. Al-Marzoug, Saudi Aramco, Dhahran 31311, Saudi Arabia

23. S. A. Messaoudi, Mathematical Sciences Department, King Fahd University of Petroleum & Minerals, Dhahran 31261, Saudi Arabia

24. F. Nekka, 1 - Faculté de Pharmacie, 2 - Centre de Recherches Mathématiques, Université de Montréal, C.P. 6128, Succ. Centre-ville, Montréal, Québec, Canada H3C 3J7

25. H. Neunzert, Fraunhofer Institute for Industrial Mathematics, Kaiserslautern, Germany

26. O. Oğuz, İstanbul Commerce University, Faculty of Engineering and Design, İstanbul, Turkey

27. M. Rahmatullaev, The Institute of Nuclear Physics, Ulughbek 702132, Tashkent

28. M. Yu. Rasulova, The Institute of Nuclear Physics, Ulughbek 702132, Tashkent

29. A. S. Al Shehri, Mathematics Department, School of Sciences, Girl's College, Dammam, Saudi Arabia

30. A. H. Siddiqi, Mathematical Sciences Department, King Fahd University of Petroleum & Minerals, Dhahran 31261, Saudi Arabia

31. N. A. Sontakke, Indian Institute of Tropical Meteorology, Dr. Homi Bhabha Road, Pashan, Pune 411008, India

32. A. Tokgozlu, Department of Geography, Faculty of Science and Letters, Süleyman Demirel University, Isparta 32260, Turkey

33. R. Tuli, Department of Mathematics, School of Basic and Applied Sciences, Guru Gobind Singh Indraprastha University, Kashmere Gate, Delhi 110006, India

34. F. D. Zaman, Mathematical Sciences Department, King Fahd University of Petroleum & Minerals, Dhahran 31261, Saudi Arabia

35. A. I. Zayed, Department of Mathematical Sciences, DePaul University, Chicago, IL 60614

Part I

Mathematics for Technology

Chapter 1

MATHEMATICS AS A TECHNOLOGY–CHALLENGES FOR THE NEXT TEN YEARS

H. Neunzert

Fraunhofer Institute for Industrial Mathematics

Abstract

The main focus of this chapter is the interlinking of mathematical models and methods to real-world systems. Six areas of technological themes which have emerged as crucial from intensive investigation in Europe, namely, Simulation of Processes and Products; Optimization, Control, and Design; Uncertainty and Risk; Management and Exploitation of Data; Virtual Material Design; and Biotechnology, Food, and Health, are elaborated. Contributions of the Fraunhofer Institute for Industrial Mathematics, Kaiserslautern, Germany in this field are highlighted.

1 Introduction

There is no doubt that mathematics has become a technology in its own right, maybe even a key technology. Technology may be defined as the application of science to the problems of commerce and industry. And science? Science may be defined as developing, testing, and improving models for the prediction of system behavior; the language used to describe these models is mathematics, and mathematics provides methods to evaluate these models. Here we are! Why has mathematics become a technology only recently? Mathematics became a technology when it received a tool to evaluate complex, "near to reality" models, and that tool was the computer. The model may be quite old. Navier–Stokes equations describe flow behavior rather well, but to solve these equations for realistic geometry and higher Reynolds numbers with sufficient precision, is even for powerful parallel computing, a real challenge. Make the models as simple as possible, as complex as necessary and then evaluate them with the help of efficient and reliable algorithms. These are genuine mathematical tasks.

Science is designed to "understand" natural phenomena; scientific technology extends the domain of the validity of scientific theories to not yet existing systems. We create a new, virtual world in which we may change and optimize much easier and quicker than in the real world. Even that is rather old. Some scholars of ancient science [9] and some philosophers [10] consider this interplay of science and technology as crucial for the birth of science during the Hellenistic period around 300 BC (with names like Euclid or Archimedes on top). But now, since we may mathematically optimize very complex virtual systems, we are able to use mathematics in order to design better machines, to minimize the risk of financial actions, and to plan optimal surgery.

This is the reason why mathematics has become a key technology. The following technology fields emerged as crucial from several investigations in Europe (see [2, 6]).

- Simulation of Processes and Products
- Optimization, Control, and Design
- Uncertainty and Risk
- Management and Exploitation of Data
- Virtual Material Design
- Biotechnology, Food, and Health

With the help of these road maps which contain examples and challenges for future mathematics gathered from all over Europe, European mathematicians shall try to influence national and international research policies in a way that may help mathematics get the weight in future programs which it has in reality already now. Mathematics was too long in an ivory tower, often used only as brain exercises for students. It needs some time and a lot of effort to catch public awareness of its new role.

In this chapter I shall show examples from different technology fields mentioned above, examples gained from our experience in the Fraunhofer Institute ITWM at Kaiserslautern. It was founded in 1996 and became a member of the Fraunhofer-Gesellschaft in 2001; the Fraunhofer-Gesellschaft is the leading German association for applied research with altogether 12,000 employees in ca. 60 institutes, an annual turnover of ca. 1.2 billion euro and branches in the US and in some European countries. Its decisive feature is that basic funding is given proportional to what is earned in industry. To make a rather complicated story simple, a Fraunhofer Institute gets 40 cents from the federal government for each euro it earns in industry. "No industrial project - no money at all and 40 % on top in order to do fundamental research related to projects"–these are the two rules which in my opinion are unique and uniquely successful worldwide.

ITWM has proved that mathematics as a technology is strong enough to follow the Fraunhofer rules. Not only that, at present it is the most successful institute of all the 15 Fraunhofer Institutes dealing with information technologies. The reason is that it has a huge market, much wider than any computer science institute. The disadvantage is that the market doesn't yet know it. The consequence is that there is a lot of space for all other really applied mathematicians and for cooperation worldwide.

But now I want to become more substantial. Here are the technology themes with examples and challenges, see references [1, 4].

2 Simulation of Processes and the Behavior of Products

Simulation means modelling-computing-visualizing. To find the right model for the behavior of car components, as simple as possible and as complicated as necessary, is, for example, a task for asymptotic analysis: identify small parameters in very complex models, study the behavior for these parameters tending to zero, and estimate the error using this "parameter = 0 - model". All this is tricky perturbation theory, sometimes advanced functional analysis. But we should never "oversimplify" in order to get an analytically treatable model; very often numerics will be necessary, and very often advanced numerical ideas are necessary. Since a realistic geometry is sometimes very complex (think of a porous medium in a microscopic view), we need, for example, new, gridfree algorithms efficiently implemented for parallel systems. And finally, long lists of numbers as a result of solving a PDE are completely useless-we have to interpret the results in terms of the original questions, and quite often we have to visualize the results as images or movies.

Simulation is now routinely used in many parts of industry all over the world to support or to replace experimentation. "It can have a dramatic effect on the design process, reducing the need for costly prototypes and increasing the speed with which new products can be brought to market [1].

There are industries where simulation has a long tradition, like aerospace or automotive industries or in oil and gas prospection. In these areas, commercial software is available and often easy to handle and efficient. It is (at least for a Fraunhofer Institute) a very hard or even impossible task to place a new algorithm to substitute this kind of software, even if this algorithm is really better than the other one. What is possible for mathematicians is to substitute some modules in software products, as, for example, the second mathematical Fraunhofer Institute SCAI does in offering an "Algebraic Multigrid Solver" for linear systems. Another possibility is postprocessing algorithms enabling the user to do an "optimal experimental design" for virtual or "numerical experiments". Industries operating with more basic technologies such as textiles, glass, or even metals just begin to use simulation. The market for commercial software seems too small, and tailor-made software is needed. How complicated this field could be will now be shown by our experience with the glass industry. ITWM has a 10-year close cooperation with Schott Glas at Mainz, where cooperation may be taken literally. The enormous knowledge of Schott scientists about materials and processes joins mathematical ideas in ITWM to find innovative solutions. (The material was provided to me by Norbert Siedow from ITWM; some parts and literature are described in the ITWM annual report 2003, page 26 ff.)

Figure 1 shows the glass making process, from the glass tank with molten glass of a temperature over $1000°C$ through a pipe to a kind of drop called gob; in this process we identified 4 mathematical tasks which are denoted by colors. Two are so-called "inverse problems" that measure the temperature in the interior of the glass flow from radiation and optimize the shape of the flanges carrying the pipe such that a given homogeneous temperature is created through electrical currents. The shape of the gob, a very viscous drop of liquid glass, has to be calculated by CFD codes able to handle free surfaces very well.

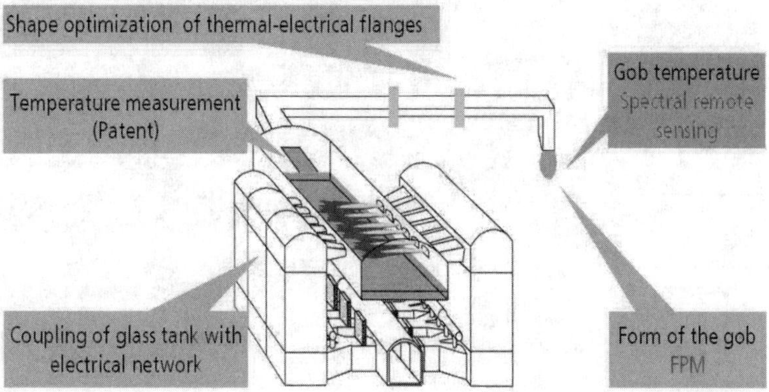

Figure 1: Mathematical Problems in Glass Industry **(Glassmaking)**

Figures 2–5 show different kinds of glass processing, Pressing of TV panels ask for the simulation of radiation. In semi-transparent media, this is a very elaborate task, since the radiation equation is a dimensional integro-differential equation with enormous computational efforts.

One uses tricky scale asymptotics (see [8]). Floatglass, an efficient production process invented by Pilkington, shows sometimes wavy patterns which have to be avoided. Whether these waves are instabilities created in a modification of the Orr-Sommerfeld equations is the subject of an ongoing PhD work. Glass fiber productions are extremely tricky processes in which the fibers interact with the air around them. Turbulent flow-fiber interaction is a topic where turbulence models are not enough, but stochastic differential equations are crucial.

Figure 3 shows classical glass processing and problems connected with the cooling of glass. I would like to mention that already around 1800 Fraunhofer who gave the name to our society produced lenses and had problems with the thermal tensions and the defects created by them.

Many of the problems here are "inverse problems" connected with heat transfer, and they are very ill-posed. Inverse problems may be counted under "optimization"; it is the combination of optimization and simulation as in inverse problems, optimal shape design, etc. which creates many mathematical challenges.

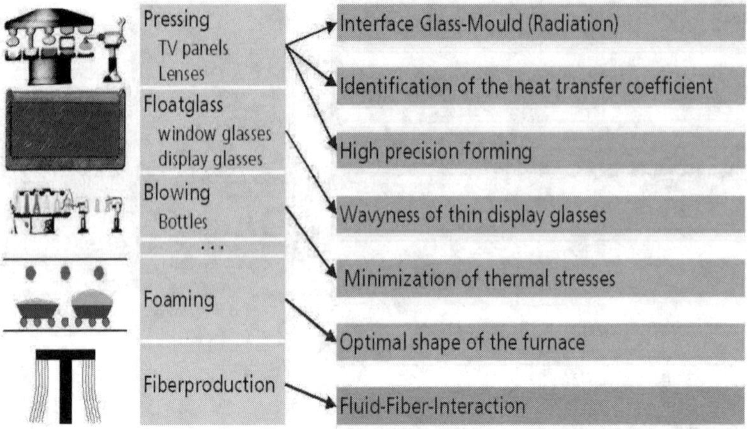

Figure 2: Mathematical Problems in Glass Industry **(Glassprocessing I)**

Let us have a closer look at a few of the problems. Figure 4 shows the details of gob forming.

The hot glass leaves the feeder when the needle opens. A drop (gob) is formed and cut off by a special cutter. J. Kuhnert (ITWM) has designed a gridfree numerical method to calculate the glass flow. It is called the "Finite Pointset Method" (FPM) and may be considered as an extension of "Smoothed Particle Hydrodynamics" (SPH) [11]. Particles are moving in the computational domain, carrying information about density, velocity, temperature, etc. This information has to be extrapolated to other positions so that derivatives of these quantities as the Laplacian of the velocity components, the temperature gradient, etc., can be calculated. These extrapolations are denoted by a tilde, and the rest is Lagrangean formalism.

The method is appropriate for fluids with free boundaries, changing even the topology, as it happens, when the gob is cut off.

A more analytical task is the question of waves at floatglass surfaces. Here is the industrial question: What is the origin of waves at the interface of glass and molten tin (the glass flows over molten tin, a classical 2-phase flow with quite different temperatures)? These waves are small defects which should be removed. What are the causes?

Let us finish the glass field by describing a very nice, very ill-posed

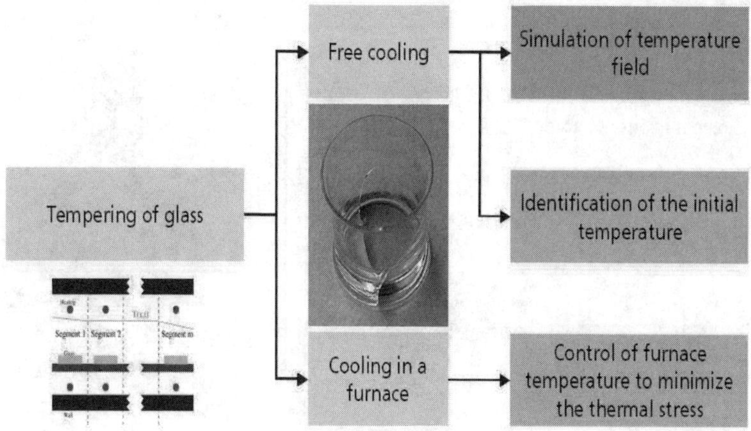

Figure 3: Mathematical Problems in Glass Industry **(Glassprocessing II)**

problem which deals with temperature measurements. The high temperature of the glass melt asks for remote measurements or at least only measurements at the boundary.

Here is the problem. We measure the temperature at parts of the boundary. Assuming that the heat transport is given by conduction and radiation and assuming that the heat flux at the boundary is known everywhere, what is the temperature inside?

The problem was solved without radiation in a very nice master's thesis by L. Justen and is with radiation the subject of a Ph.D. thesis by Pereverzyev jun. For one dimension it works, but the real world is three dimensional.

The situation is similar for melt spinning processes in textile industries; there is an intersection with the previous field when we talk about glass fibers. But, in general, we have polymer fibers, leaving nozzles as a liquid, but crystallizing when an air flow is cooling and pulling the fibers.

Here are some mathematical problems connected with the process.

Of course, there are curtains of fibers in a real process.

The industrial question belongs to "reverse engineering": these are the properties of the product we want to have (even to describe these properties is a mathematical problem). How can we create them?

The crystallization is a mathematical problem too and the subject of

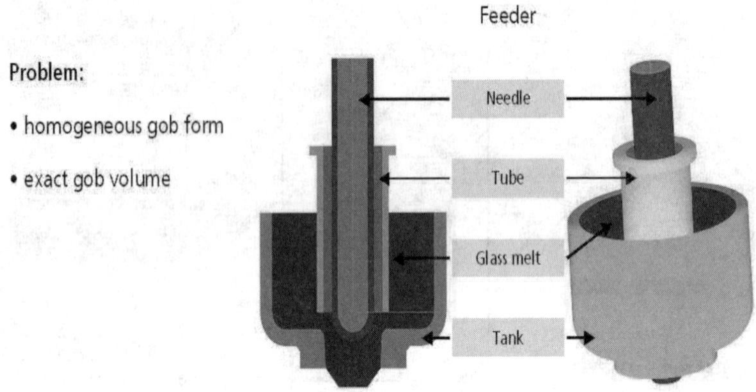

Figure 4: Mathematical Problems in Glass Industry **(Gob forming)**

a Ph.D. thesis by Renu Dhadwal . Let us have a closer look at the interaction of fibers with a turbulent flow. The main question is, How does the stochastic behavior of the turbulent air flow influence the (stochastically described) properties of the fabric? Figures 6–20 describe the work of N. Markeinkewho just finished her Ph.D.

Things may even be more complicated – see for example a quickly rotating spinneret for producing glass fibers:

Finite Pointset Method (FPM) – gridfree numerical method

$$x_i^{n+1} = x_i^n + \Delta t \cdot v_i^n$$ movement of points

$$0 = \nabla\left(\frac{1}{\rho}\nabla\left(p_{hyd}^{n+1}\right)\right);\ \tilde{p}_i^{n+1} = p_{hyd,i}^{n+1} + p_{dyn,i}^{n}$$ pressure

$$\frac{\tilde{v}_i^{n+1} - v_i^n}{\Delta t} = -\frac{1}{\rho}\nabla\left(\tilde{p}_i^{n+1}\right) + \frac{\eta}{\rho}\cdot\Delta\left(\tilde{v}_i^{n+1}\right) + g_i^{n+1}$$ impulse equation

$$\frac{T_i^{n+1} - T_i^n}{\Delta t} = \frac{1}{\rho \cdot c}\nabla\left(k \cdot \nabla T_i^{n+1}\right)$$ energy equation

$$v^{n+1} \equiv \tilde{v}^{n+1} - \Delta t \frac{1}{\rho}\nabla\left(\varepsilon^{n+1}\right);\ p_{dyn,i}^{n+1} = p_{dyn,i}^{n} - \varepsilon^{n+1}$$ correction of pressure and velocity

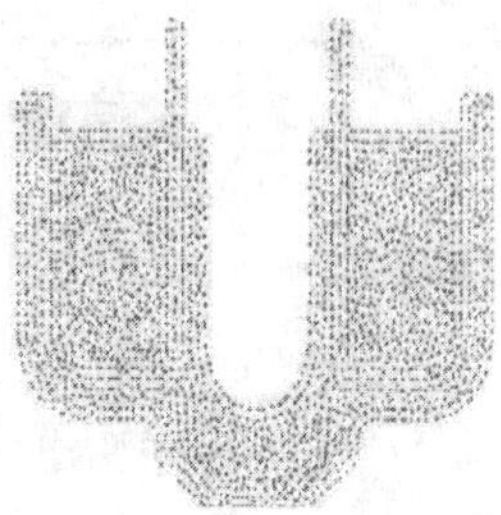

Figure 5: Mathematical Problems in Glass Industry **(Gob forming)**

Simulation

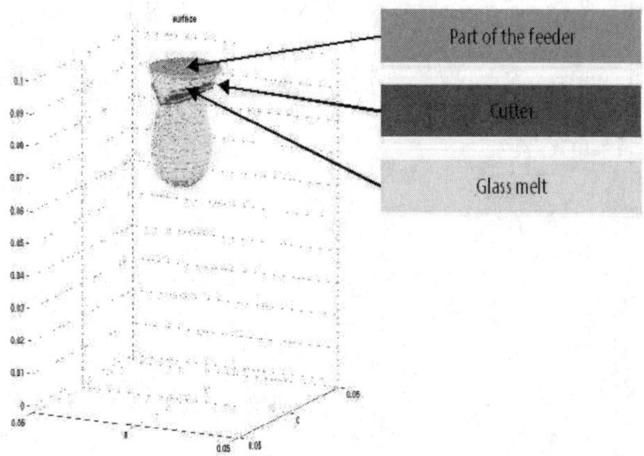

Figure 6: Mathematical Problems in Glass Industry **(Gob forming)**

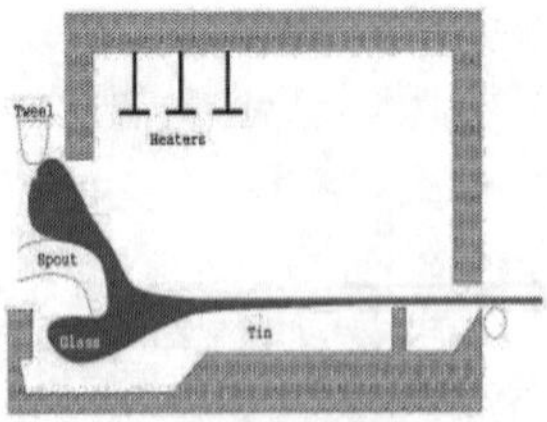

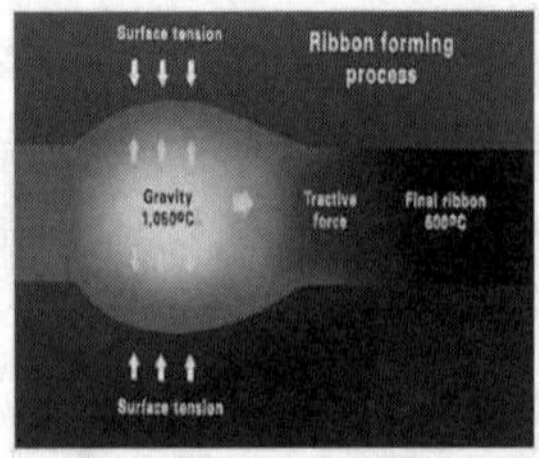

Small Waves at the Interface:

- Wavelength: $\lambda = 1 - 100\ \mu m$
- Amplitude: $a = 100 - 200$ nm

↓

LOW OPTICAL QUALITY OF THE GLASS

Causes:

- The glass is not stretched uniform (rollers)
- Mixed convection with friction (instability model)
- Non-homogeneous chemistry of the glass

Figure 7: Mathematical Problems in Glass Industry **(Floatglass)**

$$\begin{cases} \dfrac{\partial T}{\partial t} = \Delta_{\bar{r}} T - F(T), \bar{r} \in \Omega \\ \dfrac{\partial T}{\partial \bar{n}} = G(T), \bar{r} \in \partial\Omega \\ T(\bar{r},0) = \underset{\text{unknown}}{T_0(\bar{r})} \end{cases} \quad \xleftarrow{\ ?\ } \quad T(\bar{r},t) = D(\bar{r},t) \quad \begin{matrix} \bar{r} \in \partial\Omega_p \subset \partial\Omega \\ t \in [0, t_f] \end{matrix}$$

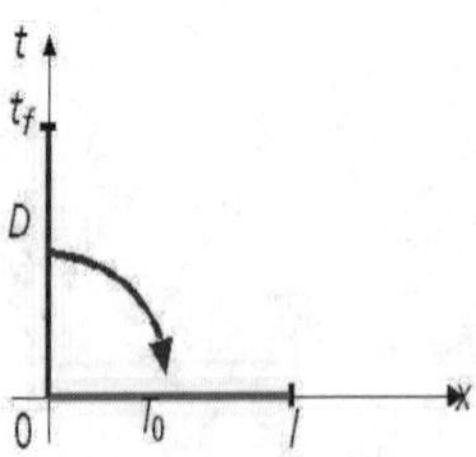

Problem is:

- non-linear ➔ *Linearization*
- ill-posed ➔ *Regularization (Tikhonov)*

Figure 8: Mathematical Problems in Glass Industry **(Reconstruciton of initial temperature)**

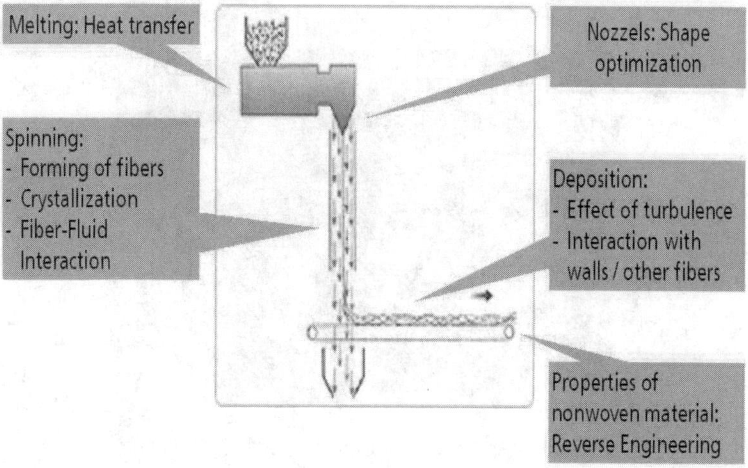

Figure 9: Mathematical Problems in Spinning Processes **(Production of nonwovens)**

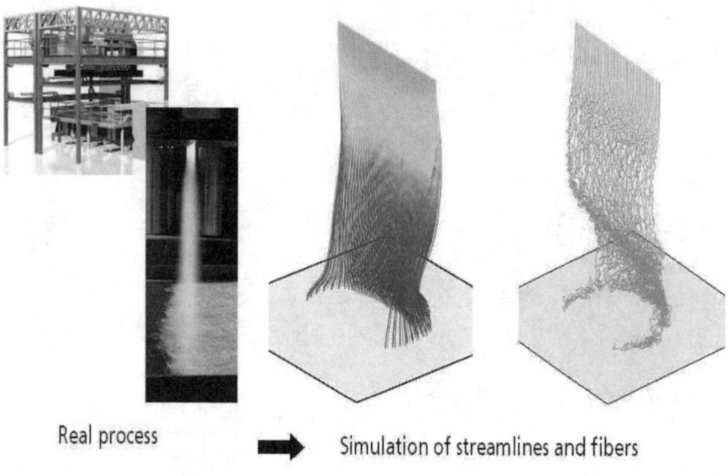

Figure 10: Mathematical Problems in Spinning Processes **(Fiber-fluid interaction: Fiber Dynamics)**

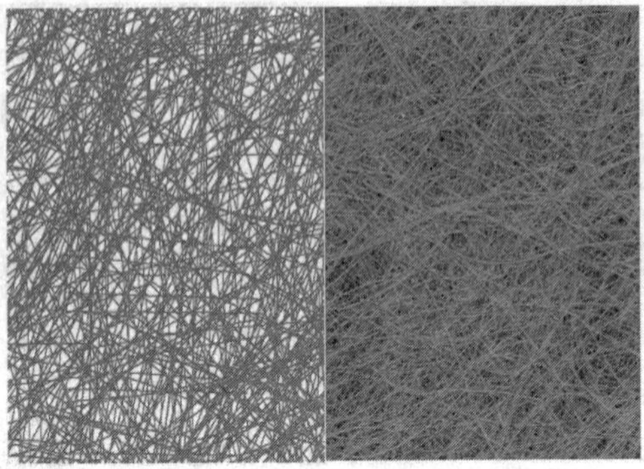

Virtual nonwoven: Check of properties, reverse engineering

Figure 11: Mathematical Problems in Spinning Processes **(Foner-fluid interaction: Nonwoven Materials)**

Asymptotic 1D-Model of fiber dynamic:

$$\rho A \partial_{tt} \mathrm{r}(s,t) = \partial_s [T(s,t)\partial_s \mathrm{r}(s,t)] - EI\partial_{ssss}\mathrm{r}(s,t) + \rho A \mathrm{g} + \mathrm{f}^{air}(s,t),$$

$$(\partial_s \mathrm{r})^2(s,t) = 1 \qquad \text{+ boundary conditions}$$

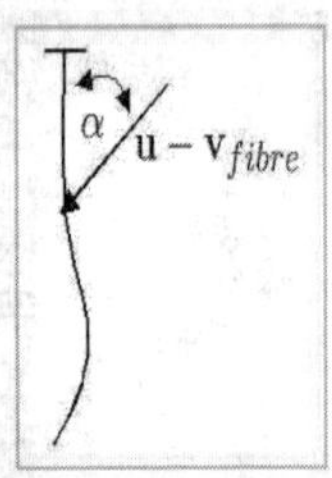

Force due to turbulent flow:

$$\mathrm{f}^{air}(s,t) = \mathrm{f}^{air}(\mathrm{u}(\mathrm{r}(s,t),t)) \quad \text{with} \quad \|\mathrm{f}^{air}\| = \frac{1}{2}\rho^{air}\, d\, \|\mathrm{u} - \mathrm{v}_{fibre}\|^2 C_\alpha$$

Turbulence-Model (k-ϵ based on RANS Equations)

$$\mathrm{u}(\mathrm{x},t) = \bar{\mathrm{u}}(\mathrm{x},t) + \mathrm{u}'(\mathrm{x},t) \Rightarrow \bar{\mathrm{u}},\ k,\ \epsilon$$

Idea: Force-Model ➔

$$\mathrm{f}^{air}(\bar{\mathrm{u}} + \mathrm{u}') \approx \underbrace{\mathrm{f}^{air}(\bar{\mathrm{u}})}_{deterministic\ \bar{\mathrm{f}}} + \underbrace{\frac{d\mathrm{f}^{air}}{d\mathrm{u}}\Big|_{\mathrm{u}=\bar{\mathrm{u}}} \mathrm{u}'}_{stochastic\ \mathrm{f}'}$$

Figure 12: Mathematical Problems in Spinning Processes **(Turbulence Effects)**

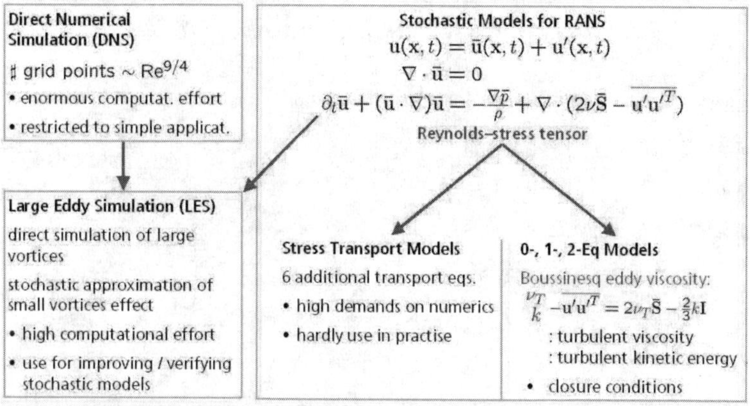

Figure 13: Mathematical Problems in Spinning Processes **(Turbulence Effects)**

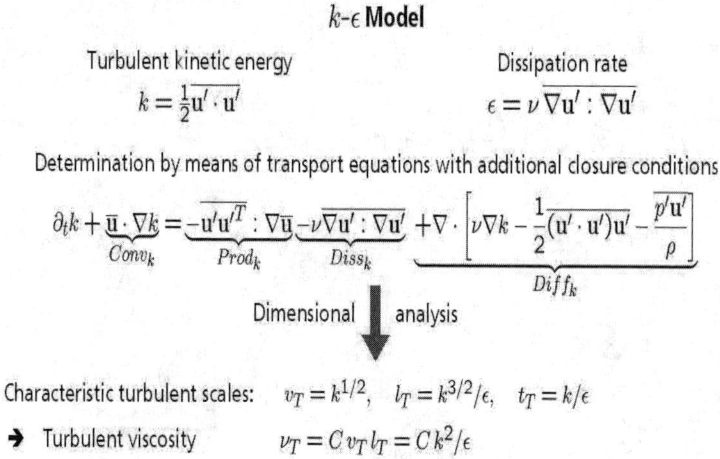

Figure 14: Mathematical Problems in Spinning Processes **(Turbulence Effects)**

Assumption: locally homogeneous, isotropic, frozen turbulence

Construct random Gaussian field for velocity fluctuations $(\mathrm{u}'_{(\mathrm{x},t)}, \mathrm{x} \in \mathbb{R}^3, t \in \mathbb{R}_0^+)$

$$\mathrm{u}'(\mathrm{x},t) = \mathrm{u}'(\mathrm{x} - \bar{\mathrm{u}}t, 0) = \mathrm{u}'(\mathrm{x} - \bar{\mathrm{u}}t)$$

with

mean $\mathbb{E}[\mathrm{u}'_\mathrm{x}] = 0$

covariance / correlation tensor

$$\mathbb{E}[\mathrm{u}'(\mathrm{x}) \otimes \mathrm{u}'(\mathrm{x}+\mathrm{r})] = \Gamma(\mathrm{r})$$

$\Gamma(\mathrm{r})$ can be expressed by 1D correlation function $c(r)$

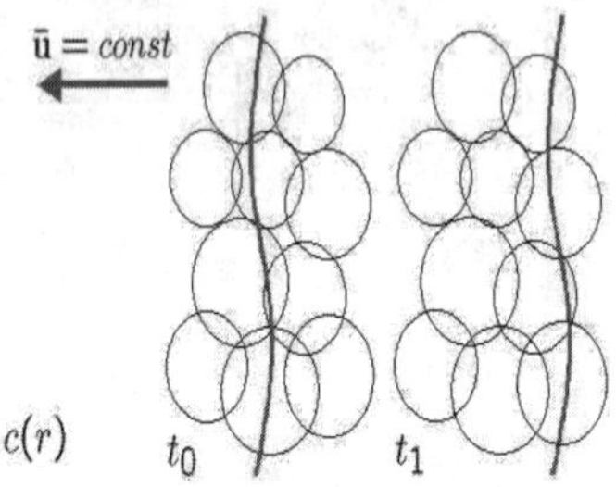

Figure 15: Mathematical Problems in Spinning Processes **(Turbulence Effects)**

Choose $c(r) := 2\int_0^\infty E(\kappa)\frac{\sin(\kappa r) - \cos(\kappa r)\kappa r}{\kappa^3 r^3}\,d\kappa$ with $E(\kappa) \in C^2(\mathbb{R}_0^+)$
+ further demands

Then $(\mathrm{u}'_{(\mathrm{x},t)}, \mathrm{x} \in \mathbb{R}^3, t \in \mathbb{R}_0^+)$ is differentiable and satisfies:

- Kolmogorovs energy spectrum
- $\overline{\mathrm{u}' \cdot \mathrm{u}'} = 2k$
- $\overline{\nabla\mathrm{u}' : \nabla\mathrm{u}'} = \epsilon/\nu$

Integrated Gaussian field for stochastic force along the fiber
Asymptotic: $\mathrm{f}'(s,t)$ white noise in space and time

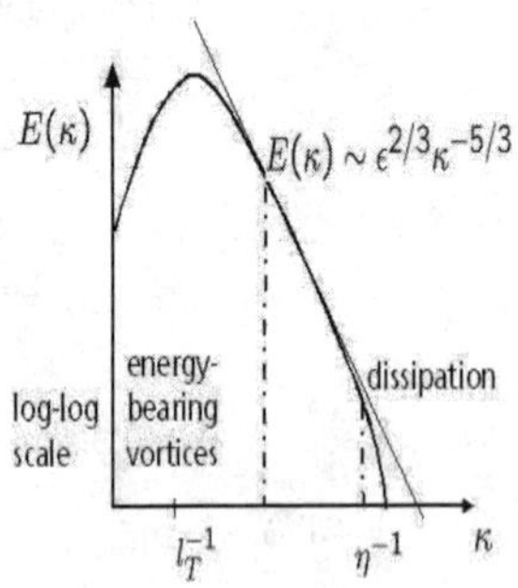

Figure 16: Mathematical Problems in Spinning Processes **(Turbulence Effects)**

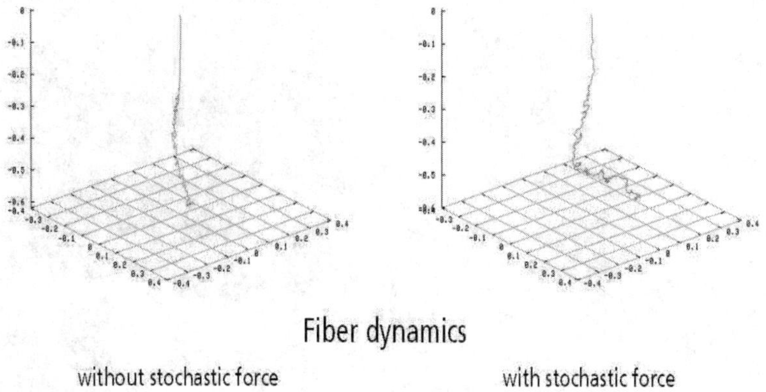

Figure 17: Mathematical Problems in Spinning Processes **(Turbulence Effects)**

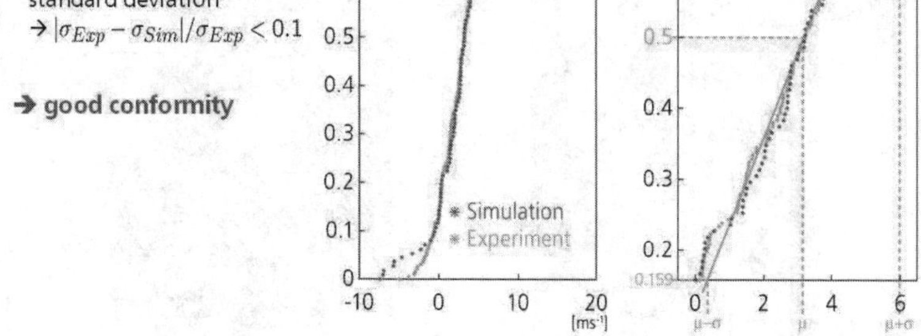

Figure 18: Mathematical Problems in Spinning Processes **(Turbulence Effects)**

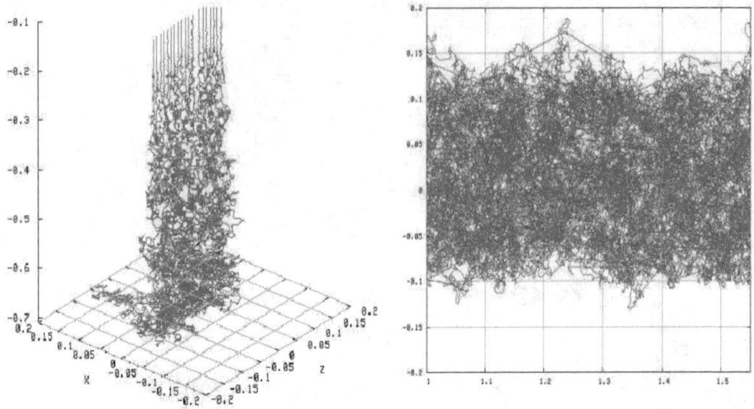

Figure 19: Mathematical Problems in Spinning Processes **(Deposition with Turbulence Effects)**

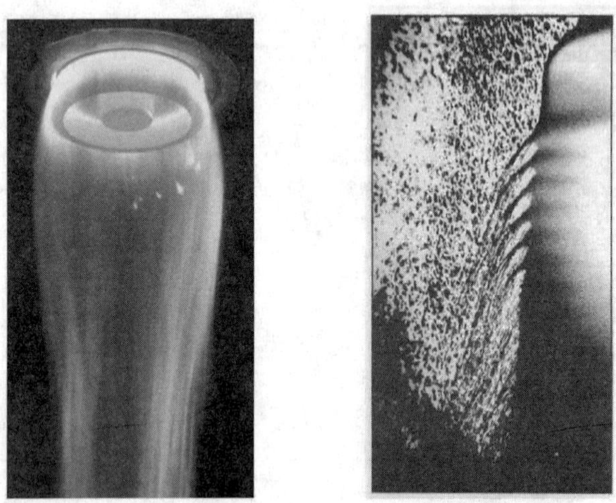

Figure 20: Mathematical Problems in Spinning Processes **(Melt-Spinning of Glass Fibers)**

3 Optimization, Control, and Design

What we finally want to achieve in our man-made world are optimal solutions: the process should be as cheap and as fast as possible, and the product should at least behave better than the products of the competitors. (Even nature seems to have a creator who is interested in optimality. That is why we have so many variational principles, and that is why animals and plants show us so many tricky solutions for their "technical" problems to be as stable, as light, as smoothly moving as possible and necessary. This is called "bionics" and there may be an interesting interplay between optimization by mathematics and optimization by evolution.) "So rather then asking how a product performs, the question is, how should the product be designed in order to perform in a specified way. Scheduling, planning and logistics also fall within that area of optimization. Optimal control is used to provide real-time control of an industrial process or a product, such as a plane or a car, in response to current operating conditions. A related area is that of inverse problems, where the parameters (or even the structure) of a model must be estimated from measurement of the system output) [1].

We have mentioned inverse problems already in (1); they appear literally everywhere. We will show two examples from our projects at ITWM; however they are very short.

There is the wide field of topological shape optimization; "topological" means that one may change the topology of a structure, for example, by admitting holes. One has to minimize an objective function (maximal stress, mean compliance, etc.) with respect to the shape.

As an example for a multicriteria optimization, we consider a project of [5].

How should we optimally control the radiation in cancer therapy such that the cancer cells are destroyed as much as possible, but at the same time organs or important healthy parts of the body remain undamaged. There are, besides optimization, a lot of simulation problems? f. e. to simulate how radiation penetrates the body, but let's concentrate on optimization assuming that the transmission of the radiation to different parts of the body given the external source, which can be controlled, is known. The goal is that a medical doctor can operate with the optimization tool, allowing more or less radiation to certain organs by "pulling" in the corresponding direction of a navigation scheme; the program then computes the different doses of different sources and different directions, getting at

Given a domain Ω and corresponding to it, the problem of linear elasticity in Ω:

$$-div\ \sigma(u_\Omega) = \rho f \quad \text{in } \Omega; \text{ where}$$

u_Ω denotes the displacements in Ω, $\sigma(u_\Omega)$ the corresponding stresses, linear depending on the strains

$$e(u_\Omega) = \left(\frac{1}{2}\left(\frac{\partial u_{\Omega_j}}{\partial x_i} + \frac{\partial u_{\Omega_i}}{\partial x_j}\right)\right), \quad \sigma = Ae.$$

The boundary conditions are:

$$\sigma(u_\Omega)n = t, \ t \text{ given surface on a part of } \partial\Omega$$
$$u_\Omega = 0 \text{ at the rest of } \partial\Omega, \text{ where the body is damped.}$$

We want to change Ω, such that a functional

$$\Omega \to J(\Omega, u_\Omega) = j(\Omega)$$

is minimal. This functional may be the mean compliance

$$\Omega \to J(\Omega, u_\Omega) = \int_\Omega Ae(u_\Omega)\cdot e(u_\Omega)dx$$

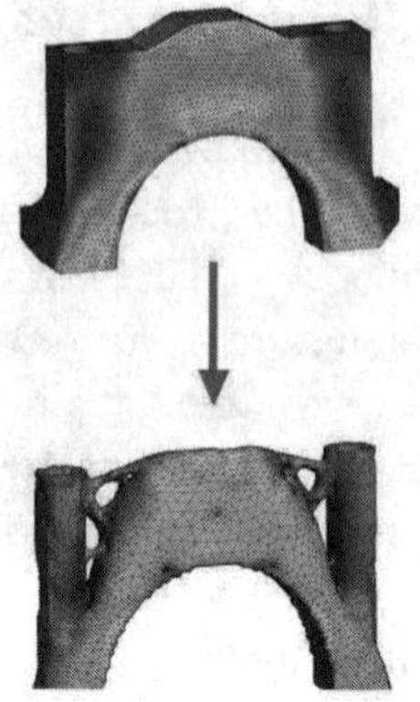

Figure 21: Topological Optimization

the end corresponding isodose levels.

To be more detailed: we have a target, the tumor and we have "risks", which should get as little as possible, but at most at given thresholds for the radiation.

One uses Pareto solutions, which are defined in the next figure:

To do this so fast, that it is finally online, and to do it so, that the doctors can easily handle it, are interesting and highly relevant mathematical tasks.

(a) Characterize Ω by level set of functions φ

$$\Omega = \{x : \varphi(x) < 0\}, \partial\Omega = \{x : \varphi(x) = 0\}$$
$$\Omega \leftrightarrow \varphi, u_\Omega = u_\varphi$$

(b) Use the concept of topological gradient: Drill a hole in Ω of radius ρ at point $x \in \Omega$:

$$\Omega_\rho^{(x)} = \Omega / \kappa_\rho(x)$$

and study, how J changes by that operations for small ρ

$$\lim_{\rho \to 0} \frac{j(\Omega_\rho(x)) - j(\Omega)}{\lambda(\rho)} = g(x) = \text{topological gradient.}$$

$\lambda(\rho)$ depends on J, but is often proportional to the volume of the hole.

$g(x) < 0 \Rightarrow$ J is improved by drilly a small hole at x.
The more negative g is, the more more you gain. $\Rightarrow$ Reduce the domain at points, where g is minimal.

Figure 22: Mathematical ideas

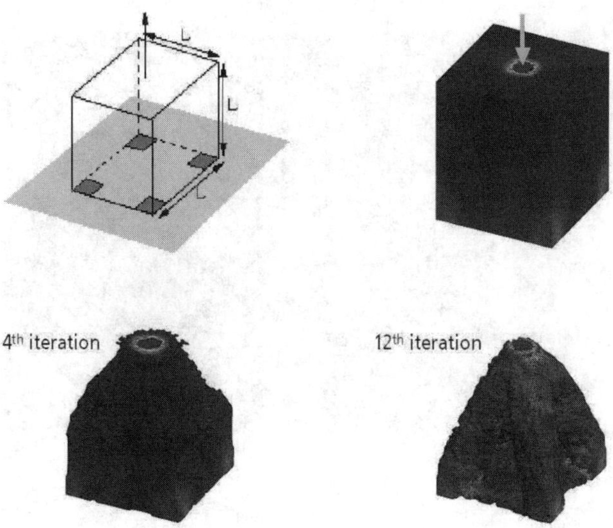

Figure 23: Cube with pointwise load: **10 % volume reduction per iteration (1)**

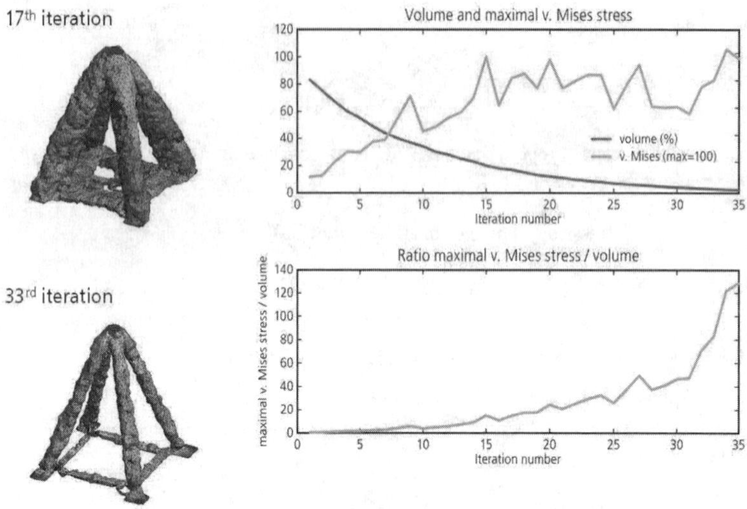

Figure 24: Cube with pointwise load: **10 % volume reduction per iteration (2)**

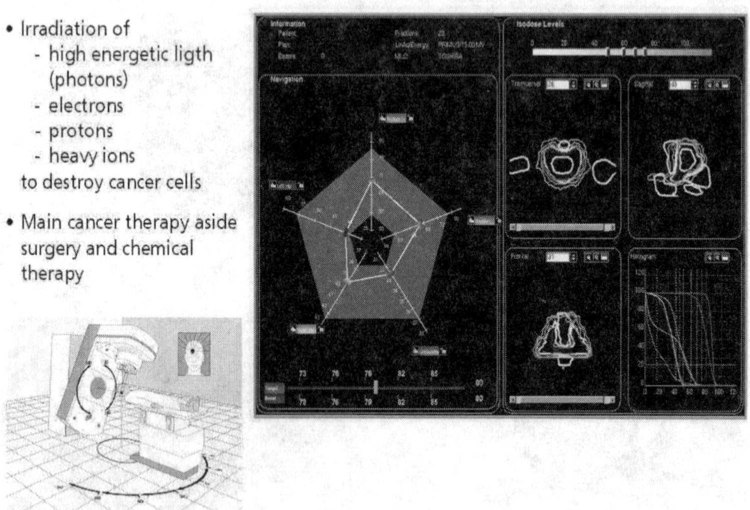

Figure 25: Optimization and Control (**Multicriteria optimization of intensity modulated radiotherapy**)

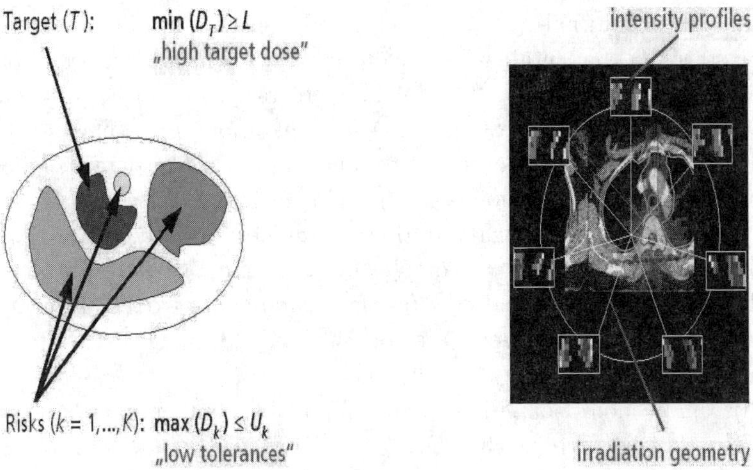

Figure 26: Ideal planning goals-not achievable

Problem:

Determine physical parameters for

Target (T):	$\min(D_T) \geq L\,(1 - F_T)$
Risks ($k = 1,\ldots,K$):	$\max(D_k) \leq U_k\,(1 + F_k)$

such that the relative deviations F_T, F_k are as small as possible.

Concept of Pareto solutions:

$$F = (F_U, F_L, F_1, \ldots, F_K) \rightarrow Min$$

F is called pareto optimal or efficient, if there is no improvement of one entry unless worsening another

Figure 27: Multicriteria approximation problem

4 Uncertainty and Risk

Many processes in nature, in economy, and even in daily life are or seem to be strongly accidental; we therefore need a stochastic theory in order to model these processes. Randomness creates uncertainty, and uncertainty creates risk, for example, in decisions about investments, about medication, and about security of technical systems like planes or power plants. Whether this randomness is genuine or just a consequence of high complexity is a philosophical question which does not influence stochastic modeling. You will find very complex systems in catastrophes like earthquakes or floods; biological systems, for example are extremely complex the human body. Experiments are not possible, and simulation is therefore highly necessary, but very difficult, too.

Also in economy, experiments are impossible, but one needs help for decisions which minimize the risk.

The law of large numbers leads often to models which are deterministic PDEs and very similar to deterministic models in natural sciences. But at a closer look they are even more complex, for example, very high dimensional (the independent variables are not geometric, but may be the values of different stocks). Therefore, even if we get at the end a treatable PDE, we have to use Monte-Carlo methods to solve them approximately, and we are back to stochastic differential equations. Now quite often derivatives of these solutions with respect to variables and parameters are needed, and to differentiate a function given by a Monte-Carlo method is not always successful. The Malliavin calculus shown in Figures 28–31, initially not invented for practical problems, is a great help [3]. Here is an example from option prizing.

Of course, there are other uncertainties and risks such as in floods and earthquakes. In technical systems, very different methods are involved.

5 Management and Exploitation of Data

We are flooded by data which, if structured, create information and finally knowledge. The extraction of this information or knowledge from data is called "data mining". Data may be given as signals or images; if we want to discover patterns, and if we want to "understand" these signals or images, we need image processing and pattern recognition methods. If we want to study and predict input-output systems for which we do not have enough theory (simple models) but many observations from the past,

Let $\Phi(t,x)$ be the price of an option at time t, where x is the present price of stocks underlying the option. Then

$$\frac{\partial \Phi}{\partial(t)} + \langle b, \nabla\Phi \rangle + \frac{1}{2}\sigma\sigma^T \Delta\Phi = 0,$$

where the final payoff of the option, $\Phi(T,x)$ is given by $f(x)$.

A Monte-Carlo method computes Φ by

$$\Phi(t,x) = E(f(X(T)/X(t) = x)$$

as an expectation value, where $X(t)$ is a solution of

$$dX(\tau) = b(X(\tau))d\tau + \sigma(X(\tau))dW(\tau).$$

f interest is often $\nabla_x \Phi(t,x)$, called "**delta**".

Figure 28: Malliavin calculus for Monte-Carlo methods (1)

we may develop "black-box" models like linear control models or neural networks. If for parts of the system a theory is available, we may talk of "grey-box" models. Data mining, signal or image processing, and black- or grey-box models are the mathematical disciplines involved here. Some of them are not as mature as PDE, optimization, or stochastics, but are certainly a field, where new ideas are needed. (There are many, especially in the field of pattern recognition: look, for example, at the articles of David Munford or Yves Meyer from the last 10 years.)

A typical input output system, where we do not have much theory, is–the human body; medicine is therefore a main application area, and we want to show only one example from our experience, the interpretation of long-term electrocardiograms. If we register only the heart beats, we get quite long sequences, $(t_i)_{i=1,\dots N}$ with $N \sim 100,000$, and have to find the information about the risk for sudden cardiac death. To do so we use Lorenz plots, sets consisting of points $\{(t_i, t_{i+1}, t_{i+2}), (t_{i+1}, t_{i+2}, t_{i+3}), \dots\}_{i=1,\dots N}$, and try to understand the structure of these sets. Of course, the beat is rather regular, if the Lorenz plot is a slim club (but too slim is again dangerous). The picture shows the clearly visible influence of drugs; to estimate the risk, one needs very tricky data mining techniques.

To compute delta, we use the "Malliavin calculus".

If Ω is a space of paths ("Wiener space") and

$F : \Omega \to \mathbb{R}$ (in 1d) is a random variable, one defines a "directional derivative"

$$D_\gamma F(\omega) := \frac{d}{d\varepsilon} F(\omega + \varepsilon\gamma) / \varepsilon = 0 \,,$$

$$\text{where } \gamma = \gamma(t) = \int_0^t g(s)ds, \qquad g \in \mathbb{L}^2.$$

If F possesses derivatives in all possible directions γ and if

there exists an $\Psi(t,w)$ such that

$$D_0 F(\omega) = \int_0^T \Psi(t,w) g(t) dt$$

then F is called "differentiable" and $\mathbb{D}_t F(\omega) := \Psi(t,w)$

Figure 29: Malliavin calculus for Monte-Carlo methods (2)

There is an "integration by parts" formula, which leads to

$$\nabla u(x) = E\left(f(X(T)\frac{1}{T}\int_0^T \sigma^{-1}(\,X(t)Y(t)\,)^* \, dW(t) \Big/ X(0) = x \right);$$

here $u(x) = \Phi(0, X)$ and Y is the "linearization around X"

$$dY(\tau) = b'(X(\tau))Y(\tau) + \sum_{i=1}^n \sigma_i'(X(\tau))Y(\tau)dW_i(\tau) \;\; ; \;\; Y(0) = E_n$$

Realize, that $\frac{1}{T}\int_0^T \sigma^{-1}(\,X(t)Y(t)\,)^* d(W(T)$ doesn't depend on f.

Figure 30: Malliavin calculus for Monte-Carlo methods (3)

A European option with one stock underlying

$$dP(t) = P(t)[r dt + \sigma dW(t)], \quad P(0) = p$$

We call $\tilde{u}(p) = E\left(e^{-rT} f(P(T))\right)$ the **option prize;**

then $$\tilde{u}'(p) = e^{-rT} E\left(f(P(T)) \cdot \frac{W(T)}{p\sigma T} \right)$$

This expactation value is simple to compute by simulation,

since again $\frac{W(T)}{p\sigma T}$ is independent of f.

However, for example for $f(x) = (x - K)_+$, the variance is very high and one needs many runs. If we do variance reduction methods, we get the following picture.

Figure 31: Malliavin calculus for Monte-Carlo methods (4)

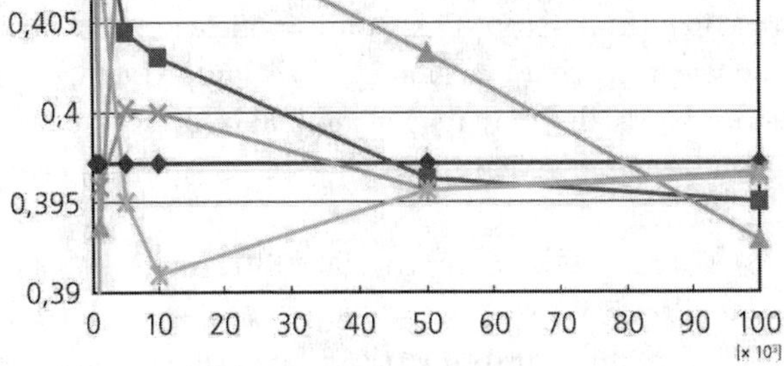

Figure 32: Comparison of computations of delta for a call

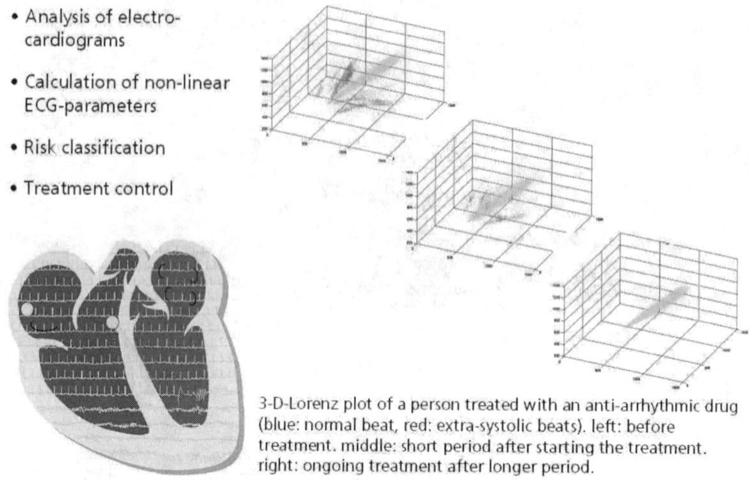

Figure 33: Risk parameters in the case of arrhythmic heartbeat

6 Virtual Material Design

One of the objectives of material science is to design new materials which have desirable properties; to do so by using simulation is called virtual material design". Mathematics is used to relate the large-scale (macroscopic) properties of materials such as stiffness, fatigue, permeability, and impedance to the small-scale (microscopic) structure of the material. The microscopic structure has to be optimized in order to guarantee the required macroscopic properties. This is an application of multi scale analysis, where we use averaging and homogenization procedures to pass from micro to macro. The scales may reach from nano to the size of constituents of composite materials. Typical materials are textiles, paper, food, drugs, and alloys.

At ITWM we try to design appropriate filter material. This is a very wide field, since filters are used everywhere: they serve different purposes and require therefore different properties. The example here deals with oil filters. The research work in its first part was done by Iliev and Laptev from ITWM.

We use a system which we get through homogenization from Navier-Stokes through a "very porous " medium: a Navier-Stokes-Brinkman sys-

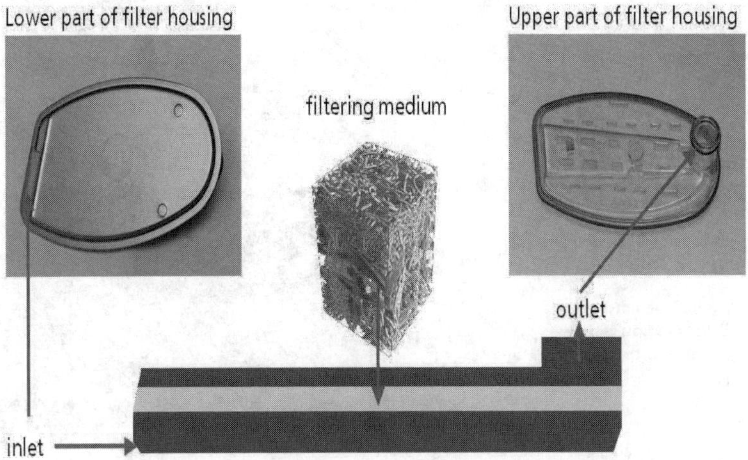

Figure 34: Simulation of 3-D flow through oil filters

tem which is a combination of incompressible, steady Navier-Stokes with a Darcy term.

The interface condition describing the behavior of the fluid on the surface of the filter material is a rather delicate issue, but in this model (with Brinkman homogenization) it is easier to handle (see the Ph.D. thesis by Laptev [7]). The flow field is given below.

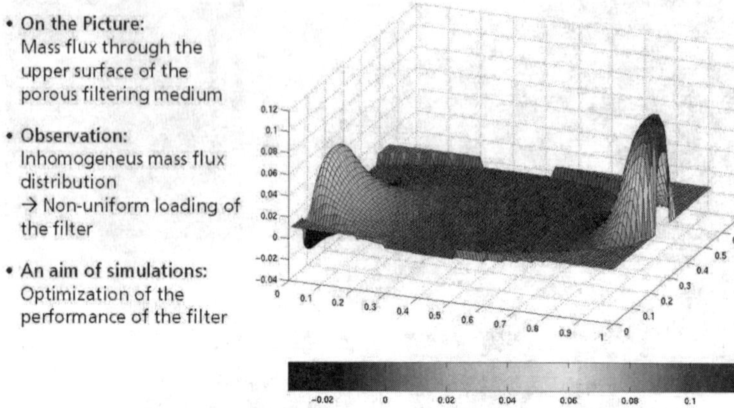

Figure 35: Simulation of Flow through a Filter Flow Rate

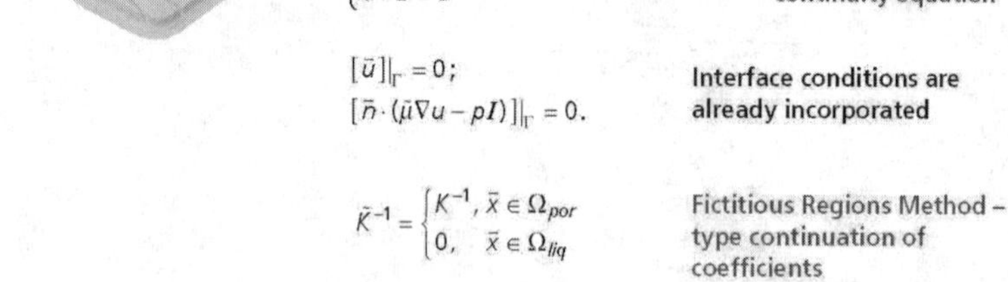

Figure 36: Simulation of 3-D flow through oil filters

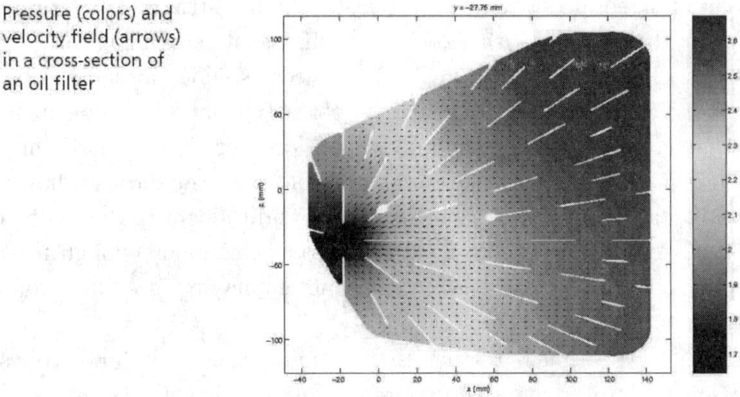

Figure 37: Simulation of 3-D flow through oil filters

The correspondence with measurements (where the pressure loss for different Reynold numbers at different temperatures with correspondingly different permeabilities) is remarkable.

I call this correspondence sometimes "prestabilized harmony": a rather crude model which is numerically approximated and gives results which correspond with nature to an extent which one really might not expect. But, of course, care is necessary. Models have their range of applicability, and their limitations should be carefully respected.

To compute the flow field of a filter is not enough to understand its efficiency. The transport of the particles, which have to be filtered out, must be simulated. Therefore, we have to model their absorption by the fibers of the filter and the motion of the particles by the fluid velocity, its friction, and the influence of diffusion. Finally, the absorption is, of course, filter and particle dependent. This is an area of exciting modelling (see, for example [8]).

7 Biotechnology, Food, and Health

This field has created new research areas which are rather interdisciplinary, for example, bio-informatics or system biology. Statistics, discrete mathematics, computer science and system and control theory, data mining, differential-algebraic systems, and parameter and structure identification are involved, together with all kinds of life sciences. Biological systems are extremely complex, involving huge molecules which interact in poorly understood ways. It is a long way to get a full understanding in terms of fundamental chemistry and physics. Moreover, it is a mathematical task to gain as much information as possible from the data we have; the classical idea to use a linear control system and to identify the coefficients does not work. We therefore need grey models, complex enough to allow prediction, but simple enough that parameters may be identified from the measurements.

Health is very much related to deterministic models for biophysical processes, a better image understanding, and efficient data mining.

Food is one of the emerging application fields of science, especially simulation. To simulate a process preparing food, for example, cooking of an omelette or frying a piece of meat in order to optimize the quality or the energy consumption, is a mathematical task of extremely high difficulty. However, the economic value is enormous for companies which offer food worldwide and for companies which produce, for example, household appliances.

The ITWM has not yet many projects in this field; however, its joint venture with Chalmers University of Technology, the Fraunhofer Chalmers Research Centre (FCC) at Gothenburg deals with bio-informatics and system biology. Figures 38–43 are taken from a presentation by Mats Jirstrand, FCC.

By metabolism we mean the processes inside living cells. These are complicated biochemical processes; even a "simple" process as glycolysis is not at all simple. We have to model biochemical pathways, i.e. chains of reactions, happening in collisions change the concentration of molecules of different types. Even simple enzymatic reactions lead to nonlinear systems. Finally, one does what every modeler has to do: we nondimensionalize and look for small parameters to apply perturbation methods. This leads to rational expression, called Michaelis-Menten dynamics in biology.

At the end we get very large, rational right-hand sides for the system of ODEs. The problem is that we do not know the parameters of the system, even the structure (which reactions should be included; do we need to include hysteresis, etc.) is not clear. Can we deduce from the behavior which structural elements the model should include? And how many parameters are we able to identify? How can we adopt the model to the knowledge we have? Some steps are done, but there is still a long way to go.

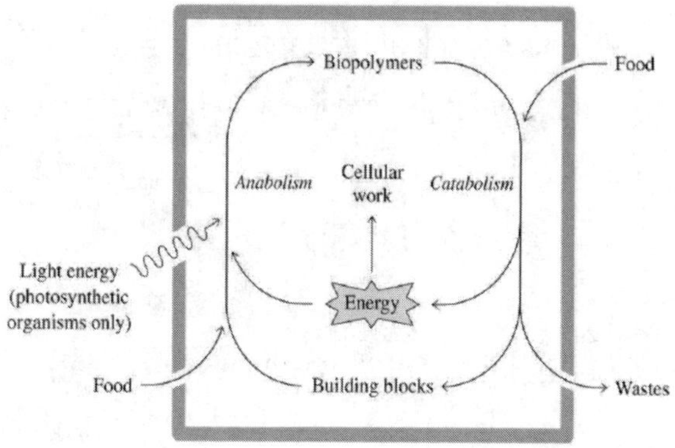

Figure 38: Metabolism

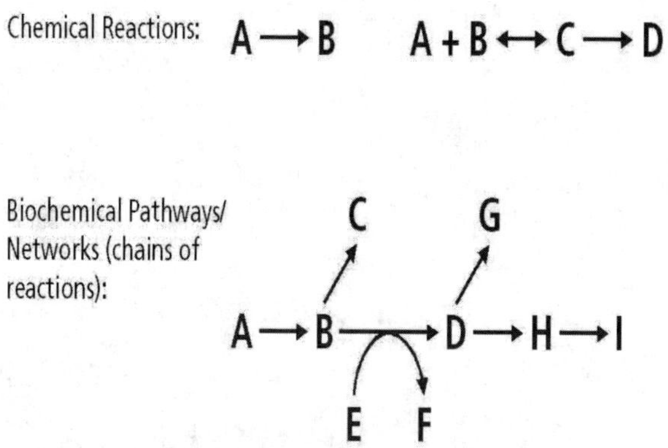

Figure 39: Modeling of Biochemical Reactions

Reaction rate in $dV \cong$ number of collisions $\cong$
number of collisions per sek of reactants in $dV \cong$
probability, that n_a molecules of type A and n_b molecules of type B enter dV simultaneously (independent events)

$$p(A)\ldots p(A)p(B)\ldots p(B) = p(A)^{n_a}\, p(B)^{n_b} \cong [A]^{n_a}[B]^{n_b}$$

Hence, $r_{dV} = k_0\ [A]^{n_a}[B]^{n_b}$ (but k_0 is volume dependent)

Double volume $\Rightarrow$ double k_0

Figure 40: Modeling of Biochemical Reactions

Enzymatic reactions: $S + E \underset{k_{-1}}{\overset{k_1}{\rightleftharpoons}} SE \xrightarrow{k_2} E + P$

$$\begin{aligned} S' &= -k_1\, S \cdot E + k_{-1}SE \\ E' &= -k_1\, S \cdot E + k_{-1}SE + k_2SE \\ SE' &= k_1\, S \cdot E - k_{-1}SE - k_2SE \\ P' &= k_2SE \end{aligned}$$

Note: $E' + SE' = 0 \ \Rightarrow \ E + SE = E_0$

Figure 41: Modeling of Biochemical Reactions

Making the model dimensionless:

$$s = S/S_0,\quad c = SE/E_0,\quad t = k_1 E_0 t \;\Rightarrow$$
$$ds/dt = -s + (s + \alpha - \beta)c$$
$$\varepsilon\, dc/dt = s - (s + \alpha)c$$

where $\alpha = (k_{-1} + k_2)/(k_1 S_0),\ \beta = k_2/(k_1 S_0),\ \varepsilon = E_0/S_0$

Now $E_0 \ll S_0 \;\Rightarrow\; \varepsilon \approx 0 \;\Rightarrow\; c = s/(\alpha + s)$

Model reduction: Singular perturbations!

Figure 42: Modeling of Biochemical Reactions

8 Conclusions

As mentioned in the beginning and shown during the description of the technology fields, one of the major drivers behind the dramatic change towards a knowledge-based economy is the advent of powerful and affordable digital computers. The rate of progress in hardware follows Moore's Law, telling us that computer power doubles every two years. Equally important, but not so widely appreciated, is the fact that there has been a similar improvement in the algorithms used to evaluate complex mathematical models. The improvement in speed, due to better algorithms, has been as significant as the improvements in hardware. All this has made computer simulation an accepted tool; in science, Computational X is dominating. Industry is already feeling the benefits of these advances, resulting in an increase in efficiency and competitiveness. This in turn makes mathematics, being at the core of all simulation, poised to become a key technology. Mathematics by its abstraction allows the transfer of ideas from one application field to another. Mathematicians are "cross thinkers". This kind of cross thinking creates creativity and leads to innovation.

To give mathematics its power, the classical "engineering mathematics"

$$ds/d\tau = -\beta s/(\alpha + s) \quad \Leftrightarrow$$
$$dS/dt = -V_{max} S/(K_m + S) \quad \text{...Michaelis-Menten dynamics}$$
$$\text{where } V_{max} = k_2 E_0, \quad K_m = (k_{-1} + k_2)/k_1$$

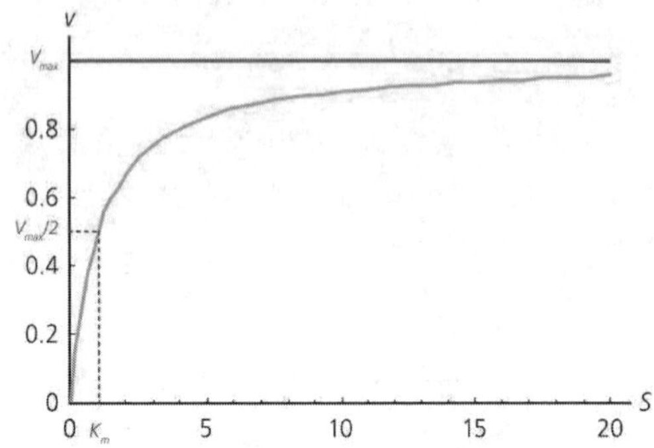

Figure 43: Modeling of Biochemical Reactions

is not sufficient. I hope I have made clear that new ideas, some from pure mathematics too, are needed in order to get good results: new function spaces, new ideas in non-linear analysis or in stochastic calculus, new ideas to deal with inverse problems and to deal with pattern recognition, etc. It is not a question of "pure or applied", there is a need for "pure and applied." Both should be in balance and they should work together; the fact that there is a widespread separation weakens both parts.

There is a need for properly educated mathematicians all over the world, too. What a proper education means for an "industrial mathematician" would be a subject in its own. The European Consortium for Mathematics in Industry (ECMI) has put a lot of effort into that issue. However, what we have to strive after is creativity and flexibility in finding proper models and more efficient algorithms. "Industrial Mathematics" or, as it is called in Europe, Technomathematics, Economathematics, or Finance Mathematics, is not a subject in its own like algebra or topology. It is more a new attitude towards the world it is the curiosity in order to understand and the drive to improve.

If we mathematicians work together, if we are courageous enough to leave the ivory tower of our science and act in the real world, I am sure we shall see a bright future for our science and for our students, too.

References

[1] A. Cliff, R. Matheij, and H. Neunzert, Mathematics: Key to the european knowledge based economy, in *MACSI-Net Roadmap for Mathematics in European Industry*, Edited by A. Cliffe, B. Matheij, and H. Neunzert, A project of European commission, Mark 2004.

[2] A. Cliffe, B. Matheij, and H. Neunzert (Eds), A project of European commission, *MACSI-Net Roadmap for Mathematics in European Industry*, Mark 2004,

[3] Fournier et al., Application of Malliavin calculus to Monte-Carlo methods in finance, *Finance and Stochastics* 3(4), 1999.

[4] Fraunhofer Institute for Industrial Mathematics, Kaiserslauntern, Germany, Annual report 2003. (info@itwm.fraunhofer.de).

[5] H. Neunzert, N. Siedow, and F. Zingsheion, Simulation temperature behavior of hot glass during cooling, In, *Mathematical Modeling*, Edited by E. Cumberbatch and A. Fitt, Cambridge University Press, 2001.

[6] H. Neunzert and U. Trottenberg (Eds), *Mathematik als Technologie*, Die Fraunhofer–Institute ITWM und SCAI, to appear.

[7] V. Laptev, *Numerical Solution of Complex Flow in Plain and Porous Media*, Dissertation, Department of Mathematics, Technical University of Kaiserlauntern, Germany, 2004.

[8] A. Latz and A. Wiegmann, Simulation of fluid particle simulation in realistic 3-dimensional fiber structures, in *Proceedings Filtech Europa*, I-353-360, 2003.

[9] L. Russo, *The Forgotten Revolution*, Springer-Verlag, Heidelberg, 2004.

[10] M. Scheler, Soziologie des Wissens, in *Die Wis-sensformen und die Gesellschaft*, Francke-Verlag, 1960.

[11] S. Tiwari and J. Kuhnert, A numerical scheme for solving incompressible and law mach number flows by finite pointset methods, in, *Meshfree Methods for Partial Differential Equations*, Edited by M. Griebel and M. Schwertzer, Volume 2, Springer-Verlag, Berlin, 2004.

Chapter 2

INDUSTRIAL MATHEMATICS – WHAT IS IT?

N. G. Barton

Sunoba Renewable Energy Systems

Abstract

This short chapter presents the author's views on the perhaps vexed issue of how to define industrial mathematics. It is argued that it is advantageous to adopt a wide-ranging definition, although the aspects of industrial mathematics that are adduced might be uncomfortably broad to some. Also presented is a list of organizational structures in which industrial mathematics is carried out, along with examples of best practice.

1 Introduction

"Industrial Mathematics" is a branch of mathematical research that has grown strongly in prevalence in the last three decades. In broad terms, industrial mathematics is the extension of applied mathematics to industrial applications. Applied mathematics covers both the development of new mathematical techniques that can be used for practical applications and the application of those techniques.

Clearly, industrial mathematics has many facets. This chapter presents the author's views on key aspects such as activities that, are embraced within the field, their outputs, and performance measures; structures that are used to undertake industrial mathematics; and examples of best practice. The key theoretical framework that is used is the classification of types of research (by type of activity, not by field of research or socioeconomic application).

This chapter is a companion article to another article by the author (Barton, 2002), which looks particularly at the role of mathematics in technological development.

2 Categories of Research

To assess where industrial mathematics fits into the broad body of activity that is mathematical research, it is useful to commence with some definitions. The Organisation for Economic Co-operation and Development (OECD) defines research as "creative work undertaken on a systematic basis in order to increase the stock of knowledge - and the use of this stock of knowledge to devise new applications".

Research has various sub-categories, for example, the following four used by the Australian Bureau of Statistics.

- **Pure basic research** is experimental and theoretical work undertaken to acquire new knowledge without looking for long-term benefits other than the advancement of knowledge.

- **Strategic basic research** is experimental and theoretical work undertaken to acquire new knowledge directed into specified broad areas in the expectation of useful discoveries. It provides the broad base of knowledge necessary for the solution of recognized practical problems.

- **Applied research** is original work undertaken primarily to acquire new knowledge with a specific application in view. It is undertaken either to determine possible uses for the findings of basic research or to determine new ways of achieving some specific and predetermined objectives.

- **Experimental development** is systematic work, using existing knowledge gained from research or practical experience, which is directed to producing new materials, products, or devices; to installing new processes, systems, and services; or to improving substantially those already produced or installed.

It might help to give an example of these sub-categories. Let us consider the development of a generic software package to compute inviscid flow around an object. These packages are widely used in engineering, particularly ship hydrodynamics and aerodynamics (a notable example being VSAERO produced by Analytical Methods Inc.), and they are grounded unambiguously on mathematical research.

The pure basic research required is the fundamental theory of partial differential equations. The strategic basic research would include the theory of irrotational flow, the fundamentals of numerical analysis, and boundary integral equations in particular. The applied research required might include mesh generation algorithms and schemes for numerical linear algebra. The required experimental development includes implementation of schemes for the numerical solution of boundary integral equations, interface of the mesh generators with CAD packages, and wrapping of the software in an interface usable by an engineer in industry.

And, there is more! These packages, and others in dozens of applications, need to be commercialised. That activity generally involves input by mathematicians as part of a team. Necessary tasks include preparation of manuals and examples, trouble-shooting, technical marketing, and ultimately user support and consulting. Those tasks just cannot be done without the input of professional mathematicians.

I hope my point is clear. The words "Industrial Mathematics" can apply to a continuum of activities. My view is that it is useful to be expansive and encompassing in definitions of industrial mathematical activities. Others will disagree, for example, by saying that the experimental development or commercialization highlighted above is not industrial mathematics. I think this is dangerous because it drives a wedge between the underlying mathematical infrastructure and the end-application. Mathematics never developed in isolation, so why now impose segregation between theoreticians and practitioners? There is also a very practical reason to adopt a broad definition – in a time when funding has never been more carefully controlled and competitive, why deny a link that can bring support?

If industrial mathematics is to apply to all phases of research, then a range of players will participate. Figure 1 is a schematic showing that universities undertake research mostly at the basic end of the spectrum, and companies are generally only involved with experimental development and commercialization, while government laboratories occupy the middle ground. (I hasten to acknowledge that there are always honourable exceptions to the preceding generalization.) Table 1 displays aspects of research and shows that there is a wide diversity between basic research and experimental development in key areas like motivation, outputs, rewards, and funding. To be very clear, there is a big gulf between pure basic research and experimental development, but I still maintain that we need a definition of industrial mathematics than spans this gulf.

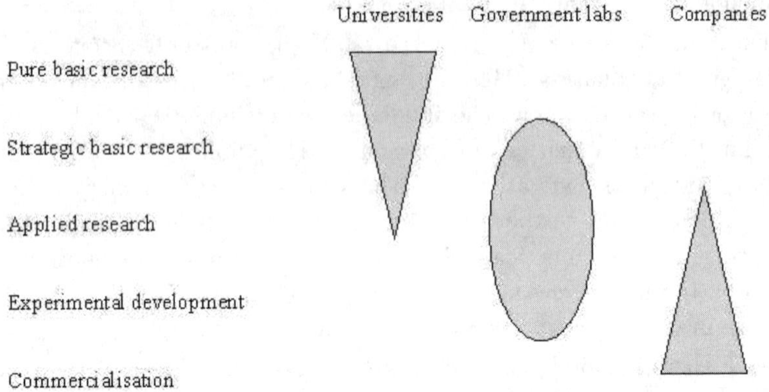

Figure 1: Illustration of the zone of activity undertaken by universities, government laboratories, and private companies.

3 What Is Industrial Mathematics?

Earlier I said that industrial mathematics is the extension of applied mathematics to industrial applications. With my previous arguments in mind, we can now compile a following list of activities that would go under the broad banner of industrial mathematics:

- an activity that bridges the gap between academic practitioners and potential industrial users (or employers of graduates);
- development of mathematics that might be used (sometime) in industry;
- mathematics that supplements industrial R & D, for example, by solving a "nearby" problem exactly or shedding light on an industrial topic;
- application of mathematics (right now) in industry; and
- contribution of mathematicians to the needs of companies to develop or improve products or processes.

The list is indeed broad, and the inclusion of the last bullet point, in particular, will not be acceptable to some. I can only repeat my arguments that such work (as in the last point) requires input by people with mathematical skills and training – practitioners if you will – and that to adopt a less encompassing list diminishes the impact of mathematics in the eyes of those who have the responsibility to make judgements about funding priorities.

Table 1: Comparison of pure basic research and experimental development in various important areas.

	Pure basic research	**Experimental development**
Freedom to publish	Public domain	Confidential
Funding	Public funded; grant applications	Privately funded; contracts, deliverables
Motivation	Motivated by spirit of enquiry, not profit	Profit motive (or perhaps national benefit)
Output	Peer reviewed journal papers	Confidential reports
Attitude to intellectual property (IP)	Exploitation of IP is not intended	Purposeful development of IP, patents are frequently used
Links to education	Very strong links to education	Few educational links, perhaps for training
Rewards	Further funding, prestige, establishment of a school of study	Commercial success
Performance measures	Fellowships, bibliometry, growth of a school, peer recognition, invitations to speak at conferences, public funding	Product or process improvement, development of a business, profit, perhaps national benefit (e.g., lower emissions)

4 Structures to Undertake Industrial Mathematics

With my broad interpretation of industrial mathematics, it is clear that a variety of organizational structures will be involved. These include:

- **University departments.** Hundreds of departments, worldwide, carry out industrial mathematics; notable examples are Oxford, RPI, Eindhoven, Linz, and Claremont. These groups and many others have strong scientific achievements, vigorous postgraduate programs, and an exemplary track record in reaching out to industry. Friedman and Lavery (1993) wrote the classic reference that describes how to initiate these activities.
- **Research Institutes**. Many countries now have a (more than one in some cases) mathematical research institute that has a strong emphasis on interacting with industrial end-users. Typically, these institutes will have a balanced program of activities, often with relatively short study programs on topics of industrial concern. Notable examples include the Newton Institute (UK), MITACS (Canada) and most recently AMSI in Australia.
- **Government laboratories**. We find that many nations have government research laboratories that are heavily involved with industrial application of mathematics. Often, these laboratories will explicitly encourage industrial work by mandating a fraction of the budget that must come from external funding sources. Examples of these government laboratories include INRIA (France), ITWM (Germany), Sintef (Norway), and CSIRO (Australia).
- **Professional associations**. The International Council for Industrial and Applied Mathematics has approximately 20 member societies (see www.iciam.org). These represent the core of the worldwide industrial and applied mathematics community. The largest and most successful of these societies is the (US-based) Society for Industrial and Applied Mathematics. These societies have a major role in influencing the culture and development of industrial and applied mathematics; they also support conferences and workshops in the field.

To supplement the above list, it must be explicitly noted that hundreds of companies around the world rely on applications of mathematics for their commercial survival, let alone profitability. At the top end, these companies include major multinationals (such as Microsoft, IBM, Shell, Toyota, and Boeing to name just five from a very long list) that have teams of mathematicians. There are also hundreds of small companies operating in particular niches that need mathematical skills. Of this group, I shall give just one shining example, Opcom Pty Ltd (see www.opcom.com.au), a company set up by former mathematics academics at the University of Queensland and now employing perhaps 50 people in public transport information systems, crew scheduling, and rostering systems; postal network modelling and optimization; and journey planning software.

Glimm (1991) and Barton (1996) have written at length on the mathematical work undertaken in companies. Their work contains examples of companies (big and small), as well as an analysis of the role of mathematics at every stage in various industry sectors.

5 Conclusion and Acknowledgment

This chapter is a written version of a presentation made by the author in a minisymposium entitled "Industrial Mathematics. What is it? And what is international best practice?" at the 5th International Congress on Industrial and Applied Mathematics, Sydney, Australia, 2003. Other speakers in the minisymposium included Arvind Gupta (MITACS, Canada), Heather Tewkesbury (The Smith Institute, UK), and Heinz Engl (Austria). The author thanks these colleagues for their contributions. The author also thanks CSIRO Mathematical and Information Sciences, his previous employer.

References

[1] Noel Barton (Ed.), *Mathematical Sciences, Adding to Australia*, National Board of Employment, Education and Training, Canberra, 1996.

[2] Noel Barton, A perspective on industrial mathematics work in recent years, in *Trends in Industrial and Applied Mathematics*, Edited by A.H. Siddiqi and M. Kocvara, Kluwer Academic Press, Dordrecht, 2002.

[3] Avner Friedman and John Lavery, *How to Start an Industrial Mathematics Program in the University*, Society for Industrial and Applied Mathematics, Philadelphia, 1993.

[4] James Glimm (Ed.), *Mathematical Sciences, Technology and Economic Competitiveness*, Board on Mathematical Sciences, National Research Council, National Academy Press, Washington D.C., 1991.

Chapter 3

MATHEMATICAL MODELS AND ALGORITHMS FOR TYPE-II SUPERCONDUCTORS

K. M. Furati and A. H. Siddiqi

King Fahd University of Petroleum & Minerals

Abstract

Superconductors are materials that exhibit zero resistivity below a critical temperature (T_c) which depends upon the material. Superconductors are classified as either type-I or type-II, with the former applying to the original element superconductors such as zinc, mercury, and aluminum and the latter referring to the modern compounds such as yttrium barium cuyprate. In 1986, the invention of high-temperature superconducting materials by Muller and Bednoroz, who won Nobel Prize of Physics in 1987 for their contributions, aroused great interest among scientists and engineers. A group of engineers started examining the possibility of replacing the traditional permanent magnet by a superconducting magnet in various types of equipment and machines, while another group of mathematicians, physicists, and engineers became engaged in mathematical modelling and numerical simulation of this phenomena to understand properly puzzling behavior of the type-II high-temperature superconductors. The real superconductor revolution everyone is awaiting is the application of high-temperature materials in the power industry after the expected manufacturing of superconducting power cables. It is predicted that it may generate markets worth many tens of billions of dollars in about 15 years. Bean and Kim et al. models of type-II superconductors are well known. In the Bean model, the current density in the superconductor cannot exceed some critical value, say J_c, which is a constant determined by the properties of high-temperature superconductors. In the Kim et al. model the critical current density depends on the magnetic field. It has been shown that the Bean model is equivalent to a parabolic variational inequality, while the Kim et al. model is equivalent to a parabolic quasi-variational inequality. It can be observed that such models also appear in safing sensor problems of metallic rings flanked with two blocks of superconductors. J.L. Lions and his collaborators, such as Glowinski and Tremolieres, have made systematic efforts to develop numerical simulation of parabolic variational inequalities. Improvement of many of their results have been studied. However, the numerical methods for parabolic quasi-variational inequalities is not well developed. In 1999, Lions also studied parallel algorithms for parabolic variational inequalities and indicated the possibility of development of such algorithms for certain classes of parabolic quasi-variational inequalities. Furati and Siddiqi studied the relevance of Lions work to Bean and Kim et al. models of superconductivity. Existence of solutions, parallel algorithms, and numerical simulation of these models will be presented in this chapter along with a brief resume of Chapman's work on superconducting fault current limiters.

1 Introduction

Superconductors are materials that exhibit zero resistivity below a critical temperature (T_c) which depends upon the material. Superconductors are classified as either type-I or type-II, with the former applying to the original elemental superconductors such as zinc, mercury, and aluminum and the latter referring to the modern compounds such as yttrium barium cuyprate.

Researches in superconductivity have fetched the 1972 Nobel Prize in Physics (BCS Theory, named for John Bardeen, Leon Cooper, and Robert Schrieffer) and the 1987 Nobel Prize in Physics (recipients were Muller and Bednorz for the discovery of lanthanum barium copper oxide exhibiting a critical temperature of 30 K). Superconductors have numerous applications of far-reaching consequences, in manufacturing the trains with fastest speed, mobile telephone systems giving great sensitivity and protection for drop-outs, ultrasensitive microwave communications, astronomy, and analysis for the pharmaceutical industry. However, the real superconductor revolution that everyone is awaiting is the application of the high-temperature materials in the power industry. Several reputed companies in different parts of the world are engaged in developing superconducting power cables. The future for all these superconducting technologies looks rosy. It is predicted that by 2020 they are expected to generate markets worth many tens of billions of dollars. However, properties of the high-temperature superconductors are a puzzle, and their proper understanding is posing serious problems. Numerical simulation and development of fast algorithms for problems of the type-II high-temperature superconductors (HTS) are challenging tasks.

A systematic effort has been initiated in the last couple of years to write mathematical models and fast algorithms for analyzing and visualizing problems of type-II superconductivity. For an updated account, we refer to Chapman [9], Prigozhin [29, 30, 31], and Barnes, McCulloch, and Dew-Hughes [2].

The mathematical formulation of safing sensor along with type-II HTS is either a variational inequality (corresponding to the Beans model) or a quasi-variational inequality (corresponding to the Kim model). Finite element method and parallel algorithms of these models have been studied in [13, 14]. This chapter is mainly based on these results. Motivation for this work was provided by Chapman [9] and Barnes, McCulloch, and Dew-Hughes [2].

Section 2 is devoted to the Bean model, while Section 3 deals with basic results of parabolic quasi-variational inequality. The quasi-variational inequality for the Kim et al. model of type-II superconductivity, existence of its solution, and parallel algorithm for this model are respectively discussed in Sections 4 to 6. We briefly mention investigation of Chapman [10] in Section 7.

It may be observed here that wavelet methods for variational inequalitiy arising from superconductivity are being studied by the authors of this chapter on the lines of Comincioli et al. [11].

2 Introduction to Parabolic Variational Inequalities

In this section we present function spaces required for formulation of problems. Basic results on parabolic variational inequality (PVI), quasi-variational inequality (QVI), and parabolic quasi-variational inequality (PQVI) which model type-II superconductivity with critical current density depending on the magnetic field in different situations are also discussed. Let H be a real Hilbert space with inner product $\langle \cdot, \cdot \rangle$ and its induced norm denoted by $\| \cdot \|$. Let H^* denote the dual space of H which is identified with H. Let K be a nonempty closed convex subset of H and without any loss of generality, and let $0 \in K$. Let $L_2(0,T;H)$ denote the space of all measurable functions $u : (0,T) \to H$, and then it is a Hilbert space with respect to the inner product

$$\langle u, v \rangle_{L_2(0,T;H)} = \int_0^T \langle u(t), v(t) \rangle_H dt. \tag{2.1}$$

Let

$$\|u\|_{L_2(0,T;H)} = \left(\int_0^T \|u(t)\|_H^2 dt \right)^{1/2}, \tag{2.2}$$

$$\|u\|_{L^\infty(0,T;H)} = \underset{0 \le t \le T}{\text{ess sup}} \|u(t)\|_H. \tag{2.3}$$

For a real Hilbert space H, we have $L_2(0,T;H) = L_2(0,T;H^*)$. Let $H^{1,p}(0,T;H)$ denote the space of functions $f \in L_p(0,T;H)$ such that their distributional derivative D_f also belongs to $L_p(0,T;H)$. The space $H^{1,p}(0,T;H)$ equipped with the norm

$$\|f\|^2_{H^{1,p}(0,T;H)} = \|f\|^2_{L_p(0,T;H)} + \|D_f\|^2_{L_p(0,T;H)}$$

is a Banach space. If $p = 2$, then $H^{1,2}(0,T;H)$ is a Hilbert space. Very often, we simply write $H^2(0,T;H)$. Let Ω be an open-bounded subset of R^3 with smooth boundary $\partial\Omega = \Gamma$.

Let $L_2(\Omega)$ and $H^1(\Omega)$ denote the space of square-integrable functions and Sobolev space of order 1, respectively. These are Hilbert spaces with respect to the inner products

$$\langle f, g \rangle_{L_2(\Omega)} = \int_\Omega f(x)\overline{g(x)}dx$$

and

$$\langle f, g\rangle_{H^1(\Omega)} = \langle f, g\rangle_{L_2(\Omega)} + \langle Df, Dg\rangle_{L_2(\Omega)}.$$

Let

$$H_0^1(\Omega) = \{v \in H^1(\Omega) | v = 0 \text{ in } R^3 \setminus \Omega\}. \tag{2.4}$$

For vector functions, let $\mathbf{L}_2 = (L_2)^3$ and $\mathbf{H}^1 = (H^1)^3$ be the Hilbert spaces with $\langle \varphi, \psi\rangle_{\mathbf{H}^1(\Omega)} = \langle \varphi, \psi\rangle_{(\mathbf{L}_2(\Omega))^3} + \langle D\varphi, D\psi\rangle_{(\mathbf{L}_2(\Omega))^3}$. We define the subspaces

$$\mathcal{H}(\text{curl}; \Omega) = \{v \in \mathbf{L}_2(\Omega) : \text{ curl } v \in \mathbf{L}_2(\Omega)\}, \tag{2.5}$$

and

$$\mathcal{H}(\text{div}; \Omega) = \{v \in \mathbf{L}_2(\Omega) : \text{ div } \in L_2(\Omega)\}. \tag{2.6}$$

For $\Box \in \{\text{curl,div}\}$, let

$$\mathcal{H}_0(\Box; \Omega) = \{v \in \mathcal{H}(\Box; \Omega); \Box v = 0 \text{ on } R^3 \setminus \Omega\}, \tag{2.7}$$

with

$$\|v\|^2_{\mathcal{H}(\Box;\Omega)} = \|v\|^2 + \|\Box v\|^2, \tag{2.8}$$

where the norms in the right-hand side are the $\mathbf{L}_2$ norms on Ω.

Note that the spaces $\mathcal{H}(\text{curl}; \Omega)$ and $\mathcal{H}(\text{div}; \Omega)$ are Hilbert spaces with the corresponding graph norm (2.8).

Define

$$\mathcal{H}_c(\Omega) = \{v \in \mathbf{L}_2(\Omega) : \text{ curl } v \in \mathbf{H}^1(\Omega), \text{ div } v \in \mathcal{H}^1(\Omega), \text{ and } v \times n|\Gamma = 0\}, \tag{2.9}$$

and

$$\mathcal{H}_d(\Omega) = \{v \in \mathbf{L}_2(\Omega) : \text{ curl } v \in \mathbf{H}^1(\Omega), \text{ div } v \in \mathcal{H}^1(\Omega), \text{ and div } v|\Gamma = 0\}, \tag{2.10}$$

where n denotes the outward unit normal vector. $\mathcal{H}_c$ and $\mathcal{H}_d$ are Hilbert spaces, and they are algebraically and topologically equivalent to $\mathbf{H}^2(\Omega)$ (see, for example, [5]). Define the subspaces

$$\mathcal{H}_{c0}(\Omega) = \{v \in \mathcal{H}_c(\Omega) : \text{ curl } v = 0 \text{ and div } v = 0 \text{ in } \Omega\}, \tag{2.11}$$

and

$$\mathcal{H}_{d0}(\Omega) = \{v \in \mathcal{H}_d(\Omega) : \text{ curl } v = 0 \text{ and div } v = 0 \text{ in } \Omega\}. \tag{2.12}$$

These two subspaces are of finite dimension ([Lemma 2.2, [5]). Let

$$\mathcal{L}_2(\Omega) = \{v : \Omega_t \to \Omega, |v(\cdot, t) \in \mathbf{L}_2(\Omega)\}. \tag{2.13}$$

For more details about Sobolov spaces of vector functions, we refer to [12].

The theory of variational inequality was invented by the famous French mathematician, scientist, and educator J.L. Lions (expired in 2000). His distinguished Ph.D. students such as Alain Bensoussan, G. Duvaut, R. Glowinski, L. Tartar, R. Temam, and R. Tremolier and collaborators from other countries such as Umberto Mosco, Stampacchia, Kinderleherer, and Vivaldi have vigorously worked on theoretical as well as numerical aspects of these two models (for update references see [34]) PQVI and PVI which correspond respectively to Kim et al. and Bean models. The Italian government has established the Stampacchia School of Mathematics, as a part of Majorana Scientific and Cultural Centre, Eriche, where annual conferences/symposia/workshops are being organized to study various aspects of variational inequalities.

Let f be given such that $f \in L_2(0,T;H^*)$. Let us consider a bilinear, continuous, coercive form $a(\cdot,\cdot)$ defined on $H \times H$, that is, $a(u,v)$ is linear, continuous and there exists $\alpha > 0$ such that

$$a(u,u) \geq \alpha \|u\|^2, \qquad \forall\, u \in H. \tag{2.14}$$

The following problem is called a PVI problem. Find $u \in L_2(0,T;H) \cap L_\infty(0,T;H)$, $\quad u(t) \in K$ a.e.

$$\left\langle \frac{\partial u}{\partial t}, v-u \right\rangle_H + a(u, v-u) \geq \langle f, v-u \rangle, \quad \forall\, v \in K$$

$$u(0) = 0. \tag{2.15}$$

When the problem is time-independent, the problem is called stationary or simply a variational inequality.

2.1 Quasi-Variational Inequalities

For every $v \in H$, let there be a given nonempty, closed convex subset $K(v)$ of H. Then the problem is find $u \in K(u)$ such that

$$a(u, v-u) \geq \langle f, v-u \rangle \quad \forall\, v \in K(u) \tag{2.16}$$

is called a QVI problem.

2.2 Existence of Solutions, Uniqueness and Numerical Analysis of QVI

One special case often considered is

$$K(v) = \{\varphi \in H^1(\Omega) | \varphi \leq M(v) \text{ a.e. in } \Omega\}, \tag{2.17}$$

where $H^1(\Omega)$ denotes the Sobolev space of order 1, and $M : \psi \to M(\psi)$ is a nonlinear operator on $L^\infty_+(\Omega) \to L^\infty_+(\Omega)$ having the following properties:

$$M \text{ is positive increasing, that is,} \tag{2.18}$$
$$0 \le \psi_1 \le \psi_2 \text{ if and only if } 0 \le M(\psi_1) \le M(\psi_2),$$

$$M(0) \ge k > 0, \quad k \quad \text{constant} , \tag{2.19}$$

where $L^\infty_+(\Omega)$ denotes the set of functions ≥ 0 a.e. in $L^\infty(\Omega)$, namely, the space of bounded functions a.e. with sup norm.

QVIs were introduced by Bensoussan and Lions in 1973. For details see Bensoussan and Lions [6], Baiocchi and Capelo [1], Mosco [26], and Lions [22].

As a special case, one may choose

$$M(\psi)(x) = k + \inf_{\substack{\xi \ge 0 \\ x, x+\xi \in \Omega}} \psi(x + \xi). \tag{2.20}$$

Laetsch [21] has proved that (2.16) has a unique solution under conditions (2.17) – (2.18), where

$$a(u, v) = \sum_{i,j} \int_\Omega a_{ij} \frac{\partial u}{\partial x_j} \frac{\partial u}{\partial x_i} + \sum \int_\Omega a_i \frac{\partial y}{\partial x_i} v dx + \int_\Omega a_0 u v dx, \tag{2.21}$$

and $f \in L^\infty_+(\Omega)$. For more results in this area, we refer to Lions [22], 161–168].

Recently, Boulbrachene et al. [8] have studied the existence and uniqueness of the solution of QVI and its finite element approximation. They have proved that the finite element approximation is optimally accurate in L^∞. The results have also been studied for the noncoercive bilinear form.

2.3 Parabolic Quasi-Variational Inequality

Find $u \in L_2(0, T; H^1(\Omega)$

$$\frac{\partial u}{\partial t} \in \mathcal{L}_2(\Omega) \tag{2.22}$$

$$u - K(u) \le 0 \quad \text{a.e. in } Q = \Omega \times (0, T), \quad u \ge 0 \text{ a.e.} \tag{2.23}$$

$$\left\langle \frac{\partial u}{\partial t}, v - u \right\rangle + a(u, v - u) \ge \langle f, v - u \rangle \quad \forall\, v \tag{2.24}$$

$$v \le K(u)(x, t) \text{ a.e. in } \Omega \tag{2.25}$$

$$\text{where } K(u)(x, t) = \inf_{\substack{\xi \ge 0 \\ x, x+\xi \in \Omega}} [k(\xi) + u(x + \xi, t)] \tag{2.26}$$

$$u(x, T) = 0 \text{ for } x \in \Omega. \tag{2.27}$$

More general PQVIs can be considered; for example, one can search solution of (2.22)-(2.27),

$$u \in K(u), \text{ where } K(v) = \{\varphi \in H^1(\Omega) : \varphi \leq K(v) \text{ a.e. in } \Omega\} \quad (2.28)$$

or $K(v)$ is any nonempty, closed subset of $H^1(\Omega)$ or any Hilbert space H. Several existence theorems under different conditions on $K(v)$ are mentioned in Section 4.

3 Fast Algorithm for the Bean Critical-State Model for Superconductivity

Perfect conductivity was discovered by Kamerlingh Onnes in 1911. When a superconducting material is cooled below a critical temperature, the resistivity drops sharply to zero, so that the material can carry an electric current without an associated electric field.

The energy penalty associated with the inclusion of the magnetic field leads to the existence of a critical magnetic field, $\mathbb{H}c(T)$, above which the superconductivity is destroyed and the sample reverts to its non-superconducting state. Superconductors with the properties are classified as type-I superconductors.

In type-II superconductors the inclusion of a magnetic field is only partial over a certain range of applied fields. The single critical field of type-I superconductors is replaced by a lower critical field $\mathbb{H}_c(T)$ and an upper critical field $\mathbb{H}_{c_1}(T)$.

For $\mathbb{H} > \mathbb{H}_{c_2}$ the material is in a normal state, while for $\mathbb{H} < \mathbb{H}_{c_1}$ the material is in a superconducting state. For $\mathbb{H}_{c_1} < \mathbb{H} < \mathbb{H}c_2$ the superconductor is in a mixed phase.

Let a superconductor occupy a three-dimensional spatial domain Ω with the boundary Γ. Let $\mathbb{H}$ denote the magnetic field intensity, $B = \mu_0 \mathbb{H}$, where μ_0 is the permeability of the vacuum. Let E denote the electric field intensity and J denote the current density, then

(i) $\frac{\partial B}{\partial t} + \text{curl } E = 0$ in Ω

(ii) $J = \text{curl } \mathbb{H}$ in Ω

(iii) $E = \rho\ J$ in Ω

(iv) $\text{curl } \mathbb{H} = J_e$ in $w = R \setminus \Omega$

Condition (iii) may be regarded as Ohm's law with an effective resistivity. However, since the resistivity is an auxiliary unknown, this relation, for a given current density, fixes only the possible direction of the electric field but not its magnitude.

According to the critical-state model, the current density in a superconductor cannot exceed some critical value, J_c. In the Bean model of the critical state, J_c is a constant determined by the properties of superconductive material [3, 4]. In the Kim et al. model [16, 17], the critical current density depends on the magnetic field, and various relations of the type $J_c = J_c(\mathbb{H})$ have been proposed [9].

The constraint on the current density may be written as

$$\|\text{curl } \mathbb{H}\| \leq J_c(\|\mathbb{H}\|) \text{ in } \Omega. \tag{3.1}$$

In the regions where the current density is less than critical, the vertices are pinned. Hence, there is no dissipation of energy, and the current is purely superconductive:

$$\|\text{curl } \mathbb{H}\| < J_c(\|\mathbb{H}\|) \Rightarrow \rho = 0. \tag{3.2}$$

To complete the model, the critical and boundary conditions must be specified

$$B|_{t=0} = B_0(x). \tag{3.3}$$

With div $B_0 = 0$ (together with (i)), the last condition ensures that div $B = 0$. On the boundary dividing the two media, the tangential component of electrical field E is continuous,

$$[E_\tau] = 0 \text{ on } \Gamma,$$

where [.] denotes the jump across the boundary. We neglect the surface current, which is much less than the total current in most applications of type-II superconductors. The tangential component of magnetic field $\mathbb{H}$ on this boundary is thus assumed to be continuous too:

$$[\mathbb{H}_\tau] = 0 \text{ on } \Gamma. \tag{3.4}$$

We also assume that

$$\|\mathbb{H}\| \to 0 \text{ as } |x| \to \infty. \tag{3.5}$$

The mathematical model obtained contains a system of equations and inequalities which is difficult to attack directly. Furthermore, in accordance with the postulates of the Bean model, the effective resistivity is not defined explicitly but only implicitly determined by (3.1)-(3.4). Prigozhin [29] has shown that the above physical phenomena, that is, the Kim et al. model, is equivalent to the mathematical model known as the PQVI.

If J_c is assumed to be an independent magnetic field, then the QVI reduces to an inequality where $K(h)$ is independent of h. This inequality is known as a parabolic or evolution variational inequality (PVI). This is the mathematical model of Bean model of type-II superconductivity.

Just a year before his death, Prof. J.L. Lions published a paper on parallel algorithms of PVI [23] and mentioned the study of such problems for QPVI as an open problem. He also referred some work in this area by P.L. Lions for equations.

Furati and Siddiqi [13, 14] have studied parallel algorithms of PVI and PQVI and have examined Lions's result in the context of the Bean model. Error estimation and the effect of external magnetic field on the magnetic field of the superconductor is studied, for detail see [13].

In brief, in the Bean model, the current density cannot exceed some critical value, say J_c, which is constant determined by the properties of superconducting material of type-II superconductors. In the Kim et al. model, the critical current density depends on the magnetic field. The Bean model is equivalent to a PVI and the Kim et al. model is equivalent to a QPVI. Numerical simulation of PVI is well developed, but we have very little knowledge about a numerical solution of PQVI.

(a) PVI modeling Bean's model of type-II superconductivity is given as find $h \in K_0$ such that

$$\left.\begin{array}{l} \left\langle \dfrac{\partial h}{\partial t} - f, \varphi - h \right\rangle \geq 0, \quad \forall\, \varphi \in K_{)} \\ f = -\dfrac{\partial h_0}{\partial t}, \; h_0 = B_0/\mu_0 - \mathbb{H}_e/t = 0 \\ \dfrac{\partial h}{\partial t} \in L_2(\Omega) = \{v : \Omega_t \to \Omega/v(\cdot, t) \in L_2(\Omega) \\ \Omega_t = \Omega \times (0,T)\}, \end{array}\right\} \tag{3.6}$$

where

$$K_0 = \{\varphi \in H^1(\Omega) : \|\text{curl } \varphi\| \leq J_c \text{ a.e. in } \Omega, \quad \text{curl } \varphi = 0 \text{ in } \omega = R^3 \setminus \Omega \text{ div } \varphi = 0 \text{ a.e. in } R^3\} \tag{3.7}$$

is a closed convex subset of $L_2(\Omega)$. $K_0 \subset H_0(\text{curl }; \Omega) \cap H(\text{div }, \Omega)$.

(b) **Theorem 3.1** *[13] The solution of (3.6) depends continuously on f with $f_1(0) = f_2(0) = 0$, with the correspsonding solutions h_1 and h_2 satisfying*

$$\|h_1 - h_2\|_{L^\infty(0,T,H^1(\Omega))} \leq \|f_1 - f_2\|_{L^\infty(0,T;H^1(\Omega))}. \tag{3.8}$$

Theorem 3.2 *[13] Let $\mu \in (0,1]$ be a mesh parameter and $\{H^\mu\}$ be a family of finite-dimensional subspaces of $H^1(\Omega)$ or $H(curl\,;\Omega)$ with the property that $\lim_{\mu\to 0} \|h - h^\mu\|_{H^1(\Omega)} = 0 \ \ \forall\ h \in H^1(\Omega)$.*

Let $K^\mu = H^\mu \cap K$. Then a semidiscrete internal approximation of (3.6) is given as find $H^\mu \in K^\mu$ such that

$$\left\langle \frac{\partial h^\mu}{\partial t} - f, \varphi^\mu - h^\mu \right\rangle \geq 0 \ \ \forall\ \varphi^\mu \in K^\mu. \tag{3.9}$$

Let $\dfrac{\partial h}{\partial t} - f \in L_2(\Omega)$. Then there exists a positive constant C independent of the subspace H^μ and of the set K^μ such that

$$\|h - h^\mu\| \leq C\left\{ \inf_{\varphi^\mu \in K^\mu} \left\{ \|h - \varphi^\mu\|^2 + \left\| \frac{\partial h}{\partial t} - f \right\|_{L_2(\Omega)} \|h - \varphi^\mu\| \right\} \right.$$
$$\left. + \left\| \frac{\partial h}{\partial t} - f \right\|_{L(\Omega)} \inf_{\varphi \in K} \|h^\mu - \varphi\| \right),$$

where h and h^μ are respectively solutions of (3.6) and (3.8).

We have discussed parallel algorithms of the Bean model and its finite element solution [13]. Iterative algorithms such as the SSORP–PCG Method are also developed in this chapter.

4 Quasi-Variational Inequality for the Kim Model of Type-II Superconductivity

Let a superconductor occupy an open-bounded domain in $\Omega \subset R^3$ with Lipschitz boundary Γ, and let $\omega = R^3 \setminus \overline{\Omega}$ be the space exterior to this domain. Let $\mathbb{H}, E$, J, and J_e, respectively, denote the magnetic field intensity, the electric field intensity, the density of current, and the density of external current. Let the following relations hold in Ω:

$$\frac{\partial B}{\partial t} + \text{curl } E = 0 \tag{4.1}$$

$$J = \text{curl } \mathbb{H} \tag{4.2}$$

$$B = \mu_0 \mathbb{H}, \tag{4.3}$$

where μ_0 is the permeability of the vacuum. In ω, we have

$$J_e = \text{curl } \mathbb{H} \tag{4.4}$$

$$\text{div } J_e = 0. \tag{4.5}$$

Let us define

$$U = H_0(\text{curl}, \Omega), \quad V = L_2(0, T; U). \tag{4.6}$$

For convenience, we choose $E = \rho J$ in Ω for some unknown nonnegative function $\rho(x, t)$. We observe that for any $\varphi \in U$, the boundary values φ_τ on both sides of Γ are defined in $H^{-1/2}(\Gamma; R^3)$ and $[\varphi_\tau] = 0$.

Let us define the external magnetic field $\mathbb{H}_e$ as a quasi-stationary magnetic field induced by the external current in the absence of the superconductor, that is, a solution of the problem

$$\left.\begin{array}{l} \text{curl } \mathbb{H}_e = J_e, \quad \text{div } \mathbb{H}_e = 0 \\ |\mathbb{H}_e| \to 0 \text{ as } |x| \to \infty. \end{array}\right\} \tag{4.7}$$

Since div $J_e = 0$, the problem has a unique solution [30], and it is the curl of convolution

$$\mathbb{H}_e = \text{curl } (\mathcal{G} * J_e), \tag{4.8}$$

where $\mathcal{G} = \frac{1}{4\pi|x|}$ is the Green function of the Laplace equation ($\mathcal{G} = (1/2\pi)\ln(1/|x|)$ for the two-dimensional problem). Let us introduce a new variable $h = \mathbb{H} - \mathbb{H}_e$ and define a set $K(h)$ as

$$K(h) = \{\varphi \in V : \|\text{curl } \varphi\| \leq J_c(|h + \mathbb{H}_e|) \text{ in } \Omega\}. \tag{4.9}$$

This set depends on h and h, itself belongs to $K(h)$.

Kim et al. [16, 17] found that generally the critical current density J_c (the current density cannot exceed some critical value, say, J_c) depends on the magnetic field. Prigozhin ([30], 241–243) has proved that $h(x, t)$ is a solution of the following QVI find $h \in K(h)$ such that

$$\langle \partial_t \{h + \mathbb{H}_e\}, \varphi - h \rangle \geq 0 \text{ for all } \varphi \in K(h), \tag{4.10}$$

$h|_{t=0} = h_0$, where $h_0 = B_0/\mu_0 - \mathbb{H}_e|_{t=0}$ ($B|_{t=0} = B_0$) and $K(h)$ defined by (3.9) are a family of closed convex sets of vectors.

It may be noted that the study of variational formulation approach of superconductivity problems was initiated by Bossavit [7]; see also [24, 25]. For a current review of the development of type-II superconductors, we refer to Chapman [9].

If one restricts to a stationary model in longitudinal geometry, then $\mathbb{H} = (0, 0, u(x))$, $x = (x_1, x_2) \in \Omega \subset R^2$, and Maxwell's equations (4.1) and (4.2) reduce to

$$\text{curl } E = 0 \tag{4.11}$$

$$\text{curl } \mathbb{H} = J, \tag{4.12}$$

with

$$J = \left(\frac{\partial u}{\partial x_2}, -\frac{\partial u}{\partial x_1}, 0 \right) = \nabla^* u.$$

We assume that the electric field inside the superconductor depends on the current density via a law in which E is a nonsmooth multivalued function defined on J, say, $E \in \gamma(J)$. This is regarded as an extension of Ohm's law $E = \rho J$, where the scalar resistivity $\rho = \rho(|J|)$ can be taken as a highly nonlinear function depending on $j = |J|$. In Bean's model, the current density cannot exceed some critical value j_c, hence enforcing $|\nabla^* u| = |\nabla u| \leq j_c$, where ∇u denotes curl u. In general, the critical value j_c may additionally depend on the magnitude of the magnetic field $|\mathbb{H}| = |u|$. This leads to a QVI [30] with the constitution relation E given by

$$E = \begin{cases} \rho_0 |\nabla u|^{p-2} \nabla^* u & \text{if } |\nabla u| < j_c(|u|) \\ (\rho_0 j_c^{p-2} + \lambda) \nabla^* u & \text{if } |\nabla u| = j_c(|u|), \end{cases} \tag{4.13}$$

where $1 < p < \infty, \rho_0 > 0$ is a constant, and $\lambda \geq 0$ can be regarded as a (unknown) Lagrange multiplier related to the inequality constraint acting only in the superconducting region $\{x \imath \Omega : \|\nabla u(x)\| = j_c(|u(x)|)\}$. The mathematical formulation of this problem is complete if we assume the boundary of the cross-section $\Omega \subset R^2$ to be Lipschitz, and we impose the boundary condition

$$u = u_1 \text{ on } \partial\Omega, \tag{4.14}$$

where u_1 is a given Lipschitz continuous function. Thus, the stationary magnetization of a superconductor described by (4.11)–(4.12) can be expressed as the following QVI for the magnetic field u [19] find $u \in K_1(u)$ such that

$$a(u, v - u) \geq 0 \quad \forall\, v \in K_1(u), \tag{4.15}$$

where

$$a(u,v) = \int_\Omega |\nabla u|^{p-2}\nabla u\cdot, \nabla v dx \tag{4.16}$$

and

$$K_1(u) = \left\{v \in H^{1,p}(\Omega) : v|_{\partial\Omega} = u_1, |\nabla v(x)| \leq j_c(|u(x)|) \text{ a.e. in } \Omega\right\}. \tag{4.17}$$

The following parabolic PQVI, which is a general version of parabolic quasi-variational inequality representing type-II superconductivity of the Kim et al. model, is studied by Prigozhin [29, 32], where $\rho_0 = 0$ and $h = u - H_e$ (see (4.10)): find $u(t) \in K(u)$ for a.e. $t \in [0,T], u(0) = u_0 \in K(u_0)$ such that

$$\left\langle \frac{\partial u}{\partial t}, v - u(t)\right\rangle + a(u, v-u(t)) \geq \langle f, v-u(t)\rangle \; \forall\, v \in K(u(t)) \text{ for a.e. } t \in [0,T], \tag{4.18}$$

where

$$a(u, v-u(t)) = \frac{\rho_0}{\mu}\int_\Omega |\nabla u(t))|^{p-2}\nabla u(t)\cdot\nabla(v-u(t))dx \tag{4.19}$$

$$\langle f, v-u(t)\rangle = \int_\Omega f(t)(v-u(t))dx, \tag{4.20}$$

$$\left\langle \frac{\partial u}{\partial t}, v - u(t)\right\rangle = \int_\Omega \frac{\partial u}{\partial t}(v-u(t))dx \tag{4.21}$$

$$K(u(t)) = \{v \in H_0^{1,p}(\Omega) : \|\nabla v\| \leq F(u), \quad F \text{ is a continuous real-valued function with property } F \geq m, \quad m > 0\}. \tag{4.22}$$

Existence theorems for such QVIs are mentioned in Section 5 (Theorems 5.4 and 5.5). For error estimation of (4.15)-(4.17) and (4.18)-(4.21), we refer to [13].

5 Existence of Solutions of the Kim Model

We have seen in the previous section that the magnetic field intensity of the type-II superconductor Kim model is a solution of QVI given by (4.10); for details, we refer to [30]. The existence of a solution of such QVIs has been studied in [15, 18, 19, 20, 21, 22, 26, 33, 35, 36] under different conditions on the set $K(h)$ of a multivalued function. We present here some of these existence theorems and discuss the existence of solutions of the Kim model in different situations applying these results.

Theorem 5.1 *[18] Let $\Gamma(\cdot) : R^m \to R^m, m \geq 1$ be a closed and convex multivalued function and $F : R^m \to R^m$ be continuously differentiable operator satisfying the following properties.*

(i) There exists a nonempty convex closed set $K \subset$ dom Γ such that

$$\Gamma(y) \subset K \text{ for all } y \in K, \text{ and } \bigcap_{y \in K} \Gamma(y) \neq \varphi. \tag{5.1}$$

(ii) F is strongly monotone on R^m, that is, there exists a number $\alpha > 0$ such that

$$\langle F(y_1) - F(y_2), y_1 - y_2 \rangle \geq \alpha \|y_1 - y_2\|^2 \text{ for } y_1, y_2 \in R^m. \tag{5.2}$$

(iii) For each $\xi \in R^m$,

$$\text{the mapping } y \to \text{Proj}_{\,\Gamma(y)}(\xi) \tag{5.3}$$

is continuous on an open set B containing K.

Then there exists at least one solution of the QVI find $y \in \Gamma(y)$ such that

$$\langle F(y), v - y \rangle \geq 0 \text{ for all } v \in \Gamma(y). \tag{5.4}$$

Theorem 5.2 *[20] Let H be a Hilbert space, and let $\Gamma(v) \subset H$ be nonempty, closed, and convex for $v \in H$. Further, assume that $f \in L_\infty([0, T; H)$ and $\Gamma(v)$ are Lipschitz continuous with Lipschitz constant L, $0 \leq L < 1$, $H = R^n$ or $H = L_2(R^n)$ $v_0 \in \Gamma(v_0)$. If $\Gamma(v(t))$ is a bounded subset of $H^1(\Omega)$, then the QVI find $v(t) \in \Gamma(v(t))$ such that*

$$\langle v'(t), w - v(t) \rangle \geq \langle f(t), w - v(t) \rangle \text{ for all} w \in \Gamma(v(t)), v(0) = v_0 \in \Gamma(v_0), \tag{5.5}$$

where $v = v(t) : [0, T] \to H, f : [0, T] \to H$.

It may be remarked that this theorem is a consequence of a remark preceding Lemma 2.1 [20] implying that condition (3.24) of Theorem 3.5 [20] is satisfied if $H = R^n$ or $H = L_2(R^n)$ and the underlying subset $\Gamma(v)$ is a bounded subset of $H^1(R^n)$.

Theorem 5.3 *[35] (i) Let C be a closed convex nonempty subset of $H^2(0, T;$ $H)$; let $K(\cdot) : C \to 2^H$ be a set-valued map and S be a mapping on C which associates a nonempty convex subset $K(z)$ of H with every $z \in C$; and*

$F(t) \in H^*$. (ii) Assume that the set-valued map $z \to K(z)$ is weakly continuous in the sense that for every $z_h \in C$ converging to some $z \in C$ weakly in H, the sequence of subsets $K(z_h)$ converges to $K(z)$ in H. (iii) $\bigcap_{h \in H} K(h) \neq \phi$ and S maps C into itself. Then the QVI: find $u \in K(u) \cap C$ such that

$$\left. \begin{array}{l} \langle \frac{\partial u}{\partial t}, v - u(t) \rangle + a(u, v - u(t)) \geq \langle F(t), v - u(t) \rangle \\ \text{for all } v \in K(u), u(t) \in K, u(0) = u^0 \in K(u^0) \end{array} \right\} \tag{5.6}$$

has at least one solution.

Theorem 5.4 *[19] Under the assumption $\|\nabla u\|_{L^\infty(\Omega)} < v \leq j_c$ for some $v > 0$, there exists at least one solution u of the QVI: find $u \in K_1(u)$ such that*

$$\int_\Omega |\nabla u|^{p-2} \nabla u \cdot \nabla(v - u) dx \geq 0 \quad \forall\, v \in K_1(u), \tag{5.7}$$

where

$$K_1(u) = \{v \in W^{1,p}(\Omega) : v|_{\partial\Omega} = u_1, |\nabla v(x)| \leq j_c(|u(x)|) \text{ a.e. in } \Omega\} \tag{5.8}$$

for a nonnegative continuous function j_c; for $p > 2$. If $1 < p < 2$ (resp. $p = 2$), then there exists at least one solution if j_c satisfies the 0 growth condition: $j(t) \leq c_0 + c_1 t^\alpha, t \geq 0$ with some $0 \leq \alpha < p/(2-p)$ (resp. $0 \leq \alpha < \infty$).

Theorem 5.5 *[33] Let Ω be a bounded open subset of R^3 with smooth boundary $\partial\Omega$, and let $T \in R^+, I = (0,T), Q_T = \Omega \times I, \Sigma = \partial\Omega \times I, \Omega_0 = \overline{\Omega} \times \{0\}$. Suppose that F, f, and h are given functions such that*

$$F \in C^0(R), \quad \exists\, m > 0 : F \geq m \tag{5.9}$$

$$f \in L^\infty(Q_T) : \quad f_t \in M(Q_T) \tag{5.10}$$

$$u \in H^{1,\infty}(\Omega), |\nabla u| \leq F(u) \text{ a.e. in } \Omega \quad \Delta_p u \in M(\Omega), \tag{5.11}$$

where $\Delta_p u = \nabla(|\nabla u|^{p-2} \nabla u)$ denotes the p-Laplacian $1 < p < \infty$ and $M(\Omega), M(Q_T)$ denote the spaces of bounded measures in Ω and Q_T, respectively. $M(\Omega) = [C^0(\overline{\Omega})]^*$ and $M(Q_T) = [C^0(\overline{Q}_T)]^*$, and $H_0^{1,\infty}(\Omega)$ denotes the space of Lipschitz functions that vanishes on $\partial\Omega$. Then, for $1 < p < \infty$, the following PQVI has at least a solution u belonging to $L^\infty(0,T; H_0^{1,\infty}(\Omega) \cap C^0(\overline{Q}_T)$, such that $\frac{\partial u}{\partial t} \in L^\infty(0,T; M(\Omega))$.

Find $u(t) \in K(u(t)) \cap L^\infty(\Omega)$ for a.e. $t \in I, u(0) = u_0$

$$\left. \begin{array}{c} \langle \frac{\partial u}{\partial t}, v - u(t) \rangle + a(u(t), v - u(t)) \geq \langle f(t), v - u(t) \rangle \\ \forall\, v \in K(u(t)) \text{ for a.e. } t \in I, \\ \text{where} \quad K(u(t)) = \{v \in H^{1,p}(\Omega) : \|\nabla v\| \leq F(u)\} \end{array} \right\} \tag{5.12}$$

and $a(\cdot,\cdot)$ and $\langle\cdot,\cdot\rangle$ are given by (4.19)-(4.21).

Remark 5.6 Uniqueness results for QVIs are rare and require additional technical assumptions, see, for example, [6, 18, 22, 27, 28]. If $K(u(t))$ is given by (2.17) satisfying (2.18)-(2.20) and $a(\cdot,\cdot)$ by (2.21), then the PQVI has a unique solution.

Remark 5.7 Theorem 5.4 has been applied to study the existence of a solution of stationary magnetization of a type-II superconductor described in (4.15)-(4.17).

Remark 5.8 Theorem 5.5 has been applied to the critical-state model of type-II superconductors in longitudinal geometry; for a more general situation than Theorem 5.4, see [33].

Remark 5.9 By Theorem 5.3, the Kim model (4.10) has at least one solution.

Verification. By using the definition of $K(h)$ and $\|\text{curl}\varphi\|$, it is easy to check that the hypothesis of Theorem 5.3 is satisfied, that is, $z_h \in K(h)$ $z_h \to z$ implies that $K(z_h) \to K(z)$ and $\bigcap_{h\in H_0(\text{curl }, \Omega)} K(h) \neq \phi$. It is clear that $K(h) \cap C \subseteq K(h)$ and so the PQVI (4.10) has at least one solution.

Remark 5.10 By Theorem 5.2, the Kim model (4.10) has a solution if the underlying space is $H_{c_0}(\Omega)$.

Verification. It is clear that $H_{c_0}(\Omega)$ is a finite-dimensional subspace, say of dimension p (Lemma 2.2, [5]), and $H_{c_0(\Omega)}$ is a subspace of $H_0(\text{curl}, \Omega)$. It is not difficult to verify that conditions of Theorem 5.2 are satisfied and hence the Kim model described by (4.10) has a solution on $H_{c_0}(\Omega)$.

6 Parallel Algorithms for the Kim Model

As seen in Section 4, the Kim model has been formulated as PQVIs and existence of their solutions has been shown in Section 5. Lions [23] has studied parallel algorithms for the solution of PVIs and indicated the possibility of such studies for PQVIs. In this section we present our work on this theme.

6.1 Formulation of Parabolic Quasi-Variational Inequalities

Let V and H be two real Hilbert spaces such that $V \subset H$, V dense in H, $V \to H$ being continuous. We identify H with its dual so that if V^* denotes the dual of V, then $V \subset H \subset V^*$. Let C be a nonempty closed convex subset of $L_2(0,T;V)$. Let $K(\cdot) : C \to 2^V$ be a set-valued mapping defined on C which associates a nonempty closed convex subset $K(v)$ of V with every $v \in C$; and $f \in L_2(0,T;V^*)$. We may assume that $0 \in K(u)$. One may choose $V = H^*$.

Let the set-valued map $z \to K(z)$ be weakly continuous on C, that is, for every sequence $z_h \in C$ converging to some $z \in C$ weakly in V, the sequence of subsets $K(z_h)$ converges to the set $K(z)$ in the sense that $K(z)$ in H.

(i) If $w_h \in K(u_h)$ and w_h converges weakly to some w in V, then $w \in K(u)$, where u is the limit of $\{u_h\}$.

(ii) For every $v \in K(u)$, there exists $v_h \in K(u_h)$ such that v_h converges strongly to v in V.

We now consider a bilinear form $(u,v) \mapsto a(u,v)$ which is continuous on $V \times V$. The form is assumed to be coercive (not necessarily symmetric), that is,

$$a(u,v) \geq \alpha \|u\|_V^2, \quad \alpha > 0, \ \forall\, u \in V. \tag{6.1}$$

We assume that the sets $K(v)$ for all $v \in C$ have a nonempty intersection, that is, there is at least one element, say, $u^0 \in \bigcap_{h \in V} K(h)$, and S maps C into itself, where

$$S : C \to V, w = S(z) \text{ for each fixed } z \in C, \tag{6.2}$$

and w is a unique solution of the quasi-variational problem: find $u \in K(u) \cap C$ such that

$$\left\langle \frac{\partial u}{\partial t}, v - u(t) \right\rangle + a(u, v - u(t)) \geq \langle f(t), v - u(t) \rangle$$
$$\forall\, v \in K(u), u(t) \in K(u) \cap C, u(0) = u^0 \in K(u^0).$$

We are interested in the study of parallel algorithms for solutions of PQVIs in their most general form: find u such that

$$u \in L_2(0,T;V) \cap L^\infty(0,T;H), u(t) \in K(u(t)) \text{ a.e.}$$

$$\left\langle \frac{\partial u}{\partial t}, v - u \right\rangle_H + a(u, v-u) \geq \langle f, v - u \rangle \qquad \forall\, v \in K(u) \tag{6.3}$$

$$u(0) = 0. \tag{6.4}$$

If we consider

$$v \in L_2(0,T;V), \frac{\partial v}{\partial t} \in L_2(0,T;V) \tag{6.5}$$

$$v(0) = 0 \tag{6.6}$$

on the lines of Lions [23], we can define a weak solution of (6.3)-(6.4) as a function u such that

$$u \in L_2(0,T;V), u(t) \in K(u(t)) \text{ a.e.} \tag{6.7}$$

which satisfies

$$\int_0^T \left[\left\langle \frac{\partial v}{\partial t}, v-u \right\rangle + a(u, v-u) - \langle f, v-u \rangle \right] dt \geq 0, \tag{6.8}$$

for v satisfying (3.76) and (3.77).

6.2 Decomposition Method

We introduce N couples of Hilbert spaces V_i and H_i such that

$$V_i \subset H_i \subset V_i^*, \quad i = 1, 2, \ldots, N. \tag{6.9}$$

Let $K_i(u)$ be a nonempty closed convex subset of V_i, and r_i a sequence of linear operators such that

$$r_i \in \mathcal{L}(H; H_i) \cap \mathcal{L}(V, V_i) \tag{6.10}$$

which maps $K(u)$ into $K_i(u), i = 1, 2, 3, \ldots, N$.

Let us have a family of subspaces H_{ij} such that

$$H_{ij} = H_{ji} \qquad \forall\, i, j \in [1, 2, , 3, \ldots, N] \tag{6.11}$$

and a family of operators r_{ij} such that

$$r_{ij} \in \mathcal{L}(H_j, H_{ij}). \tag{6.12}$$

The following hypotheses are made:

$$r_j r_{ji} \varphi = r_i r_{ij} \varphi \qquad \forall\, \varphi \in V, \tag{6.13}$$

and if N elements $u_i \in K_i(u)$ are given such that

$$r_{ij} u_j = r_{ji} u_i \quad \forall\, i, j, \tag{6.14}$$

then there exists $u \in K$ such that

$$u_i = r_i u, \text{ and moreover } \|u\|_V^2 \leq c \left(\sum_{i=1}^{N} \|u\|_{V_i}^2 \right). \tag{6.15}$$

Remark 6.1 (a) The special cases of $K(u)$ can be taken as in (2.17)-(2.19); in particular M of (2.17) can be taken as (2.20).

(b) Case $K_i(u) = K$ is discussed by Lions [23].

We introduce the following bilinear forms:

$$c_i(u_i, v_i) \text{ is continuous, symmetric on } H_i \times H_i,$$

which satisfies

$$c_i(u_i, v_i) \geq \gamma_i \|u_i\|_{H_i}^2, \quad \gamma_i > 0, \quad \forall\, u_i \in H_i. \tag{6.16}$$

$a_i(u_i, v_i)$ is continuous (not necessarily symmetric) on $V_1 \times V_i$ and it satisfies

$$a_i(u_i, u_i) \geq \alpha_i \|u_i\|_{V_i}^2, \alpha_i > 0 \quad \forall\, u_i \in V_i. \tag{6.17}$$

We assume that

$$\sum_{i=1}^{N} c_i(r_i u, r_i u) = \langle u, v\rangle_H \quad \forall\, u, v \in H, \tag{6.18}$$

$$\sum_{i=1}^{N} a_i(r_i u, r_i \widehat{u}) = a(u, v) \quad \forall\, u, v \in V. \tag{6.19}$$

We further assume that f is decomposed as follows: given $f_i \in L_2(0, T; V_i^*)$, we have

$$\sum_{i=1}^{N} \langle f_i, r_i \widehat{u}\rangle = \langle f, \widehat{u}\rangle \quad \forall\, \widehat{u} \in V. \tag{6.20}$$

6.3 Decomposition Approximation

Find $u_i, i = 1, 2, \ldots, N$ such that

$$c_i\left(\frac{\partial u_i}{\partial t}, v_i - u_i\right) + a_i(u_i, \widehat{u}_i - u_i) + \frac{1}{\epsilon}\sum_j \langle r_{ji} u_i$$

$$-r_{ij} u_j, r_{ji}(v_i - u_i)\rangle_{H_{ij}} \geq \langle f_i, v_i - u_i\rangle \quad \forall\, v_i \in K_i(u_i) \tag{6.21}$$

$$u_i \in L_2(0, T; V_i), u_i(t) \in K_i(u_i(t)) \text{ a.e. }, \quad u_i(0) = 0. \tag{6.22}$$

Theorem 6.2 *Let $K(\cdot)$ be as in Theorem 5.3, then the set of (decomposed) PQVIs (6.21)-(6.22) admits a solution $u_i = u_i^\epsilon (i = 1, 2, \ldots, N)$. Further, as $\epsilon \to 0$, we get*

$$u_i^\epsilon \to u_i \text{ in } L_2(0, T; V_i) \text{ locally} \tag{6.23}$$

$$\text{and} \quad u_i = r_i u, \tag{6.24}$$

where u is a solution of (6.3) or (6.8).

Theorem 6.3 *Let $K(u)$ be given by (2.17)–(2.19), then the set of (decomposed) PQVIs (6.21)–(6.22), where $a(\cdot,\cdot)$ is given by (2.21), $f \in L_+^\infty(\Omega)$, and $H = V$ is $H_0^1(\Omega)$ or $H^1(\Omega)$, admits a unique solution $u_i = u_i^\epsilon, i = 1, 2, \ldots, N$. Further, as $\epsilon \to 0$, one has*

$$u_i^\epsilon \to u_i \text{ in } L_2(0,T;V_i) \quad \text{weakly and } u_i = r_i u,$$

where u is the solution of PQVI (6.3) for this special type of $a(\cdot,\cdot)$ and $K(v)$.

Theorem 6.4 *Let $K(u)$ be given by (2.17)–(2.19), then the set of (decomposed) PQVIs (6.21)–(6.22), where $a(\cdot,\cdot)$ is given by (2.21), $f \in L_+^\infty(\Omega)$, and $H = V$ is $H_0^1(\Omega)$ or $H^1(\Omega)$, admits a unique solution $u_i = u_i^\epsilon, i = 1, 2, \ldots, N$. Further, as $\epsilon \to 0$, one has*

$$u_i^\epsilon \to u_i \text{ in } L_2(0,T;V_i) \quad \text{weakly and } u_i = r_i u,$$

where u is the solution of PQVI (6.3) for this special type of $a(\cdot,\cdot)$ and $K(v)$.

7 Superconducting Fault Current Limiters

There is currently much interest in constructing fault current limiter from high-temperature superconducting materials. These devices are used in high-power electronic systems with rapidly dangerous "fault" level currents and are used to prevent damage to equipment within the system in the time before conventional circuit breakers can operate, or before the fault subsides.

For ease of fabrication, such fault current limiters are usually made of films, where the superconductor is deposited onto a substrate.

Chapman [10] has derived a model for such a thin-film fault current limiters, and has studied its simulation.

Acknowledgment. K.M. Furati and A.H. Siddiqi would like to express their gratitude to the King Fahd University of Petroleum & Minerals for financial support via Project No. # MS/Safing Sensor/234.

References

[1] C. Baiocchi and A. Capelo, *Variational and Quasivariational Inequalities,* John Wiley and Sons Ltd., New York, (1984).

[2] G. Barnes, M. McCulloh, and D. Dew-Hughes, Computer modeling of type-II superconductors in appplications, *Supercond. Sci. Technol.,* **12** (1999), 518–522.

[3] C.P. Bean, Magnetization of hard superconductors, *Phys. Rev. Lett.,* **8** (1962), 250–253.

[4] C.P. Bean, Magnetization of high-field superconductors, *Rev. Mod. Phys.,* **36** (1964), 31–39.

[5] A. Bendali, J.M. Dominguez, and S. Gallic, A variational approach for the vector potential formulation of the Stokes and Navier-Stokes problems in three-dimensional domains, *J. Math. Anal. Appl.,* **107** (1985), 537–560.

[6] A. Bensoussan and J.L. Lions, *Impulse Control and Quasi-variational Inequalities,* Gauthier-Villars, Paris (1982).

[7] A Bossavit, Numerical modeling of superconductors in three dimensions: A model and finite element method, *IEEE Trans. Magn.,* **30** (1994), 3363–3366.

[8] M. Boulbrachene, P. Cortey-Dumont, and J.C. Miellou, L_∞-error estimates for a class of semilinear variational inequalities, *Int. J. Math. Math. Sci.,* **27–6** (2001), 309–319.

[9] S.J. Chapman, A hierarchy of models for Type-II superconductor, *SIAM Rev.,* **42** (2000), 555–598.

[10] J. Chapman, Superconductor fault current limiters, 5th ICIAM, p. 247, Sydney, 7–11, July 2003.

[11] V. Comincioli, T. Scapolla, G. Naldi, and P. Venini, A wavelet like Galerkin method for numerical solution of variational inequalities arising in elastoplasticity, *Comm. Numr. Math. Engng.,* **16** (2000), 133–144.

[12] R. Dautray and J.L. Lions, *Mathematical Analysis and Numerical Methods for Science and Technology,* Springer-Verlag, Heidelberg (1990).

[13] K.M. Furati and A.H. Siddiqi, Fast algorithms for Bean model of type-II superconductivity, *J. Numer. Funct. Anal. & Optim.,* in press.

[14] K.M. Furati and A.H. Siddiqi, Parallel for quasi-variational inequality arising in type-II superconductivity model, *J. Numer. Funct. Anal. & Optim.,* in process.

[15] B. Hanouzet and J.L. Joly, Convergence uniform des itérés definessant La solution d'inéquations quasi-variationales et application á low, *Num. Funct. Anal. and Optimi.,* **1** (1979), 399–414.

[16] Y.B. Kim, C.F. Hempstead, and A.R. Strand, Critical persistent currents in hard superconductors, *Phys. Rev. Lett.,* **9** (1962), 306–309.

[17] Y.B. Kim, C.F. Hempstead, and A.R. Strand, Magnetization and critical current, *Phys. Rev. Lett.,* **129** (1963), 528–535.

[18] M. Kocvara and J.V. Outrata, On a class of quasi-variational inequalities, *Optimi. Meth. Software,* **5** (1995), 275–295.

[19] M. Kunze and J.F. Rodrigue, An elliptic quasi-variational inequality with gradient constraints and some of its applications, **23** (2000), 897–908.

[20] M.M. Kunze and M.D.P. Marques, On parabolic quasi-varational inequalities and state dependent sweeping process, *Top. Meth. Nonlinear Anal.*, **12** (1998), 178–191.

[21] T.H. Laetsch, Uniqueness theorem for elliptic quasi-variational inequality, *Funct. Anal.*, **18** (1975), 286–288.

[22] J.L. Lions, Sur quelques questions d'analyse, de mécanique et de contrôle optimal, *Les Presses de L'université de Montréal,* (1976).

[23] J.L. Lions, Parallel algorithms for the solution of variational inequalities, *Interfaces and Free Boundaries,* **1** (1999), 3–16.

[24] M. Maslouh, F. Bouillault, A. Bossavit, and J.C. Vérité. From Bean's model to the H-M characteristic of a superconductor: Some numerical experiments, *IEEE Trans. Appl. Superconductivity,* **7** (1997), 3797–3801.

[25] M. Maslouh, F. Bouillault, A. Bossavit, and J.C. Vérité, Numerical modeling of superconductor materials using an anisotropic Kim law, *IEEE Trans. Magn.*, **34** (1998), 3064–3067.

[26] U. Mosco. Some introductory remarks on implicit variational poroblems. In A.H. Siddiqi, editor, *Recent Developments in Applicable Mathematics,* Macmillan India Limited, New Delhi-Bangalore-Madras, (1994), 1–45.

[27] J.V. Outrata and J. Zowe, A Newton method for a class of quasivariational inequalities, *Comput. Optimi. Appl.*, **4,** (1995), 5–21.

[28] J.V. Outrata and J. Zowe, Nonsmooth approach to optimization problems with equilibrium constraints, *Math. Prog.,* **68** (1995), 105–130.

[29] L. Prigozhin, The Bean model in superconductivity: Variational formulation and numerical solution, *J. Comput. Phys.,* **129** (1996), 190–200.

[30] L. Prigozhin, On the Bean critical state model in superconductivity, *European J. Appl. Math.,* **7** (1996), 237-248.

[31] L. Prigozhin, Variational model of sandpile growth, *European J. Appl. Math.,* **7** (1996), 225–235.

[32] L. Prigozhin, Analysis of critical state problems in Type-II superconductivity, *IEEE Trans. Appl. Superconductivity,* **7** (1997), 3866–3873.

[33] J.F. Rodrigues and L. Santos, A parabolic quasi-variational inequality arising in a superconductivity model, *Ann. Scuola Norm. Pisa Cl. Sci.,* **29** (2000), 153–169.

[34] A.H. Siddiqi, *Applied Functional Analysis*, Marcel Dekker, New York (2004).

[35] A.H. Siddiqi and P. Manchanda, Certain remarks on a class of evolution quasi-variational inequalities, *Int. J. Math. Sci.*, **24** (2000), 851–855.

[36] B. Ton, Time-dependent quasi-variational inequalities and nash equilibrium, *Nonlinear Anal.,* **41** (2000), 1057–1081.

Part II

Wavelet Methods for Real-World Problems

Chapter 4

WAVELET FRAMES AND MULTIRESOLUTION ANALYSIS

O. Christensen

Technical University of Denmark

Abstract

Multiresolution analysis was introduced around 1985 as a tool to construct wavelet orthonormal bases for $L^2(\mathbb{R})$ of the type $\{2^{j/2}\psi(2^jx - k)\}_{j,k\in\mathbb{Z}}$. Much later, several authors observed that some of the limitations on the possible wavelet constructions could be removed by considering frames instead of orthonormal bases. We review some of the most important results for frames constructed via various multi scale methods; in order to put the results in the right perspective we also discuss general wavelet frames.

1 Introduction

Multiresolution analysis was introduced around 1985 as a tool to construct orthonormal bases for $L^2(\mathbb{R})$ of the type $\{2^{j/2}\psi(2^jx-k)\}_{j,k\in\mathbb{Z}}$; one speaks about a *wavelet basis*, and ψ is called a *wavelet*. Already today, basic wavelet theory is well known among analysts, and a very large number of researchers work on constructing wavelets with desirable properties.

It was observed early that some of the limitations of the original theory could be removed by considering Riesz bases of the type $\{2^{j/2}\psi(2^jx - k)\}_{j,k\in\mathbb{Z}}$ rather than orthonormal bases. The purpose of this chapter is to go one step further and study the advantages of constructing *frames* rather than bases. We give the formal definition in the next section. For now, we just mention that a frame is some kind of overcomplete basis: via a frame for $L^2(\mathbb{R})$, one can express each function $f \in L^2(\mathbb{R})$ as an unconditionally convergent infinite linear combination of the frame elements, exactly as we are used to for bases. However, the coefficients in the expansion are not necessarily unique. Theoretically, this opens the opportunity to *choose* the most convenient coefficients, and we will actually demonstrate that this is particulary useful for wavelet frames.

The overcompleteness of frames has already proved useful in the context of noise compression, and its use is currently investigated in several areas of signal processing. In this chapter we concentrate on the mathematical properties. The main part is devoted to construction of frames via various extensions of the classical multiresolution analysis scheme. We believe that the reader will appreciate the results more by knowing the results valid for general wavelet frames, and for this reason we begin by a review of those results (and a few general frame results) in Sections 2 and 3; Section 2 considers results for wavelet frames with general translation and dilation parameters, while Section 3 takes a step forward to the multiresolution constructions by dealing with dyadic frames. After that we discuss frame multiresolution analysis, which is an important intermediate step between the "classical theory" and the frame constructions in the later sections. Finally, the theory founded by Ron and Shen for general frame constructions is presented in Sections 5–7.

2 Classical Frame and Wavelet Theory

2.1 General Frame Theory

Let $\mathcal{H}$ be a separable Hilbert space with the inner product $\langle \cdot, \cdot \rangle$ linear in the first entry. A countable family of elements $\{f_k\}_{k \in I}$ in $\mathcal{H}$ is a

(i) *Bessel sequence* if there exists a constant $B > 0$ such that

$$\sum_{k \in I} |\langle f, f_k \rangle|^2 \leq B||f||^2, \ \forall f \in \mathcal{H}.$$

(ii) *frame* for $\mathcal{H}$ if there exist constants $A, B > 0$ such that

$$A||f||^2 \leq \sum_{k \in I} |\langle f, f_k \rangle|^2 \leq B||f||^2, \ \forall f \in \mathcal{H}. \tag{2.1}$$

The numbers A, B in (2.1) are called *frame bounds.*

(iii) *Riesz basis* for $\mathcal{H}$ if $\overline{\text{span}}\{f_k\}_{k\in I} = \mathcal{H}$ and there exist constants $A, B > 0$ such that

$$A\sum |c_k|^2 \leq \left|\left|\sum c_k f_k\right|\right|^2 \leq B\sum |c_k|^2 \qquad (2.2)$$

for all finite sequences $\{c_k\}$.

Every orthonormal basis is a Riesz basis, and every Riesz basis is a frame (the bounds A, B in (2.2) are frame bounds). That is, Riesz bases and frames are natural tools to gain more flexibility than possible with an orthonormal basis. For an overview of the general theory for frames and Riesz bases, we refer to [9]; a deeper treatment is given in [11]. Here, we just mention that the difference between a Riesz basis and a frame is that the elements in a frame might be dependent. More precisely, a frame $\{f_k\}_{k\in I}$ is a Riesz basis if and only if

$$\sum_{k\in I} c_k f_k = 0, \ \{c_k\} \in \ell^2(I) \Rightarrow c_k = 0, \ \forall k \in I.$$

Given a frame $\{f_k\}_{k\in I}$, the associated *frame operator* is a bounded invertible operator on $\mathcal{H}$, defined by

$$Sf = \sum_{k\in I} \langle f, f_k\rangle f_k.$$

The series defining the frame operator converges unconditionally for all $f \in \mathcal{H}$. Via the frame operator we obtain the *frame decomposition*, representing each $f \in \mathcal{H}$ as an infinite linear combination of the frame elements:

$$f = SS^{-1}f = \sum_{k\in I} \langle f, S^{-1}f_k\rangle f_k. \qquad (2.3)$$

The family $\{S^{-1}f_k\}_{k\in I}$ is itself a frame, called the *canonical dual frame.* In case $\{f_k\}_{k\in I}$ is a frame but not a Riesz basis, there exist other frames $\{g_k\}_{k\in I}$ which satisfy

$$f = \sum_{k\in I} \langle f, g_k\rangle f_k, \ \forall f \in \mathcal{H}; \qquad (2.4)$$

each family $\{g_k\}_{k\in I}$ with this property is called a *dual frame.* All dual frames associated to a given frame have been characterized by Li [33].

Theorem 2.1 *Let $\{f_k\}_{k\in I}$ be a frame for $\mathcal{H}$. The dual frames of $\{f_k\}_{k\in I}$ are precisely the families*

$$\{g_k\}_{k\in I} = \left\{ S^{-1}f_k + h_k - \sum_{j\in I} \langle S^{-1}f_k, f_j\rangle h_j \right\}_{k\in I}, \tag{2.5}$$

where $\{h_k\}_{k\in I}$ is a Bessel sequence.

Note that if $\{f_k\}_{k\in I}$ is a Riesz basis, then $g_k = S^{-1}f_k$ for all choices of $\{h_k\}$.

Formula (2.3) is the main reason for considering frames, but it also immediately reveals one of the fundamental problems with frames. In fact, in order for (2.3) to be practically useful, one has to invert the frame operator, which is difficult when $\mathcal{H}$ is infinite-dimensional. One way to avoid this difficulty is to consider *tight* frames, i.e., frames $\{f_k\}_{k\in I}$ for which

$$\sum_{k\in I} |\langle f, f_k\rangle|^2 = A||f||^2, \ \forall f \in \mathcal{H} \tag{2.6}$$

for some $A > 0$. For a tight frame, $\langle Sf, f\rangle = A||f||^2$, which implies that $S = AI$, and therefore

$$f = \frac{1}{A} \sum_{k\in I} \langle f, f_k\rangle f_k, \ \forall f \in \mathcal{H}. \tag{2.7}$$

We shall not go into details with the general frame theory, but only mention a stability result, which will be used frequently; it appears in [8].

Theorem 2.2 *Let $\{f_k\}_{k\in I}$ be a frame for $\mathcal{H}$ with bounds A, B. Let $\{g_k\}_{k\in I}$ be a sequence in $\mathcal{H}$, and assume that there exist constants $\lambda, \mu \geq 0$ such that $\lambda + \frac{\mu}{\sqrt{A}} < 1$ and*

$$\left|\left|\sum c_k(f_k - g_k)\right|\right| \leq \lambda \left|\left|\sum c_k f_k\right|\right| + \mu \left(\sum |c_k|^2\right)^{1/2} \tag{2.8}$$

for all finite scalar sequences $\{c_k\}$. Then $\{g_k\}_{k\in I}$ is a frame with bounds

$$A\left(1 - (\lambda + \frac{\mu}{\sqrt{A}})\right)^2, \quad B\left(1 + \lambda + \frac{\mu}{\sqrt{B}}\right)^2.$$

Moreover, if $\{f_k\}_{k\in I}$ is a Riesz basis, then $\{g_k\}_{k\in I}$ is a Riesz basis.

2.2 Wavelet Frames

Given a function $\psi \in L^2(\mathbb{R})$ and parameters $a > 1, b > 0$, the associated *wavelet system* is the collection of functions $\{a^{j/2}\psi(a^j x - kb)\}_{j,k\in\mathbb{Z}}$. A frame of this type is called a *wavelet frame.* The definition shows that all the functions in the wavelet frame are generated by certain scalings and translations of the single function ψ, a feature which is very important in computations. A slight generalization is to consider frames generated by scaling and translating of a *finite* collection of functions $\psi_1, \ldots, \psi_n$; a frame $\{a^{j/2}\psi_\ell(a^j x - kb)\}_{j,k\in\mathbb{Z},\ell=1,\ldots,n}$ is called a *multiwavelet frame.*

Wavelet frames are typically defined in terms of properties of the Fourier transform of the generating functions. Our convention for the Fourier transform of $f \in L^1(\mathbb{R})$ is

$$\mathcal{F}f(\gamma) = \hat{f}(\gamma) = \int_{-\infty}^{\infty} f(x)e^{-2\pi i x\gamma}dx,$$

with the usual extension to $f \in L^2(\mathbb{R})$.

An immediate generalization of a result from [7], [9] to multiwavelet frames is given below.

Theorem 2.3 *Let $a > 1, b > 0$ and $\psi_1, \ldots, \psi_n \in L^2(\mathbb{R})$ be given. Suppose that*

$$B := \sup_{|\gamma|\in[1,a]} \sum_{\ell=1}^{n} \sum_{j,k\in\mathbb{Z}} |\hat{\psi_\ell}(a^j\gamma)\hat{\psi_\ell}(a^j\gamma + \frac{k}{b})| < \infty. \tag{2.9}$$

Then $\{a^{j/2}\psi_\ell(a^j x - kb)\}_{j,k\in\mathbb{Z},\ell=1,\ldots,n}$ is a Bessel sequence. If furthermore

$$A := \inf_{|\gamma|\in[1,a]} \left[\sum_{\ell=1}^{n} \sum_{j\in\mathbb{Z}} |\hat{\psi_\ell}(a^j\gamma)|^2 - \sum_{\ell=1}^{n} \sum_{k\neq 0} \sum_{j\in\mathbb{Z}} |\hat{\psi_\ell}(a^j\gamma)\hat{\psi_\ell}(a^j\gamma + \frac{k}{b})| \right] > 0, \tag{2.10}$$

then $\{a^{j/2}\psi_\ell(a^j x - kb)\}_{j,k\in\mathbb{Z},\ell=1,\ldots,n}$ is a frame for $L^2(\mathbb{R})$ with bounds $\frac{A}{b}, \frac{B}{b}$.

Let us for a moment consider wavelet systems generated by just one function ψ. If we are allowed to vary the parameters a, b, it is folklore that a large class of functions ψ generate wavelet frames, simply by choosing b sufficiently small for a given value of $a > 1$.

Proposition 2.4 *Let $\psi \in L^2(\mathbb{R})$ and $a > 1$ be given. Assume that* *(i)* $\inf_{|\gamma|\in[1,a]} \sum_{j\in\mathbb{Z}} |\hat{\psi}(a^j\gamma)|^2 > 0$,

(ii) there exists a constant $C > 0$ *such that*

$$|\hat{\psi}(\gamma)| \leq C\frac{|\gamma|}{(1+|\gamma|^2)^{3/2}} \quad a.e. \tag{2.10}$$

Then $\{a^{j/2}\psi(a^jx - kb)\}_{j,k\in\mathbb{Z}}$ *is a frame for* $L^2(\mathbb{R})$ *for all sufficiently small* $b > 0$.

Example 2.5 Let $a = 2$, and consider the function

$$\psi(x) = \frac{2}{\sqrt{3}}\pi^{-1/4}(1-x^2)e^{-\frac{1}{2}x^2}.$$

Due to its shape, ψ is called the Mexican hat. Its Fourier transform is

$$\hat{\psi}(\gamma) = -8\sqrt{\frac{2}{3}}\pi^{9/4}\gamma^2 e^{-2\pi^2\gamma^2}.$$

A numerical calculation shows that

$$\inf_{|\gamma|\in[1,2]} \sum_{j\in\mathbb{Z}} |\hat{\psi}(2^j\gamma)|^2 > 3.27.$$

Numerical calculations based on the expressions for A, B in Theorem 2.3 give that $\{2^{j/2}\psi(2^jx - kb)\}_{j,k\in\mathbb{Z}}$ is a frame if $b < 1.97$. The frame bounds A, B for some selected values for b are as follows:

b	0.25	0.5	0.75	1	1.25	1.5	1.75	1.97
A	13.1	6.55	4.36	3.26	2.33	1.25	0.422	0.0069
B	14.2	7.1	4.73	3.57	3.09	3.13	3.5	3.54

For small values of b, the frame bounds are almost identical to the values obtained in [22] via a different criterion. For large values of b the bounds above are sharper (for $b = 1.5$, the bounds given in [22] are $A = 0.325, B = 4.221$). Furthermore, the criterion used in [22] suggests that the frame property breaks down already before $b = 1.75$.

The dependence on b can be illustrated by looking at the functions

$$G_0(\gamma, b) = \frac{1}{b}\sum_{j\in\mathbb{Z}} |\hat{\psi}(a^j\gamma)|^2, \; G_1(\gamma, b) = \frac{1}{b}\sum_{k\neq 0}\sum_{j\in\mathbb{Z}} |\hat{\psi}(a^j\gamma)\hat{\psi}(a^j\gamma + k/b)|. \tag{2.12}$$

In terms of these functions, Theorem 2.3 says that $\{2^{j/2}\psi(2^jx-kb)\}_{j,k\in\mathbb{Z}}$ is a frame with lower frame bound A if

$$A := \inf_{\gamma\in[1,2]} (G_0(\gamma,b) - G_1(\gamma,b)) > 0. \tag{2.11}$$

As upper frame bound we can use

$$B = \sup_{\gamma\in[1,2]} (G_0(\gamma,b) + G_1(\gamma,b))\,.$$

Plots of these functions for various values of b are shown in [11].

2.3 Irregular Wavelet Frames

This section can be considered as a small detour and is not needed for our purpose of studying multiscale wavelet frames. We include it because it describes a topic which is mathematically interesting and challenging.

Till now we have exclusively considered translations with integer-multiples of the parameter b and dilations by $a^j, j \in \mathbb{Z}$. A more general and considerably more complicated question is:

Which conditions on a sequence $\{(\lambda_j, \mu_j)\}_{j\in I}$ in $\mathbb{R}^+ \times \mathbb{R}$ and a function $\psi \in L^2(\mathbb{R})$ imply that

$$\{\lambda_j^{1/2}\psi(\lambda_j x - \mu_j)\}_{j\in I} \text{ is a frame for } L^2(\mathbb{R})?$$

A frame of this type is called an *irregular wavelet frame*. Very few results are known about irregular wavelet frames, mainly because of their complicated structure. For example, the proof of Theorem 2.3 does not generalizes to this very general case. But if we assume that the translates are still of the type $b\mathbb{Z}$, essentially the same result with a^j replaced by λ_j holds, cf. [12]. This type of result is typical for what is known; most results are in fact, obtained by assuming that either translation or dilation is like that in the traditional wavelet setting.

Let us consider a wavelet system of the form $\{\lambda_j^{\frac{1}{2}}\psi(\lambda_j\gamma - kb)\}_{j,k\in\mathbb{Z}}$, where $b > 0$. In order for this system to be a frame, $\{\lambda_j\}_{j\in\mathbb{Z}}$ has to satisfy certain separateness conditions.

Following [13], we say that a sequence $\{\lambda_j\}_{j\in\mathbb{Z}}$ of positive numbers is *logarithmically separated by* $\lambda > 1$ if

$$|\log\lambda_j - \log\lambda_k| \geq \log\lambda, \quad \forall k \neq j.$$

If $\{\lambda_j\}$ is ordered increasingly, this is equivalent to $\frac{\lambda_{j+1}}{\lambda_j} \geq \lambda,\ \forall j \in \mathbb{Z}$.

Proposition 2.6 *Let $\psi \in L^2(\mathbb{R})$ and $\{\lambda_j\}_{j\in\mathbb{Z}}$ be a sequence in $\mathbb{R}^+$. Suppose that $\{\lambda_j^{\frac{1}{2}}\psi(\lambda_j\gamma - kb)\}_{j,k\in\mathbb{Z}}$ is a frame and that $\hat{\psi}$ is continuous in a point γ_0 where $\hat{\psi}(\gamma_0) \neq 0$. Then $\{\lambda_j\}_{j\in\mathbb{Z}}$ is a finite union of logarithmically separated sets.*

Recently, Sun and Zhou [41] obtained one of the first useful results concerning wavelet frames where both the dilation and the scaling are allowed to be irregular.

Theorem 2.7 *Let $\psi \in L^2(\mathbb{R})$ be a real-valued function for which $x\psi(x)$, $\psi'(x)$, $x\psi''(x)$ all are in $L^2(\mathbb{R})$. Assume that $\hat{\psi}(0) = 0$. Then there exist constants $a > 1, b > 0$ such that*

$$\left\{ s_{j,k}^{-1/2}\psi\left(\frac{x-\mu_{j,k}}{s_{j,k}}\right)\right\}_{j,k\in\mathbb{Z}}$$

is a frame for $L^2(\mathbb{R})$ for all sequences $\{(s_{j,k},\mu_{j,k})\}_{j,k\in\mathbb{Z}}$ for which

$$(s_{j,k},\mu_{j,k}) \in [a^j, a^{j+1}] \times [a^j bk, a^j b(k+1)], \ \ j,k \in \mathbb{Z}. \tag{2.12}$$

Theorem 2.7 can naturally be considered as a perturbation result; in fact, the constants a, b are chosen such that $\{a^{-j/2}\psi(\frac{x-a^jbk}{a^j})\}_{j,k\in\mathbb{Z}}$ is a frame, and the condition (2.12) is strong enough to guarantee that Theorem 2.2 can be applied on $\{s_{j,k}^{-1/2}\psi(\frac{x-\mu_{j,k}}{s_{j,k}})\}_{j,k\in\mathbb{Z}}$. We refer to [41] for the details.

At the moment no results about the structure of irregular wavelet frames are known. They are not associated with shift-invariant spaces, and we do not know anything about the duals (except that the dual frame exists by general frame theory).

3 Dyadic Wavelet Frames

In the rest of the chapter we will concentrate on wavelet frames with scaling parameter $a = 2$ and translation parameter $b = 1$. Several results will be formulated in terms of the translation and scaling operators defined on $L^2(\mathbb{R})$ by

$$(T_k f)(x) = f(x-k), \ k \in \mathbb{Z}, \quad (Df)(x) = 2^{1/2}f(2x).$$

A frame $\{2^{j/2}\psi(2^jx - k)\}_{j,k\in\mathbb{Z}}$ is said to be *dyadic*; in terms of the operators D^j, T_k it can be written as $\{D^jT_k\psi\}_{j,k\in\mathbb{Z}}$. We will frequently use the short notation $\{\psi_{j,k}\}_{j,k\in\mathbb{Z}}$. Correspondingly, a multiwavelet will be written $\{D^jT_k\psi_\ell\}_{j,k\in\mathbb{Z},\ell=1,\dots,n}$ or $\{\psi_{\ell;j,k}\}_{j,k\in\mathbb{Z},\ell=1,\dots,n}$.

3.1 The Duals of a Wavelet Frame

It is not difficult to construct wavelet frames via Theorem 2.3, but they are sometimes inconvenient to use. Letting S denote the frame operator associated with a frame $\{\psi_{j,k}\}_{j,k\in\mathbb{Z}}$, we have already noticed that the frame decomposition

$$f = \sum_{j,k\in\mathbb{Z}} \langle f, S^{-1}\psi_{j,k}\rangle \psi_{j,k} \tag{3.1}$$

is difficult to apply due to the presence of the operator S^{-1}. Another annoying fact is that the canonical dual frame $\{S^{-1}\psi_{j,k}\}_{j,k\in\mathbb{Z}}$ usually does not have the wavelet structure; see the example below, which appears in [22]:

Example 3.1 Let $\{\psi_{j,k}\}_{j,k\in\mathbb{Z}}$ be a wavelet orthonormal basis for $L^2(\mathbb{R})$. Given $\epsilon \in]0,1[$, we define a function θ by

$$\theta = \psi + \epsilon D\psi.$$

We want to prove that $\{\theta_{j,k}\}_{j,k\in\mathbb{Z}}$ is a Riesz basis and find the dual Riesz basis. First, direct calculation shows that

$$\psi_{j,k} - \theta_{j,k} = -\epsilon D^j T_k D\psi = -\epsilon D^{j+1} T_{2k}\psi; \tag{3.2}$$

thus, given any finite scalar sequence $\{c_{j,k}\}$,

$$\left|\left|\sum_{j,k} c_{j,k}(\psi_{j,k} - \theta_{j,k})\right|\right|^2 = \epsilon^2 \left|\left|\sum_{j,k} c_{j,k} D^{j+1} T_{2k}\psi\right|\right|^2 = \epsilon^2 \sum_{j,k} |c_{j,k}|^2.$$

Via the general perturbation result stated in Theorem 2.2 we see that $\{\theta_{j,k}\}_{j,k\in\mathbb{Z}}$ is a Riesz basis. Thus, we can define a bounded invertible operator U on $L^2(\mathbb{R})$ by $U\psi_{j,k} := \theta_{j,k}$, and the frame operator for $\{\theta_{j,k}\}_{j,k\in\mathbb{Z}}$ is $S = UU^*$. The canonical dual associated to $\{\theta_{j,k}\}_{j,k\in\mathbb{Z}}$ is

$$\{S^{-1}\theta_{j,k}\}_{j,k\in\mathbb{Z}} = \{(U^*)^{-1} U^{-1}\theta_{j,k}\}_{j,k\in\mathbb{Z}} = \{\left(U^{-1}\right)^* \psi_{j,k}\}_{j,k\in\mathbb{Z}}. \tag{3.3}$$

We want to obtain a more concrete expression for the dual. In terms of the operator U, (3.2) means that

$$(I - U)\psi_{j,k} = -\epsilon D^j D T_{2k}\psi = -\epsilon\psi_{j+1,2k};$$

expanding an arbitrary $f \in L^2(\mathbb{R})$ in the orthonormal basis $\{\psi_{j,k}\}_{j,k\in\mathbb{Z}}$ it follows that

$$(I-U)f = -\epsilon \sum_{j,k\in\mathbb{Z}} \langle f, \psi_{j,k}\rangle \psi_{j+1,2k}.$$

Direct calculation of the adjoint operator shows that

$$(I-U)^*g = (I-U^*)g = -\epsilon \sum_{j,k\in\mathbb{Z}} \langle g, \psi_{j+1,2k}\rangle \psi_{j,k}, \quad g \in L^2(\mathbb{R}). \tag{3.4}$$

In particular, $||I-U^*|| \le \epsilon < 1$, which implies that $(U^*)^{-1}$ can be expanded in a Neumann series,

$$(U^*)^{-1} = \sum_{n=0}^{\infty} (I-U^*)^n\,;$$

now, (3.3) implies that the dual Riesz basis of $\{\theta_{j,k}\}_{j,k\in\mathbb{Z}}$ is

$$\{S^{-1}\theta_{j,k}\}_{j,k\in\mathbb{Z}} = \{\sum_{n=0}^{\infty} (I-U^*)^n\, \psi_{j,k}\}_{j,k\in\mathbb{Z}}. \tag{3.5}$$

Using expression (3.4) we can go one step further. In fact, since $\{\psi_{j,k}\}_{j,k\in\mathbb{Z}}$ is an orthonormal system, the action of $(I-U)^*$ on the functions $\psi_{j,k}$, $j,k \in \mathbb{Z}$, is given by

$$(I-U^*)\psi_{j,2k} = -\epsilon\psi_{j-1,k}, \quad \text{while} \quad (I-U^*)\psi_{j,2k+1} = 0.$$

In particular,

$$S^{-1}\theta_{j,2k+1} = \psi_{j,2k+1} \text{ for all } j,k \in \mathbb{Z}.$$

Also, for $k \neq 0$ there exists a value of $n \in \mathbb{N}$ for which

$$(I-U^*)^n\, \psi_{j,k} = 0,$$

so $S^{-1}\theta_{j,k}$ is a finite linear combination of functions $\{\psi_{j,k}\}_{j,k\in\mathbb{Z}}$. For $k=0$, we have

$$S^{-1}\psi_{j,0} = \sum_{n=0}^{\infty} (-\epsilon)^n \psi_{j-n,0}. \tag{3.6}$$

In particular, the dual frame of $\{\theta_{j,k}\}_{j,k\in\mathbb{Z}}$ does *not* have the wavelet structure!

We also note that the above calculations show that there are other properties which are not inherited by the dual. For example, if we assume that the function ψ has compact support, then also θ has compact support, and all the functions $\{\theta_{j,k}\}_{j,k\in\mathbb{Z}}$ have compact support. If we look at the dual $\{S^{-1}\theta_{j,k}\}_{j,k\in\mathbb{Z}}$, then we obtain functions with compact support when $k \neq 0$ because the elements in the dual frame are finite linear combinations of the functions in $\{\psi_{j,k}\}_{j,k\in\mathbb{Z}}$ in this case. However, for $k = 0$ expression (3.6) shows that $S^{-1}\theta_{j,0}$ is *not* compactly supported.

Example 3.1 provides us with a good reason to consider tight wavelet frames, because the dual frame automatically has the right structure in this case. However, for general frames this is the point where it is natural to exploit the potential overcompleteness of frames. In other words, if $\{\psi_{j,k}\}_{j,k\in\mathbb{Z}}$ is a frame but not a Riesz basis, we know that for given $f \in L^2(\mathbb{R})$ there exist coefficients $\{c_{j,k}\}_{j,k\in\mathbb{Z}} \neq \{\langle f, S^{-1}\psi_{j,k}\rangle\}_{j,k\in\mathbb{Z}}$ such that $f = \sum_{j,k\in\mathbb{Z}} c_{j,k}\psi_{j,k}$. Thus, it is natural to ask if we can find a function $\tilde{\psi}$ such that

$$f = \sum_{j,k\in\mathbb{Z}} \langle f, \tilde{\psi}_{j,k}\rangle \psi_{j,k}, \ \forall f \in L^2(\mathbb{R}).$$

Definition 3.2 *Consider two sequences of functions $\psi_1, \ldots, \psi_n \in L^2(\mathbb{R})$, respectively, $\tilde{\psi}_1, \ldots, \tilde{\psi}_n \in L^2(\mathbb{R})$. Then $\{D^jT_k\psi_\ell\}_{j,k\in\mathbb{Z},\ell=1,\ldots,n}$ and $\{D^jT_k\tilde{\psi}_\ell\}_{j,k\in\mathbb{Z},\ell=1,\ldots,n}$ are called a pair of dual wavelet frames if both are Bessel sequences and*

$$f = \sum_{\ell=1}^{N} \sum_{j,k\in\mathbb{Z}} \langle f, D^jT_k\psi_\ell\rangle D^jT_k\tilde{\psi}_\ell, \ \forall f \in L^2(\mathbb{R}). \tag{3.7}$$

That Bessel sequences $\{D^jT_k\psi_\ell\}_{j,k\in\mathbb{Z},\ell=1,\ldots,n}$ and $\{D^jT_k\tilde{\psi}_\ell\}_{j,k\in\mathbb{Z},\ell=1,\ldots,n}$ are frames if they satisfy (3.7) follows from Cauchy-Schwartz' inequality applied to (3.7): the lower frame bound for one of the sets of functions is implied by the upper bound for the other family. A pair of dual wavelet frames is also called *sibling frames* or *bi-frames*.

Daubechies and Han give in [24] an example of a wavelet system $\{\psi_{j,k}\}_{j,k\in\mathbb{Z}}$ for which the canonical dual does not have the wavelet structure; however, there exist infinitely many functions $\tilde{\psi}$ for which $\{\tilde{\psi}_{j,k}\}_{j,k\in\mathbb{Z}}$ is a dual frame. The generator of this specific frame has the property that $\hat{\psi} = \chi_k$ for a compact subset K of $\mathbb{R}$. Sets having this property are called *wavelet frame sets* and have been investigated by, e.g., Han [30] and Dai et al. [20].

A characterization of all pairs of dual wavelet frames was obtained by Frazier et al. [29]; for convenience we remove the specification of the indices $j, k \in \mathbb{Z}, \ell = 1, \dots, n$.

Theorem 3.3 *Let $\psi_1, \dots, \psi_n, \tilde{\psi}_1, \dots, \tilde{\psi}_n \in L^2(\mathbb{R})$, and assume that $\{D^j T_k \psi_\ell\}$ and $\{D^j T_k \tilde{\psi}_\ell\}$ are Bessel sequences. Then $\{D^j T_k \psi_\ell\}$ and $\{D^j T_k \tilde{\psi}_\ell\}$ are a pair of dual wavelet frames if and only if*

$$\sum_{\ell=1}^{n} \sum_{j \in \mathbb{Z}} \hat{\psi}_\ell(2^j \gamma) \overline{\hat{\tilde{\psi}}_\ell(2^j \gamma)} = 1$$

and

$$\sum_{\ell=1}^{n} \sum_{j=0}^{\infty} \hat{\psi}_\ell(2^j \gamma) \overline{\hat{\tilde{\psi}}_\ell(2^j (\gamma + q))} = 0 \text{ for all odd integers } q.$$

3.2 Multiresolution Analysis

Orthonormal bases for $L^2(\mathbb{R})$ of the type $\{\psi_{j,k}\}_{j,k \in \mathbb{Z}}$ are traditionally constructed via *multiresolution analysis.*

Definition 3.4 *A multiresolution analysis for $L^2(\mathbb{R})$ consists of a sequence of closed subspaces $\{V_j\}_{j \in \mathbb{Z}}$ of $L^2(\mathbb{R})$ and a function $\phi \in V_0$ such that*

(i) $\cdots V_{-1} \subset V_0 \subset V_1 \cdots$.

(ii) $\cap_j V_j = \{0\}$ *and* $\overline{\cup_j V_j} = L^2(\mathbb{R})$.

(iii) $f \in V_j \Leftrightarrow Df \in V_{j+1}$.

(iv) $f \in V_0 \Rightarrow T_k f \in V_0, \ \forall k \in \mathbb{Z}$.

(v) $\{T_k \phi\}_{k \in \mathbb{Z}}$ *is an orthonormal basis for* V_0.

The property (i) is formulated by saying that the spaces V_j are nested. This is a very convenient property in, for example, approximation theory, especially when there is an easy recipe for moving around between the spaces V_j. This last property is also guaranteed by Definition 3.4 because (iii) implies that $V_j = D^j V_0$, i.e., that all the spaces V_j are scaled versions of V_0.

If we want to approximate a function $f \in L^2(\mathbb{R})$ via a multiresolution analysis, the natural starting point is to search for an approximation within a certain V_j-space. In case no element in this space approximates f well enough, we choose a larger j-value; then we obtain a better approximation, and it is taken from a space which is just a scaled version of the previous space.

Definition 3.4 is central in numerous constructions of orthonormal bases, and the topic is already well covered with many excellent books (see the references and the review [2] by Benedetto). For this reason we will not discuss this subject in detail, but only mention that the main idea is to let W_j denote the orthogonal complement of V_j in V_{j+1}; then

$$L^2(\mathbb{R}) = \bigoplus_{j\in\mathbb{Z}} W_j,$$

and one can prove that the spaces W_j satisfy the same scaling relationship as V_j, i.e., $W_j = D^j W_0$. Now one constructs a function $\psi \in W_0$ for which $\{T_k\psi\}_{k\in\mathbb{Z}}$ is an orthonormal basis for W_0, and it follows that $\{D^j T_k \psi\}_{k\in\mathbb{Z}}$ is an orthonormal basis for $L^2(\mathbb{R})$.

The classical example of a wavelet generated by a multiresolution analysis is the *Haar wavelet*,

$$\psi(x) = \begin{cases} 1 & \text{if } x \in [0, \frac{1}{2}[\\ -1 & \text{if } x \in [\frac{1}{2}, 1[\\ 0 & \text{otherwise.} \end{cases}$$

However, for applications the question is not only what is possible mathematically, but also what one needs to focus on constructions which are convenient to use. One of the shortcommings of the Haar wavelet is the missing regularity. Some of the properties which are relevant for a basis $\{2^{j/2}\psi(2^j x - k)\}_{j,k\in\mathbb{Z}}$ are

- that ψ has a computationally convenient form, for example, that ψ is a piecewise polynomial (a spline);
- regularity of ψ;
- symmetry (or anti-symmetry) of ψ, i.e., that

$$\psi(x) = \psi(-x) \text{ or } \psi(x) = -\psi(-x);$$

- compact support of ψ; and
- that ψ has *vanishing moments*, i.e., that for a certain $m \in \mathbb{N}$,

$$\int_{-\infty}^{\infty} x^\ell \psi(x)dx = 0 \text{ for } \ell = 0, 1, \ldots, m.$$

The role of vanishing moments is not immediately clear, but the following proposition shows that a large number of vanishing moments is important if we want to obtain smooth wavelets. For the proof we refer to [22].

Proposition 3.5 *Assume that $\psi \in L^2(\mathbb{R})$ is m times continuously differentiable with bounded derivatives, that $\{2^{j/2}\psi(2^j x - k)\}_{j,k\in\mathbb{Z}}$ is an orthonormal system, and that there exist constants $C, \epsilon > 0$ such that*

$$|\psi(x)| \leq \frac{C}{(1+|x|)^{1+m+\epsilon}}, \quad \forall x \in \mathbb{R}. \tag{3.8}$$

Then

$$\int_{-\infty}^{\infty} x^\ell \psi(x)dx = 0 \text{ for all } \ell = 0, 1, \ldots, m.$$

The decay condition (3.8) is automatically satisfied if ψ is a bounded function with compact support. So in this case vanishing moments are unavoidable for ψ to generate a wavelet basis and be differentiable at the same time. This is also the reason that vanishing moments play a role in Daubechies' construction of compactly supported m times continuously differentiable wavelets.

Vanishing moments are also essential in the context of compression, which we now sketch. Assuming that $\{\psi_{j,k}\}_{j,k\in\mathbb{Z}}$ is an orthonormal basis for $L^2(\mathbb{R})$, every $f \in L^2(\mathbb{R})$ has the representation

$$f = \sum_{j,k\in\mathbb{Z}} \langle f, \psi_{j,k}\rangle \psi_{j,k}. \tag{3.9}$$

All information about f is stored in the coefficients $\{\langle f, \psi_{j,k}\rangle\}_{j,k\in\mathbb{Z}}$, and (3.9) tells us how to reconstruct f based on knowledge of the coefficients. However, in practice one cannot store an infinite sequence of non-zero numbers, so one has to select a finite number of the coefficients to keep. This is usually done by *thresholding*: one chooses a certain $\epsilon > 0$ and keeps only the coefficients $\langle f, \psi_{j,k}\rangle$ for which $|\langle f, \psi_{j,k}\rangle| \geq \epsilon$. Here the vanishing moments come in again: one can prove that if ψ has a large number of vanishing moments, then only relatively few coefficients $\langle f, \psi_{j,k}\rangle$ will be large, i.e., we have obtained an efficient compression of the signal f. We refer to the paper by Beylkin, Coifman, and Rokhlin [1] for more details.

Compact support (or at least fast decay) of ψ is essential for the use of computer-based methods, where a function with unbounded support always has to be truncated. For the same reason we often want the support to be small. The condition of ψ being symmetric is less important (or even irrelevant) in many contexts, but there are cases where it is a helpful property; without going into detail, this is the case in image compression, where a nonsymmetric wavelet will generate nonsymmetric errors, which are more disturbing to the human eye than symmetric errors.

The next result, which is also proved in [22], shows that it is difficult to combine the classical multiresolution analysis with the desire of having a symmetric wavelet ψ.

Proposition 3.6 *Assume that a real-valued and compactly supported function $\phi \in L^2(\mathbb{R})$ generates a multiresolution analysis. If the associated wavelet ψ is real-valued and compactly supported and has either a symmetry axis or an anti-symmetry axis, then ψ is necessarily the Haar wavelet.*

Thus, under the above assumptions we are back at the function we want to avoid! Proposition 3.6 was one of the reasons for Cohen, Daubechies, and Feauveau to introduce *biorthogonal multiresolution analysis* [19], where one constructs a Riesz basis $\{\psi_{j,k}\}_{j,k\in\mathbb{Z}}$ and its dual $\{\tilde{\psi}_{j,k}\}_{j,k\in\mathbb{Z}}$ for $L^2(\mathbb{R})$ instead of an orthonormal basis. Note that the dual also has the wavelet structure in this particular case.

The construction in [19] allows the functions $\psi, \tilde{\psi}$ to be symmetric and compactly supported, but only one of them can be a spline. A related result by Chui and Wang [18] allows as well ψ as $\tilde{\psi}$ to be symmetric splines, but only one can have compact support. Another important step is the idea of using multiwavelets. In [27], Donovan, Geronimo, and Hardin prove that one can construct orthonormal bases of the type $\{\psi_{\ell;j,k}\}_{j,k\in\mathbb{Z},\ell=1,\dots,n}$, where the functions $\psi_1, \dots, \psi_n$ are symmetric splines with compact support.

4 Frame Multiresolution Analysis

We are now ready to extend the classical multiresolution scheme to construction of frames. There are several good reasons to do so. Let us just mention that no C^∞-function with exponential decay and bounded derivatives can generate an orthonormal basis. However, a function with those properties can very well generate a frame, as we saw in Example 2.5. Thus, it is desirable also to be able to construct overcomplete frames via multiresolution schemes.

Since we describe a generalization of multiresolution analysis, it is important to get an understanding of the relationship between the conditions in Definition 3.4. First, we note that property (v) implies that $V_0 = \overline{\text{span}}\{T_k\phi\}_{k\in\mathbb{Z}}$; and therefore property (iii) forces that

$$V_j = \overline{\text{span}}\{D^j T_k\phi\}_{k\in\mathbb{Z}},\ j \in \mathbb{Z}. \tag{4.1}$$

With this observation we can use the following fundamental result by Boor, DeVore, and Ron [6] to conclude that the first condition property in (ii) automatically is satisfied and that the second follows by a weak assumption on ϕ.

Lemma 4.1 *Let $\phi \in L^2(\mathbb{R})$, and define the spaces V_j by (4.1). Then the following holds.*

(i) $\cap_j V_j = \{0\}$.

(ii) If the spaces V_j are nested and $|\hat{\phi}| > 0$ on a neighborhood of zero, then $\cup_j V_j$ is dense in $L^2(\mathbb{R})$.

Benedetto and Li [4] defined *frame multiresolution analysis* by a straightforward extension of Definition 3.4, simply by replacing property (v) by the condition that $\{T_k\phi\}_{k\in\mathbb{Z}}$ is a frame for V_0. By our discussion of the redundancy in Definition 3.4, we can give a shorter definition.

Definition 4.2 *A function $\phi \in L^2(\mathbb{R})$ generates a frame multiresolution analysis if $\{T_k\phi\}_{k\in\mathbb{Z}}$ is a frame sequence and the spaces $\{V_j\}_{j\in\mathbb{Z}}$ defined by (4.1) satisfy the conditions*

(i) $\cdots V_{-1} \subset V_0 \subset V_1 \cdots$

(ii) $\overline{\cup_j V_j} = L^2(\mathbb{R})$.

Frame multiresolution analysis is not the most general way to obtain frames via multiscale techniques, but it provides us with a natural link from the classical constructions to the more advanced theory presented in the following sections.

Ron and Shen gave in [39] a sufficient condition for the spaces V_j in (4.1) being nested. Let $L^2(\mathbb{T})$ denote the 1-periodic functions whose restriction to $]-\frac{1}{2}, \frac{1}{2}[$ belongs to L^2, and let $L^\infty(\mathbb{T})$ denote the 1-periodic bounded functions.

Lemma 4.3 *Assume that $\phi \in L^2(\mathbb{R})$, and define the spaces V_j by (4.1). Then the following holds.*

(i) *If* $\psi \in L^2(\mathbb{R})$ *and there exists a function* $F \in L^\infty(\mathbb{T})$ *such that* $\hat{\psi}(2\gamma) = F(\gamma)\hat{\phi}(\gamma)$, *then* $\psi \in V_1$.

(ii) *If there exists a function* $H_0 \in L^\infty(\mathbb{T})$ *such that*

$$\hat{\phi}(2\gamma) = H_0(\gamma)\hat{\phi}(\gamma), \tag{4.2}$$

then $V_0 \subseteq V_1$.

An equation of the type (4.2) is called a *refinement equation*, and H_0 is a *two-scale symbol*. A sufficient condition for ϕ generating a frame multiresolution analysis was given in [4]. The corresponding result for multiresolution analysis is well known.

Theorem 4.4 *Suppose that* $\phi \in L^2(\mathbb{R})$, *that* $\{T_k\phi\}_{k\in\mathbb{Z}}$ *is a frame for* V_0, *and that* $|\hat{\phi}| > 0$ *on a neighborhood of zero. If* ϕ *satisfies a refinement equation with* $H_0 \in L^\infty(\mathbb{T})$, *then* ϕ *generates a frame multiresolution analysis.*

The principle for constructing a frame via a frame multiresolution analysis follows the classical approach: we let W_j denote the orthogonal complement of V_j in V_{j+1} and search for a function $\psi \in W_0$ such that $\{T_k\psi\}_{k\in\mathbb{Z}}$ is a frame for W_0. Combining results in [5] we have the following proposition.

Proposition 4.5 *Assume that* $\phi \in L^2(\mathbb{R})$ *generates a frame multiresolution analysis with a two-scale symbol* $H_0 \in L^\infty(\mathbb{T})$. *Let* $F \in L^\infty(\mathbb{T})$, *and define* $\psi \in V_1$ *by* $\hat{\psi}(2\gamma) = F(\gamma)\hat{\phi}(\gamma)$. *If there exist functions* $G_0, G_1 \in L^\infty(\mathbb{T})$ *such that*

$$\overline{H_0}F\Phi + T_{1/2}(\overline{H_0}F\Phi) = 0 \tag{4.3}$$

$$H_0G_0\Phi + FG_1\Phi = \Phi \tag{4.4}$$

$$T_{1/2}(H_0\Phi)G_0 + T_{1/2}(F\Phi)G_1 = 0, \tag{4.5}$$

then $\{T_k\psi\}_{k\in\mathbb{Z}}$ *is a frame for* W_0.

In the rest of the analysis, the set

$$\Gamma := \{\gamma \in]-\frac{1}{2},\frac{1}{2}[\ : \Phi(2\gamma) = 0, \Phi(\gamma) > 0, \Phi(\gamma + \frac{1}{2}) > 0\} \tag{4.6}$$

plays the key role: in case Γ is a null set we are actually able to solve (4.3), (4.4), and (4.5), and if Γ has positive Lebesgue measure no solution exists.

Theorem 4.6 *Assume that $\phi \in L^2(\mathbb{R})$ generates a frame multiresolution analysis, and define the set Γ by (4.6). Then the following holds.*

(i) If Γ has positive Lebesgue measure, there does not exist a function $\psi \in W_0$ such that $\{T_k\psi\}_{k\in\mathbb{Z}}$ is a frame for W_0.

(ii) If Γ has vanishing Lebesgue measure, then there exists a function $\psi \in W_0$ such that $\{T_k\psi\}_{k\in\mathbb{Z}}$ is a frame for W_0 and $\{D^jT_k\psi\}_{j,k\in\mathbb{Z}}$ is a frame for $L^2(\mathbb{R})$.

In order to base a frame construction on Proposition 4.5 we have to find F such that (4.3), (4.4), and (4.5) can be solved. Sometimes, different choices for F will lead to the same frame construction, but in other cases the freedom is very useful (see [10] for a detailed discussion). One consequence of the freedom is that one can construct a tight frame $\{D^jT_k\psi\}_{j,k\in\mathbb{Z}}$ via an appropriate choice of F.

Corollary 4.7 *Assume that $\phi \in L^2(\mathbb{R})$ generates a frame multiresolution analysis and that $|\Gamma| = 0$. Then there exists a function $\psi \in W_0$ such that $\{D^jT_k\psi\}_{k\in\mathbb{Z}}$ is a tight frame for $L^2(\mathbb{R})$ with the frame bound equal to one.*

A proof using an extension of Daubechies' orthonormalization trick is given in [36]; a different proof is in [10].

In light of the negative result in Theorem 4.6(i), it is interesting that we can always associate a multiwavelet frame to a frame multiresolution analysis. This was first proven by Kim and Lim in [31]; a different proof is in [10].

Theorem 4.8 *Assume that $\phi \in L^2(\mathbb{R})$ generates a frame multiresolution analysis, and let Q_j denote the orthogonal projection onto W_j. Then*

$$\{D^jT_kQ_0D\phi\}_{j,k\in\mathbb{Z}} \cup \{D^jT_kQ_0DT_1\phi\}_{j,k\in\mathbb{Z}}$$

is a multiwavelet frame for $L^2(\mathbb{R})$.

5 The Unitary Extension Principle

In this section we present results by Ron and Shen [38–40], which enable us to construct tight multiwavelet frames (an extension to general wavelet frames is in Section 7). The generators $\psi_1, \dots, \psi_n$ will be constructed on the basis of a function which satisfies a refinement equation, and since we will work with all those functions simultaneously, it is convenient to change our previous notation slightly and denote the refinable function by ψ_0.

General setup: Let $\psi_0 \in L^2(\mathbb{R})$. Assume that $\lim_{\gamma\to 0} \hat{\psi_0}(\gamma) = 1$ and that there exists a function $H_0 \in L^\infty(\mathbb{T})$ such that

$$\hat{\psi_0}(2\gamma) = H_0(\gamma)\hat{\psi_0}(\gamma). \tag{5.1}$$

Let $H_1, \ldots, H_n \in L^\infty(\mathbb{T})$, and define $\psi_1, \ldots, \psi_n \in L^2(\mathbb{R})$ by

$$\hat{\psi_\ell}(2\gamma) = H_\ell(\gamma)\hat{\psi_0}(\gamma), \; \ell = 1, \ldots, n. \tag{5.2}$$

Finally, let H denote the $(n+1) \times 2$ matrix-valued function defined by

$$H(\gamma) = \begin{pmatrix} H_0(\gamma) & T_{1/2}H_0(\gamma) \\ H_1(\gamma) & T_{1/2}H_1(\gamma) \\ \cdot & \cdot \\ \cdot & \cdot \\ H_n(\gamma) & T_{1/2}H_n(\gamma) \end{pmatrix}. \tag{5.3}$$

We will frequently suppress the dependence on γ and simply speak about the matrix H.

The purpose is to find $H_1, \ldots, H_n$ such that

$$\{D^jT_k\psi_1\}_{j,k\in\mathbb{Z}} \cup \{D^jT_k\psi_2\}_{j,k\in\mathbb{Z}} \cup \cdots \cup \{D^jT_k\psi_n\}_{j,k\in\mathbb{Z}} \tag{5.4}$$

constitutes a tight multiwavelet frame. The above explanation does not immediately reveal the relationship to multiresolution analysis; however, by defining

$$V_j = D^j\overline{\text{span}}\{T_k\psi_0\}_{k\in\mathbb{Z}}, \tag{5.5}$$

it follows from the general setup and Lemma 4.1 and Lemma 4.3 that the conditions in Definition 3.4 are satisfied, except maybe the condition (v). Thus, the general setup also leads to a multiscale analysis, with all the associated computational benefits.

The *unitary extension principle* by Ron and Shen shows that a condition on the matrix H will imply that the multiwavelet system in (5.4) is a tight frame for $L^2(\mathbb{R})$.

Theorem 5.1 *Let $\{\psi_\ell, H_\ell\}_{\ell=0,\ldots,n}$ be as in the general setup, and assume that the 2×2 matrix $H(\gamma)^*H(\gamma)$ is the identity for a.e. γ. Then the multi wavelet system $\{D^jT_k\psi_\ell\}_{j,k\in\mathbb{Z},\ell=1,\ldots,n}$ constitutes a tight frame for $L^2(\mathbb{R})$ with the frame bound equal to one.*

As an application, we show how one can construct compactly supported tight spline frames.

Example 5.2 Fix any $m = 1, 2, \ldots$, and consider the function ψ_0 defined by

$$\hat{\psi_0}(\gamma) = \frac{\sin^{2m}(\pi\gamma)}{(\pi\gamma)^{2m}}.$$

ψ_0 is known as the *B-spline of order* $2m$, and

$$\psi_0 = \chi_{[-\frac{1}{2},\frac{1}{2}]} * \chi_{[-\frac{1}{2},\frac{1}{2}]} * \cdots * \chi_{[-\frac{1}{2},\frac{1}{2}]} \quad (2m \text{ factors}).$$

It is clear that $\lim_{\gamma\to 0} \hat{\psi_0}(\gamma) = 1$, and by direct calculation,

$$\hat{\psi_0}(2\pi\gamma) = \cos^{2m}(\pi\gamma)\hat{\psi_0}(\gamma).$$

Thus, ψ_0 satisfies the refinement equation with

$$H_0(\gamma) = \cos^{2m}(\pi\gamma).$$

Let $\binom{2m}{\ell}$ denote the binomial coefficients $\frac{(2m)!}{(2m-\ell)!\ell!}$, and define the 1-periodic bounded functions $H_1, H_2, \ldots, H_{2m}$ by

$$H_\ell(\gamma) = \sqrt{\binom{2m}{\ell}} \sin^\ell(\pi\gamma)\cos^{2m-\ell}(\pi\gamma).$$

Then

$$H(\gamma) = \begin{pmatrix} H_0(\gamma) & T_{1/2}H_0(\gamma) \\ H_1(\gamma) & T_{1/2}H_1(\gamma) \\ \cdot & \cdot \\ \cdot & \cdot \\ H_n(\gamma) & T_{1/2}H_n(\gamma) \end{pmatrix}$$

$$= \begin{pmatrix} \cos^{2m}(\pi\gamma) & \sin^{2m}(\pi\gamma) \\ \sqrt{\binom{2m}{1}}\sin(\pi\gamma)\cos^{2m-1}(\pi\gamma) & -\sqrt{\binom{2m}{1}}\cos(\pi\gamma)\sin^{2m-1}(\pi\gamma) \\ \sqrt{\binom{2m}{2}}\sin^2(\pi\gamma)\cos^{2m-2}(\pi\gamma) & \sqrt{\binom{2m}{2}}\cos^2(\pi\gamma)\sin^{2m-2}(\pi\gamma) \\ \cdot & \cdot \\ \cdot & \cdot \\ \sqrt{\binom{2m}{2m}}\sin^{2m}(\pi\gamma) & \sqrt{\binom{2m}{2m}}\cos^{2m}(\pi\gamma) \end{pmatrix}.$$

Now consider the 2×2 matrix $M := H(\gamma)^* H(\gamma)$. Using the binomial formula

$$(x+y)^{2m} = \sum_{\ell=0}^{2m} \binom{2m}{\ell} x^\ell y^{2m-\ell},$$

we see that the first entry in the first row of M is

$$M_{1,1} = \sum_{\ell=0}^{2m} \binom{2m}{\ell} \sin^{2\ell}(\pi\gamma)\cos^{2(2m-\ell)}(\pi\gamma) = 1.$$

A similar argument gives that $M_{2,2} = 1$. Also,

$$\begin{aligned} M_{1,2} &= \sin^{2m}(\pi\gamma)\cos^{2m}(\pi\gamma)\left(1 - \binom{2m}{1} + \binom{2m}{2} - \cdots + \binom{2m}{2m}\right) \\ &= \sin^{2m}(\pi\gamma)\cos^{2m}(\pi\gamma)(1-1)^{2m} = 0. \end{aligned}$$

Thus, M is the identity on $\mathbb{C}^2$ for all γ; by Theorem 5.1 this implies that the $2m$ functions $\psi_1, \ldots, \psi_{2m}$ defined by

$$\begin{aligned} \hat{\psi}_\ell(\gamma) &= H_\ell(\gamma/2)\hat{\psi}_0(\gamma/2) \\ &= \sqrt{\binom{2m}{\ell}} \frac{\sin^{2m+\ell}(\pi\gamma/2)\cos^{2m-\ell}(\pi\gamma/2)}{(\pi\gamma/2)^{2m}} \end{aligned}$$

generate a multiwavelet frame for $L^2(\mathbb{R})$.

Frequently, one takes a slightly different choice of H_ℓ, namely,

$$H_\ell(\gamma) = i^\ell \sqrt{\binom{2m}{\ell}} \sin^\ell(\pi\gamma)\cos^{2m-\ell}(\pi\gamma).$$

Inserting this expression in $\hat{\psi}_\ell(\gamma) = H_\ell(\gamma/2)\hat{\psi}_\ell(\gamma/2)$ and using the commutator relations for the operators $\mathcal{F}, D, T_k$ show that ψ_ℓ is a finite linear combination with real coefficients of the functions

$$DT_k\psi_0, \; k = -m, \ldots, m.$$

It follows that ψ_ℓ is a real-valued spline with support in $[-m, m]$, degree $2m-1$, smoothness class C^{2m-2}, and knots at $\mathbb{Z}/2$. Note in particular that we obtain smoother generators by starting with higher order splines, but that the price to pay is that the number of generators increases as well.

Several interesting constructions can be based on the unitary extension principle, but there are some general limitations on the properties they can have. Let us discuss this in some detail. Assume that a tight frame $\{\psi_{\ell;j,k}\}_{j,k\in\mathbb{Z},\ell=1,\dots,n}$ is constructed via the unitary extension principle based on a refinable function ψ_0. We say that ψ_0 *provides approximation order* s if for all f in the Sobolev space $H^s(\mathbb{R})$,

$$dist(f, V_j) = O(2^{-js}).$$

Because $\{\psi_{\ell;j,k}\}_{j,k\in\mathbb{Z},\ell=1,\dots,n}$ is a tight frame, we know that for all $f \in L^2(\mathbb{R})$,

$$f = \sum_{\ell=1}^{n} \sum_{j,k\in\mathbb{Z}} \langle f, \psi_{\ell;j,k}\rangle \psi_{\ell;j,k}.$$

As an approximation of f we can thus use

$$Q_J f := \sum_{\ell=1}^{n} \sum_{j<J} \sum_{k\in\mathbb{Z}} \langle f, \psi_{\ell;j,k}\rangle \psi_{\ell;j,k}$$

for a reasonably large value of $J \in \mathbb{Z}$. We say that *the frame* $\{\psi_{\ell;j,k}\}_{j,k\in\mathbb{Z},\ell=1,\dots,n}$ *provides approximation order* s if for all $f \in H^s(\mathbb{R})$,

$$||f - Q_J f|| = O(2^{-sJ}).$$

One can prove that $Q_J \in V_J$ for all $J \in \mathbb{Z}$. Therefore, the approximation order of the frame $\{\psi_{\ell;j,k}\}_{j,k\in\mathbb{Z},\ell=1,\dots,n}$ cannot exceed the approximation order m of the underlying refinable function ψ_0. Note that in the case of a classical multiresolution analysis, where a refinable function leads to the construction of an orthonormal basis $\{D^jT_k\psi\}_{j,k\in\mathbb{Z}}$ for $L^2(\mathbb{R})$, the operator Q_J is the orthogonal projection onto V_j and the two types of approximation orders coincide; in general they might be different.

Since every implementation has to be done with a finite collection of vectors, the approximation order of $\{D^jT_k\psi_\ell\}_{j,k\in\mathbb{Z},\ell=1,\dots,n}$ is clearly important in applications: we want it to be as large as possible. If $\{D^jT_k\psi_\ell\}_{j,k\in\mathbb{Z},\ell=1,\dots,n}$ is a tight frame constructed via the unitary extension principle, one can prove (see [26]) that if the refinable function ψ_0 provides approximation order m and all functions ψ_ℓ have vanishing moments of order m_0, then the approximation order of $\{D^jT_k\psi_\ell\}_{j,k\in\mathbb{Z},\ell=1,\dots,n}$ is equal to $\min(m, 2m_0)$.

Thus, it is desirable that the functions $\{\psi_\ell\}_{\ell=1}^n$ have a reasonably large number of vanishing moments. However, there are restrictions on the number of vanishing moments one can obtain via the unitary extension principle.

Example 5.3 We return to the B-spline ψ_0 of order $2m$ considered in Example 5.2. Here, $H_0(\gamma) = \cos^{2m}(\pi\gamma)$. If we want to construct a frame via the unitary extension principle, we have to find $\{H_\ell\}_{\ell=1}^n$ such that $1 = \sum_{\ell=0}^n |H_\ell(\gamma)|^2$, i.e., such that

$$\sum_{\ell=1}^{n} |H_\ell(\gamma)|^2 = 1 - \cos^{2m}(\pi\gamma). \tag{5.6}$$

The order of the zero at $\gamma = 0$ for the function $1 - \cos^{2m}(\pi\gamma)$ is two, so also on the left-hand side of (5.6) we can only factor γ^2 out; this implies that at least one of the functions $|H_\ell|^2$ can at most have a zero at $\gamma = 0$ of order two, and, therefore, at least one of the functions ψ_ℓ can at most have one vanishing moment. Thus, the approximation order of a frame $\{D^j T_k \psi_\ell\}_{j,k\in\mathbb{Z},\ell=1,\ldots,n}$ constructed via the unitary extension principle is two.

The method presented in the following section will make it possible to obtain better approximation orders. Note that this will not be apparent from our examples, where we only consider splines of order two. A reader interested in this point is encouraged to consult [26].

6 The Oblique Extension Principle

Due to restrictions like the one just mentioned, there are certain frame constructions that seem impossible when working with a set of functions $\{\psi_0, H_0\}$ satisfying the general setup. However, sometimes another choice of those functions could lead to the unexcepted construction! An important reformulation of Theorem 5.1 was obtained by Daubechies et al. in [26]. It removes some of the constraints, in the sense that it gives a more natural recipe for construction of frames than Theorem 5.1; for example, the switch from one general setup to another is possible. It is called the *oblique extension principle:*

Theorem 6.1 *Let $\{\psi_\ell, H_\ell\}_{\ell=0}^n$ be as in the general setup. Assume that there exists a strictly positive function $\theta \in L^\infty(\mathbb{T})$ for which* $\lim_{\gamma\to 0} \theta(\gamma) =$

1 *and such that for a.e.* γ,

$$H_0(\gamma)\overline{H_0(\gamma+\nu)}\theta(2\gamma)+\sum_{\ell=1}^{n}H_\ell(\gamma)\overline{H_\ell(\gamma+\nu)}=\begin{cases}\theta(\gamma) & \text{if } \nu=0\\ 0 & \text{if } \nu=\frac{1}{2}\end{cases}. \tag{6.1}$$

Then the functions $\{D^jT_k\psi_\ell\}_{j,k\in\mathbb{Z},\ell=1,\dots,n}$ *constitute a tight frame for* $L^2(\mathbb{R})$ *with the frame bound equal to one.*

Proof. Assume that the conditions in Theorem 6.1 are satisfied, and define the function $\tilde{\psi}_0\in L^2(\mathbb{R})$ by

$$\hat{\tilde{\psi}}_0(\gamma)=\sqrt{\theta(\gamma)}\hat{\psi}_0(\gamma). \tag{6.2}$$

Define the 1-periodic functions $\tilde{H}_0,\dots,\tilde{H}_n$ by

$$\tilde{H}_0(\gamma)=\sqrt{\frac{\theta(2\gamma)}{\theta(\gamma)}}H_0(\gamma),\quad \tilde{H}_\ell(\gamma)=\sqrt{\frac{1}{\theta(\gamma)}}H_\ell(\gamma),\ \ell=1,\dots,n. \tag{6.3}$$

The idea in the proof is to apply the unitary extension principle to $\tilde{\psi}_0, \tilde{H}_0, \dots, \tilde{H}_n$ and thereby obtain a tight frame $\{D^jT_k\tilde{\psi}_\ell\}_{j,k\in\mathbb{Z},\ell=1,\dots,n}$; finally, it turns out that $\tilde{\psi}_\ell=\psi_\ell, \ell=1,\dots,n$.

We now prove that $\tilde{\psi}_0, \tilde{H}_0,\dots,\tilde{H}_n$ satisfy the conditions in the general setup. First,

$$\hat{\tilde{\psi}}_0(2\gamma)=\sqrt{\theta(2\gamma)}\hat{\psi}_0(2\gamma)=\tilde{H}_0(\gamma)\hat{\tilde{\psi}}_0(\gamma),$$

and

$$\lim_{\gamma\to 0}\hat{\tilde{\psi}}_0(\gamma)=\lim_{\gamma\to 0}\left(\sqrt{\theta(\gamma)}\hat{\psi}_0(\gamma)\right)=1.$$

Via (6.1) with $\nu=0$,

$$\sum_{\ell=0}^{n}|\tilde{H}_\ell(\gamma)|^2=\frac{\theta(2\gamma)}{\theta(\gamma)}|H_0(\gamma)|^2+\sum_{\ell=1}^{n}\frac{|H_\ell(\gamma)|^2}{\theta(\gamma)}=1,$$

so $\tilde{H}_0,\dots,\tilde{H}_n\in L^\infty(\mathbb{T})$. Because $\theta(2(\gamma+\frac{1}{2}))=\theta(2\gamma)$, we also have

$$\sum_{\ell=0}^{n}\tilde{H}_\ell(\gamma)\overline{\tilde{H}_\ell(\gamma+\frac{1}{2})}=\frac{\theta(2\gamma)}{\sqrt{\theta(\gamma)\theta(\gamma+\frac{1}{2})}}H_0(\gamma)\overline{H_0(\gamma+\frac{1}{2})}$$
$$+\frac{1}{\sqrt{\theta(\gamma)\theta(\gamma+\frac{1}{2})}}\sum_{\ell=1}^{n}H_\ell(\gamma)\overline{H_\ell(\gamma+\frac{1}{2})}=0.$$

Defining the functions $\tilde{\psi}_1, \dots, \tilde{\psi}_n$ by

$$\hat{\tilde{\psi}}_\ell(2\gamma) = \tilde{H}_\ell(\gamma)\hat{\tilde{\psi}}_0(\gamma), \ \ell = 1, \dots, n \tag{6.4}$$

it follows from Theorem 5.1 that $\{D^j T_k \tilde{\psi}_\ell\}_{j,k \in \mathbb{Z}, \ell=1,\dots,n}$ constitutes a tight frame for $L^2(\mathbb{R})$ with the frame bound equal to one. The proof is now completed by the observation that for $\ell = 1, \dots, n$,

$$\hat{\psi}_\ell(2\gamma) = H_\ell(\gamma)\hat{\psi}_0(\gamma) = \sqrt{\theta(\gamma)}\tilde{H}_\ell(\gamma)\frac{1}{\sqrt{\theta(\gamma)}}\hat{\tilde{\psi}}_0(\gamma) = \hat{\tilde{\psi}}_\ell(2\gamma),$$

which shows that $\psi_\ell = \tilde{\psi}_\ell$. ■

By taking $\theta = 1$ in Theorem 6.1 we obtain Theorem 5.1. From the extra freedom in Theorem 6.1 concerning the choice of θ, one could expect it to be a more general result than Theorem 5.1, but the proof shows that the class of frames which can be constructed is the same for the two theorems. However, in practice Theorem 6.1 gives more flexibility because it naturally leads to some constructions one would not expect from Theorem 5.1. Let us explain this in more detail. Suppose that ψ_0 is a compactly supported function satisfying (5.1) for some 1-periodic function $H_0 \in L^\infty(\mathbb{T})$ and that the functions H_ℓ, θ are chosen as trigonometric polynomials satisfying the conditions in Theorem 6.1. Writing $H_\ell(\gamma) = \sum_k c_{k\ell} e^{2\pi i k \gamma}$ (a finite sum), we see that

$$\hat{\psi}_\ell(2\gamma) = H_\ell(\gamma)\hat{\psi}_0(\gamma) = \mathcal{F}\sum_k c_{k\ell} T_{-k}\psi_0(\gamma).$$

This shows that the frame $\{D^j T_k \psi_\ell\}$ is generated by functions having compact support. The proof of Theorem 6.1 shows that the same frame can be constructed via Theorem 5.1, based on the function $\tilde{\psi}_0$ which might not be compactly supported; the fact that the resulting frame $\{D^j T_k \psi_\ell\}$ is generated by compactly supported functions is somewhat miraculous and could certainly not be predicted in advance. In short, this shows that there are constructions which appear naturally via Theorem 6.1, but one would not even think about constructing them via Theorem 5.1.

In order to apply the oblique extension principle one needs to choose the functions θ and $H_1, \dots, H_n$ simultaneously such that (6.1) is satisfied. It is not clear how to do this in general, but an extra condition on the choice of θ will make it easy to construct frames.

Corollary 6.2 *Let ψ_0 and H_0 be as in the general setup on page 91. Let θ be a strictly positive 1-periodic function, chosen such that the function*

$$\eta(\gamma) := \theta(\gamma) - \theta(2\gamma)\left(|H_0(\gamma)|^2 + |H_0(\gamma + \frac{1}{2})|^2\right) \qquad (6.5)$$

is positive as well. Let $\{G_\ell\}_{\ell=2}^n$ be trigonometric polynomials for which

$$\sum_{\ell=2}^{n} |G_\ell(\gamma)|^2 = 1, \text{ and } \sum_{\ell=2}^{n} G_\ell(\gamma)\overline{G_\ell(\gamma + \frac{1}{2})} = 0. \qquad (6.6)$$

Let ρ, σ be 1-periodic functions such that

$$|\rho(\gamma)|^2 = \theta(\gamma), \quad |\sigma(\gamma)|^2 = \eta(\gamma), \qquad (6.7)$$

and define $\{H_\ell\}_{\ell=1}^n$ by

$$H_1(\gamma) = e^{2\pi i\gamma}\rho(2\gamma)\overline{H_0(\gamma + \frac{1}{2})}, \quad H_\ell(\gamma) = G_\ell(\gamma)\sigma(\gamma), \; \ell = 2, \ldots, n.$$

Then the functions $\{\psi_\ell\}_{\ell=1}^n$ given by (5.2) generate a tight frame $\{D^jT_k\psi_\ell\}$ for $L^2(\mathbb{R})$.

In practice one would like to have as few generators as possible, and Corollary 6.2 makes it easy to obtain frames with three generators. In fact, (6.6) is satisfied with

$$G_2(\gamma) = \frac{1}{\sqrt{2}}, \quad G_3(\gamma) = \frac{1}{\sqrt{2}}e^{2\pi i\gamma}. \qquad (6.8)$$

The flexibility of the oblique extension principle is demonstrated in [26], where tight frames are obtained via some kind of interpolation between B-splines and the functions used by Daubechies in her construction of orthonormal wavelet bases with compact support.

The assumption (6.5) on the function η even implies that we can construct a frame generated by two functions.

Corollary 6.3 *Let ψ_0 and H_0 be as in the general setup on page 91. Let θ be a strictly positive 1-periodic function, chosen such that the function η in (6.5) is positive as well. Define the functions ρ, σ as in (6.7), and let*

$$H_1(\gamma) = e^{2\pi i\gamma}\rho(2\gamma)\overline{H_0(\gamma + \frac{1}{2})}, \quad H_2(\gamma) = H_0(\gamma)\sigma(2\gamma). \qquad (6.9)$$

Then the functions $\{\psi_\ell\}_{\ell=1}^2$ given by (5.2) generate a tight frame $\{D^jT_k\psi_\ell\}_{j,k\in\mathbb{Z},\ell=1,2}$ for $L^2(\mathbb{R})$.

Other constructions of multiwavelet frames with few generators were, in fact, previously given by Chui and He [15] and Petukhov [37]. In particular they proved that the general setup together with the assumption

$$|H_0(\gamma)|^2 + |H_0(\gamma + \frac{1}{2})|^2 \leq 1$$

always make it possible to construct a frame with two generators; also, if H_0 is a polynomial of degree m, one can choose H_1 and H_2 as polynomials of degree at most m. Petukhov also describes all solutions to the matrix equation $H(\gamma)^* H(\gamma) = I$.

An interesting result by Chui, He, and Stöckler [16] excludes the option of having a single generator in many cases; for example, there does not exist dual wavelet frame pairs $\{D^j T_k \psi\}_{j,k \in \mathbb{Z}}, \{D^j T_k \tilde{\psi}\}_{j,k \in \mathbb{Z}}$ for which ψ is a finite linear combination of splines $B_m(2x - k),\ k \in \mathbb{Z}$. Multiwavelet frames with generators of this type were obtained in Example 5.2.

Note that if θ and H_0 are trigonometric polynomials, then η defined in (6.5) is also a trigonometric polynomial. The assumption that θ and η are positive implies by spectral factorization (see [22]) that we can choose ρ, σ to be trigonometric polynomials. Thus, the generators ψ_ℓ in the above corollaries are finite linear combinations of functions $DT_k\psi_0,\ k \in \mathbb{Z}$.

Let us return to B-splines.

Theorem 6.4 *Let B_m denote the spline of order $m \in \mathbb{N}$ and $H_0(\gamma) = \cos^m(\pi\gamma)$ the associated two-scale symbol. Then there exists a trigonometric polynomium such that the function η in* (6.5) *is positive.*

Theorem 6.4 is proved in [26]. It follows that we can construct multiwavelet frames with two or more generators based on *any* B-spline B_m. Let us, for example, consider Corollary 6.3. If we choose the functions ρ, σ in (6.7) to be trigonometric polynomials, then the functions H_1, H_2 in (6.9) are trigonometric polynomials, which imply that the associated frame generators ψ_1, ψ_2 are finite linear combinations of terms $B_m(2x-k), k \in \mathbb{Z}$. By choosing m large enough, we can thus obtain generators belonging to any prescribed smoothness class $C^N(\mathbb{R})$. In contrast with what we obtained for applications of the unitary extension principle in Example 5.2, the number of generators is *not* forced to grow with the desired smoothness.

We give an example on frame constructions via Theorem 5.1 and Theorem 6.1.

Example 6.5 Let $\psi_0 = T_1 B_2$, where B_2 is the B-spline of order two considered in Example 6.5; it is refinable with the two-scale symbol

$$H_0(\gamma) = \frac{(1+e^{-2\pi i\gamma})^2}{4} = e^{-2\pi i\gamma}\cos^2(\pi\gamma).$$

(i) Defining

$$H_1(\gamma) = ie^{-2\pi i\gamma}\sqrt{2}\sin(\pi\gamma)\cos(\pi\gamma) = \frac{\sqrt{2}}{4}(1-e^{-4\pi i\gamma}),$$

$$H_2(\gamma) = -e^{-2\pi i\gamma}\sin^2(\pi\gamma) = \frac{(1-e^{-2\pi i\gamma})^2}{4}, \tag{6.10}$$

the unitary extension principle shows that the associated functions $\psi_1^{(i)} := \psi_1$ and ψ_2 generate a tight frame for $L^2(\mathbb{R})$; direct calculation gives that

$$\psi_1^{(i)}(x) = \frac{1}{\sqrt{2}}(B_2(2x-1) - B_2(2x-3)) \tag{6.11}$$

$$\psi_2(x) = \frac{1}{2}\left(B_2(2x-1) - 2B_2(2x-2) + B_2(2x-3)\right). \tag{6.12}$$

(ii) An alternative construction can be obtained via the oblique extension principle. Let

$$\theta(\gamma) := \frac{4-\cos(2\pi\gamma)}{3}. \tag{6.13}$$

In this example we keep the choice of H_2 in (6.10); thus, if we want to use the oblique extension principle, we have to choose H_1 such that the two conditions in (6.1) are satisfied. That is, we require that

$$|H_1(\gamma)|^2 = \theta(\gamma) - |H_0(\gamma)|^2\theta(2\gamma) - |H_2(\gamma)|^2,$$

$$H_1(\gamma)\overline{H_1(\gamma+\tfrac{1}{2})} = -H_0(\gamma)\overline{H_0(\gamma+\tfrac{1}{2})}\theta(2\gamma) - H_2(\gamma)\overline{H_2(\gamma+\tfrac{1}{2})}.$$

Direct computation gives the equations

$$|H_1(\gamma)|^2 = \frac{1}{6}(\cos(2\pi\gamma)+2)^2(\cos(2\pi\gamma)-1)^2,$$

$$H_1(\gamma)\overline{H_1(\gamma+\tfrac{1}{2})} = \frac{1}{6}(\cos(2\pi\gamma)+2)(\cos(2\pi\gamma)-2) \times(\cos(2\pi\gamma)-1)(\cos(2\pi\gamma)+1).$$

Those equations are satisfied if we let

$$\begin{aligned}H_1(\gamma) &= \frac{1}{\sqrt{6}}(\cos(2\pi\gamma)+2)(\cos(2\pi\gamma)-1)\\ &= \frac{1}{4\sqrt{6}}(e^{4\pi i\gamma}+e^{-4\pi i\gamma}+2e^{2\pi i\gamma}+2e^{-2\pi i\gamma}-6).\end{aligned}$$

This leads to the function ψ_1 given by

$$\begin{aligned}\hat{\psi}_1(\gamma) &= H_1(\gamma/2)\hat{\psi}_0(\gamma/2)\\ &= \frac{1}{4\sqrt{3}}\mathcal{F}\left(T_{3/2}+T_{-1/2}+2T_1+2-6T_{1/2}\right)DB_2(\gamma).\end{aligned}$$

Thus,

$$\begin{aligned}\psi_1(\gamma) = &\frac{1}{2\sqrt{6}}\left(B_2(2\gamma-3)+B_2(2\gamma+1)+2B_2(2\gamma-2)\right)\\ &+\frac{1}{2\sqrt{6}}\left(2B_2(2\gamma)-6B_2(2\gamma-1)\right).\end{aligned}$$

This function has support on $[-1,2]$. Instead of taking this generator, we take

$$\begin{aligned}\psi_1^{(ii)}(\gamma) &:= \psi_1(\gamma-1)\\ &= \frac{1}{2\sqrt{6}}\left(B_2(2\gamma-5)+B_2(2\gamma-1)+2B_2(2\gamma-4)\right)\\ &\quad+\frac{1}{2\sqrt{6}}\left(2B_2(2\gamma-2)-6B_2(2\gamma-3)\right),\end{aligned}$$

which generates the same wavelet system and has support on $[0,3]$.

(iii) Systematic constructions with two generators can be given via Corollary 6.3. Define again θ by (6.13). Then, the function η in (6.5) is

$$\eta(\gamma) = \frac{2}{3}(8\cos^4(\pi\gamma)+1)(\cos(\pi\gamma)-1)^2(\cos(\pi\gamma)+1)^2. \tag{6.14}$$

Since $\eta(\gamma)\geq 0$ for all γ, the conditions in Corollary 6.3 are satisfied. The remaining work consists of extracting square roots ρ, resp. σ of θ resp. η. This is rather cumbersome and will not be done here.

7 Construction of Dual Wavelet Pairs

So far the constructions via the extension principles have concerned tight frames. However, the technique is more far reaching, and one can actually extend the results and construct dual wavelet pairs. We cite a result from [26], which was (in a slightly different form) also obtained by Chui, He, and Stöckler [16]:

Theorem 7.1 *Let $\{H_\ell, \psi_\ell\}_{\ell=0}^n$ and $\{\tilde{H}_\ell, \tilde{\psi}_\ell\}_{\ell=0}^n$ be two sets of functions satisfying the conditions in the general setup on page 91, and such that for some $C>0$ and $\rho > \frac{1}{2}$,*

$$|\hat{\psi}_0(\gamma)|, |\hat{\tilde{\psi}}_0(\gamma)| \leq \frac{C}{(1+|\gamma|)^\rho}, \quad a.e. \tag{7.1}$$

Assume that there exists a function $\theta \in L^\infty(\mathbb{T})$ such that $\lim_{\gamma \to 0} \theta(\gamma) = 1$ and

$$H_0(\gamma)\overline{\tilde{H}_0(\gamma+\nu)}\theta(2\gamma) + \sum_{\ell=1}^n H_\ell(\gamma)\overline{\tilde{H}_\ell(\gamma+\nu)} = \begin{cases} \theta(\gamma) & \text{if } \nu = 0 \\ 0 & \text{if } \nu = \frac{1}{2}. \end{cases} \tag{7.2}$$

Then $\{D^jT_k\psi_\ell\}_{j,k\in\mathbb{Z},\ell=1,\dots,n}$ and $\{D^jT_k\tilde{\psi}_\ell\}_{j,k\in\mathbb{Z},\ell=1,\dots,n}$ are a pair of dual wavelet frames.

The decay condition (7.1) is stronger than necessary, but on the other hand, it is weak enough to be satisfied for almost all interesting constructions. To find a pair of dual frames via Theorem 7.1 is much easier than to construct tight frames via the oblique extension principle because the function θ is not assumed to be positive. It also has great advantages compared to frame constructions via Theorem 2.3: we avoid the cumbersome inversion of the frame operator and the problem that the canonical dual of a wavelet frame might not have the wavelet structure.

Similar to what we saw for the oblique extension principle, we can use Theorem 7.1 to construct frames with three generators. For a given real-valued function $\theta \in L^\infty(\mathbb{T})$, we define the function

$$\eta(\gamma) := \theta(\gamma) - \theta(2\gamma)\left(H_0(\gamma)\overline{\tilde{H}_0(\gamma)} + H_0(\gamma+\frac{1}{2})\overline{\tilde{H}_0(\gamma+\frac{1}{2})}\right).$$

We will assume that η is real-valued and has a zero of order at least two at the origin. We are not assuming positivity of θ and η, so we cannot just copy our choices of H_ℓ in Corollary 6.2. Instead, we choose two real-valued functions η_1, η_2 with period 1 such that

$$\eta = 2\eta_1\eta_2, \text{ and } \eta_1(0) = \eta_2(0) = 0. \tag{7.3}$$

Also, noting that $\gamma \to \theta(2\gamma)$ has period $\frac{1}{2}$, we choose two $\frac{1}{2}$-periodic and real-valued functions θ_1, θ_2 such that

$$\theta(2\gamma) = \theta_1(\gamma)\theta_2(\gamma). \tag{7.4}$$

Corollary 7.2 *Let $\{\psi_0, H_0\}$ and $\{\tilde{\psi}_0, \tilde{H}_0\}$ be as in the general setup. Assume that θ satisfies the above conditions, and define $\{H_\ell\}_{\ell=1}^3$ and $\{\tilde{H}_\ell\}_{\ell=1}^3$ by*

$$\begin{aligned} H_1(\gamma) &= e^{2\pi i\gamma}\theta_1(\gamma)\overline{\tilde{H}_0(\gamma+\frac{1}{2})}, & \tilde{H}_1(\gamma) &= e^{2\pi i\gamma}\tilde{\theta}_1(\gamma)\overline{H_0(\gamma+\frac{1}{2})} \\ H_2(\gamma) &= \eta_1(\gamma), & \tilde{H}_2(\gamma) &= \eta_2(\gamma) \\ H_3(\gamma) &= e^{2\pi i\gamma}\eta_1(\gamma), & \tilde{H}_3(\gamma) &= e^{2\pi i\gamma}\eta_2(\gamma). \end{aligned}$$

Then, if the associated functions $\{D^jT_k\psi_\ell\}_{j,k\in\mathbb{Z},\ell=1,2,3}$ and $\{D^jT_k\tilde{\psi}_\ell\}_{j,k\in\mathbb{Z},\ell=1,2,3}$ are Bessel sequences, they constitute a pair of dual wavelet frames.

As for the oblique extension principle, the number of generators can be reduced to two by defining

$$\begin{aligned} H_1(\gamma) &= e^{2\pi i\gamma}\theta_1(\gamma)\overline{\tilde{H}_0(\gamma+\frac{1}{2})}, & \tilde{H}_1(\gamma) &= e^{2\pi i\gamma}\tilde{\theta}_1(\gamma)\overline{H_0(\gamma+\frac{1}{2})} \\ H_2(\gamma) &= H_0(\gamma)\eta_1(2\gamma), & \tilde{H}_2(\gamma) &= \eta_2(2\gamma)\tilde{H_0}(\gamma). \end{aligned}$$

We have assumed the factorizations of $\theta(2\cdot)$ and η to be real-valued. This is not strictly necessary. However, a standard application is to the case where $\theta, H_0, \tilde{H}_0$ are trigonometric polynomial, and in this case we will also choose η_1, η_2 and θ_1, θ_2 to be trigonometric polynomials. Assuming that they are real-valued implies that the frame generators $\{\psi_\ell\}_{\ell=1}^3$ and $\{\tilde{\psi}_\ell\}_{\ell=1}^3$ are symmetric if the refinable functions ψ_0 and $\tilde{\psi}_0$ are symmetric real-valued functions. Thus, the above process will lead to symmetric dual wavelet pairs when applied to even order B-splines.

That it is much easier to construct general frames than their tight counterparts is evident from Example 6.5(iii). In fact, if we base the choices of H_1, H_2 and $\tilde{H}_1, \tilde{H}_2$ on the same refinable function, namely, $\psi_0 = \tilde{\psi}_0 = T_1B_2$, and we again define θ by (6.13), then we need to find functions $\eta_1, \eta_2, \theta_1, \theta_2$ satisfying (7.3) and (7.4) with η given by (6.14), but this is easy. We leave it to the reader to give concrete constructions.

Let us finally mention that Daubechies and Han in [24] proved that based on any two refinable functions with compact support, one can construct a pair of dual wavelet frame having generators with compact support.

Acknowledgment. The author would like to thank Professor Abul Hasan Siddiqi for the invitation to include this chapter.

References

[1] G. Beylkin, R. Coifman, and V. Rokhlin, Fast wavelet transforms and numerical algorithms, *Comm. Pure Appl. Math.*, **44** (1991), 141–183.

[2] J. Benedetto, Ten books on wavelets, *SIAM Rev.*, **42**(1)1 (2000), 127–138.

[3] J. Benedetto and S. Li, Subband coding and noise reduction in frame multiresolution analysis, In *Proceedings of SPIE Conference on Mathematical Imaging*, San Diego, July 1994.

[4] J. Benedetto and S. Li, The theory of multiresolution analysis frames and applications to filter banks, *Appl. Comp. Harm. Anal.*, **5** (1998), 389–427.

[5] J. Benedetto and O. Treiber, Wavelet frames: Multiresolution analysis and extension principles, Chapter 1, In *Wavelet Transforms and Time-Frequency Signal Analysis*, Edited by L. Debnath, Birkhäuser, Boston, 2001, pp. 33–36.

[6] C. Boor, R. DeVore, and A. Ron, On the construction of multivariate (pre)wavelets, *Constr. Approx.*, **9** (1993), 123–166.

[7] P.G. Casazza and O. Christensen, Weyl-Heisenberg frames for subspaces of $L^2(\mathbb{R})$, *Proc. Amer. Math. Soc.*, **129** (2001), 145–154.

[8] O. Christensen, A Paley-Wiener theorem for frames, *Proc. Amer. Math. Soc.*, **123** (1995), 2199–2202.

[9] O. Christensen, Frames, bases, and discrete Gabor/wavelet expansions, *Bull. Amer. Math. Soc.*, **38** no.3 (2001), 273–291.

[10] O. Christensen, On frame multiresolution analysis, *Arabian J. Sci. Engg.*, **28**(1C) (2003), 59–72 (special issue on wavelets).

[11] O. Christensen, *An Introduction to Frames and Riesz Bases*, Birkhäuser, Boston, 2003.

[12] O. Christensen, S. Favier, and F. Zó, Irregular wavelet frames and Gabor frames, *Appr. Theory Appl.*, **17**(3) (2001), 90–101.

[13] O. Christensen and A. Lindner, Lower bounds for finite Gabor and wavelet systems, *Appr. Theory Appl.*, **17**(1) (2001), 18–29.

[14] C. Chui, *Wavelets - A Tutorial in Theory and Practice*, Academic Press, New York, 1992.

[15] C. Chui and W. He, Compactly supported tight frames associated with refinable functions, *Appl. Comp. Harm. Anal.*, **8** (2000), 293–319.

[16] C. Chui, W. He, and J. Stöckler, Compactly supported tight and sibling frames with maximum vanishing moments, *Appl. Comp. Harm. Anal.*, **13** (2002), 224–262.

[17] C. Chui and X. Shi, Bessel sequences and affine frames, *Appl. Comp. Harm. Anal.*, **1** (1993), 29–49.

[18] C. Chui and J. Wang, On compactly supported spline wavelets and a duality principle, *Trans. Amer. Math. Soc.*, **330** (1992), 903–915.

[19] A. Cohen, I. Daubechies, and J.-C. Feauveau, Biorthogonal bases of compactly supported wavelets, *Comm. Pure Appl. Math.*, (1993), 485-560.
[20] X. Dai, Y. Diao, and Q. Gu, Frame wavelet sets in $\mathbb{R}$, *Proc. Amer. Math. Soc.*, **129**(7) (2000), 2045–2055.
[21] I. Daubechies, The wavelet transformation, time-frequency localization and signal analysis, *IEEE Trans. Inform. Theory*, **36** (1990), 961–1005.
[22] I. Daubechies, *Ten Lectures on Wavelets*, SIAM Conf. Series in Applied Math, Boston, 1992.
[23] I. Daubechies, A. Grossmann, and Y. Meyer, Painless nonorthogonal expansions, *J. Math. Phys.*, **27** (1986), 1271–1283.
[24] I. Daubechies and B. Han, The canonical dual of a wavelet frame, *Appl. Comp. Harm. Anal.*, **12**(3) (2002), 269–285.
[25] I. Daubechies and B. Han, Pairs of dual wavelet frames from any two refinable functions, *Preprint*, 2001.
[26] I. Daubechies, B. Han, A. Ron, and Z. Shen, Framelets: MRA-based constructions of wavelet frames, *Appl. Comp. Harm. Anal.*, **14**(1) (2003), 1–46.
[27] G. Donovan, J.S. Geronimo, and D.P. Hardin, Intertwining multiresolution analyses and the construction of piecewise-polynomial splines, *SIAM J. Math. Anal.*, **27**(6) (1996), 1791–1815.
[28] M. Frazier, *An Introduction to Wavelets Through Linear Algebra*, Springer-Verlag, Heidelberg, 2001.
[29] M. Frazier, G. Garrigos, K. Wang, and G. Weiss, A characterization of functions that generate wavelet and related expansion, *J. Fourier Anal. Appl.*, **3** (1997), 883–906.
[30] B. Han, On dual wavelet tight frames, *Appl. Comp. Harm. Anal.*, **4**(4) (1997), 380–413.
[31] H.O. Kim and J.K. Lim, On frame wavelets associated with frame multiresolution analysis, *Appl. Comp. Harm. Anal.*, **10**(1) (2001), 61–70.
[32] H.O. Kim and J.K. Lim, Frame multiresolution analysis, *Comm. Korean Math. Soc.*, **15** (2000), 285–308.
[33] S. Li, On general frame decompositions, *Numer. Funct. Anal. Optimiz.*, **16**(9 & 10) (1995), 1181–1191.
[34] S. Mallat, *A Wavelet Tour of Signal Processing*, Academic Press, New York, 1999.
[35] M. Meyer, *Wavelets and Operators*, Herman, Paris.
[36] M. Paluszynski, H. Sikic, G. Weiss, and S. Xiao, Generalized low pass filters and MRA frame wavelets, *J. Geometric Anal.*, **11**(2) (2001), 311–342.
[37] A. Petukhov, Explicit construction of framelets, *Appl. Comp. Harm. Anal.*, **11** (2001), 313–327.
[38] A. Ron and Z. Shen, Affine systems in $L_2(\mathbb{R}^d)$: The analysis of the analysis operator, *J. Funct. Anal.*, **148** (1997), 408–447.
[39] A. Ron and Z. Shen, Affine systems in $L_2(R^d)$ II: Dual systems, *J. Fourier Anal. Appl.*, **3** (1997), 617–637.
[40] A. Ron and Z. Shen, Compactly supported tight affine spline frames in $L_2(R^d)$, *Math. Comp.*, **67** (1998), 191–207.

[41] W. Sun and X. Zhou, Irregular wavelet/Gabor frames, *Appl. Comp. Harm. Anal.*, **13**(1) (2002), 63–76.

[42] D. Walnut, *An Introduction to Wavelet Analysis,* Birkhäuser, Boston, 2001.

[43] P. Wojtaszczyk, *A Mathematical Introduction to Wavelets*, Cambridge University Press, London; New York, 1999.

Chapter 5

COMPARISON OF A WAVELET-GALERKIN PROCEDURE WITH A CRANK-NICOLSON-GALERKIN PROCEDURE FOR THE DIFFUSION EQUATION SUBJECT TO THE SPECIFICATION OF MASS

S.H. Behiry
Mansoura University

J.R. Cannon
University of Central Florida

H. Hashish
Mansoura University

A. I. Zayed
DePaul University

Abstract

A Wavelet-Galerkin procedure on a bounded interval is compared with a Crank-Nicolson-Galerkin procedure for the initial-boundary value problem for the diffusion equation subject to the specification of mass. The Wavelet-Galerkin procedure is derived, and computations of several example problems are reported and compared to the results of the computations with a Crank-Nicholson-Galerkin procedure in a previous paper of one of the authors. Overall, the Wavelet-Galerkin procedure competes well with the Crank-Nicolson-Galerkin procedure.

1 Introduction

In [19], Cannon, Perez-Esteva, and van der Hoek considered a Galerkin procedure for obtaining numerical approximations to the solution $U = U(x,t)$ which satisfies

$$U_t = U_{xx}, \quad 0 < x < 1, \quad 0 < t \le T, \tag{1.1a}$$

$$U(x,0) = f(x), \quad 0 < x < 1, \tag{1.1b}$$

$$U_x(1,t) = g(t), \quad 0 < t \le T \tag{1.1c}$$

$$M(t) = \int_0^b U(x,t)dx, \quad 0 < b < 1, \tag{1.1d}$$

where f, g, and M were assumed sufficiently smooth to guarantee a unique smooth solution U. The subscripts x and t denote the respective partial differentiation, see [12, 13, 14]. A continuous Galerkin approximation was derived. Namely, for arbitrary $\psi(x,t)$ in a finite dimensional vector space S_N, we solve for $W(x,t) \in S_N$ so that

$$\begin{aligned}(\psi, W_t) + (\psi_x, W_x) = {} & \psi(1,t)g(t) + \psi(0,t)M'(t) - \psi(0,t)D(t) \\ & + \psi(0,t)\int_0^t 2\frac{\partial^2\theta}{\partial x^2}(b, t-\tau)W(0,\tau)d\tau \\ & - \psi(0,t)\int_0^t 2\frac{\partial^2\theta}{\partial x^2}(b-1, t-\tau)W(1,\tau)d\tau \text{ for } \psi \in S_N,\end{aligned} \tag{1.2}$$

where

$$(\psi, \varphi) = \int_0^1 \psi(x,t)\varphi(x,t)dx, \tag{1.3}$$

$$\begin{aligned}D(t) = {} & 2\theta(b,t)f(0) - \{\theta(b-1,t) + \theta(b+1,t)\}f(1) \\ & + \int_0^1 \{\theta(b-\xi,t) + \theta(b+\xi,t)\}f'(\xi)d\xi\end{aligned} \tag{1.4}$$

$$\theta(x,t) = \sum_{n=-\infty}^{\infty} K(x+2m,t), \quad t > 0, \tag{1.5}$$

and

$$K(x,t) = \frac{1}{2\sqrt{\pi t}} \exp\left(\frac{-x^2}{4t}\right), \quad t > 0. \tag{1.6}$$

Error estimates were obtained for the continuous Galerkin procedure. A Crank-Nicolson-Galerkin approximation scheme was derived from the continuous Galerkin method, and error estimates for it were obtained. Finally, some numerical experiments were performed to test the Crank-Nicolson-Galerkin procedure.

In this chapter we derive a Wavelet-Galerkin procedure, where the variational equation is simply

$$(\psi, W_t)+(\psi_x, W_x) = \psi(1,t)g(t)+\psi(0,t)M'(t)-\psi(0,t)W_x(b,t) \text{ for } \psi \in S_N, \tag{1.7}$$

and compare its performance against that of the Crank-Nicolson-Galerkin procedure in [19]. Section 2 is devoted to the derivation of the Wavelet-Galerkin method. In Section 3, we consider the algebraic problem generated by the wavelet basis. Finally, in Section 4, we compare the results for the Wavelet-Galerkin method against the Crank-Nicolson-Galerkin procedure utilizing the same numerical examples that were considered in [19]. In conclusion, we discuss the results of the comparison against the theoretical approximation results of the wavelet basis. The demonstration of the convergence of the solution of the Wavelet-Galerkin method to the solution U of (1.1) remains on open problem.

2 Wavelet-Galerkin Method

This section is devoted to applying the Wavelet-Galerkin method for the diffusion equation subject to the specification of mass described in (1.1). But first let us give a brief introduction of wavelets, while for greater insight into wavelets, their properties, and their construction, the reader is referred to [1–3] or any other introductory materials in wavelets.

Wavelets are a family of functions generated by translation and dilation of a single function $\psi(x)$. This family of functions, denoted by $\psi_{j,k}(x)$, is given by

$$\psi_{j,k}(x) = 2^{j/2}\psi(2^j x - k), \quad j,k \in \mathbb{Z},$$

where $\mathbb{Z}$ denotes the integers. The mother wavelet function $\psi(x)$ has a companion, the scaling function $\varphi(x)$. In this chapter we will be dealing with Daubechies' compactly supported wavelets only. In such a case, the functions φ and ψ satisfy the following relations

$$\varphi(x) = \sum_{j=0}^{N-1} h_j \varphi(2x-j),$$

$$\psi(x) = \sum_{j=0}^{N-1} (-1)^j h_{N-j-1} \varphi(2x-j),$$

where $h_j, j = 0, 1, ..., N-1$ is a set of N (an even integer) coefficients. These coefficients, which are called the wavelet filter coefficients, were derived by Daubechies. For Daubechies' wavelets, the support of the scaling function $\varphi(x)$ is the interval $[0, N-1]$, while that of the corresponding wavelet $\psi(x)$ is the interval $\left[1-\frac{N}{2}, \frac{N}{2}\right]$. For a fixed N, ψ is called the Daubechies N wavelet. Any $L^2(\mathbb{R})$ function $f(x)$ may be approximated at scale j by

$$P_j f(x) = \sum_k c_{j,k}\varphi_{j,k}(x), \quad k \in \mathbb{Z},$$

where, using the Daubechies notation, $P_j f$ represents the projection of the function f onto the space of scaling functions at level (scale) j, the scale level, which is spanned by the basis functions

$$\varphi_{j,k}(x) = 2^{j/2}\varphi(2^j x - k), \quad k \in \mathbb{Z}. \tag{2.1}$$

The Wavelet-Galerkin method entails representing the solution of the differential equation as an expansion of scaling function at a particular level J. The Wavelet-Galerkin solution involves the evaluation of connection coefficients (see [4] and [5]) to approximate derivatives. The connection coefficients and associated computation developed in [4–6] (for instance) are essentially based on an unbounded domain. Therefore, most applications of the Wavelet-Galerkin method are limited to the cases where the problem domain is unbounded or the boundary condition is periodic [7]. To apply the Wavelet-Galerkin method to the solution of finite domain problems, [8] derived algorithms for computing some finite integrals of products of wavelets and their derivatives.

This work is based on [8] to solve problem (1.1). The solution of such a problem by the Wavelet-Galerkin method has its own difficulties. It will be shown in the following section how the authors overcame these difficulties.

Let the Wavelet-Galerkin approximation to the solution of (1.1) at scale J be

$$U(x,t) \simeq \sum_{k=2-N}^{2^J-1} u_{J,k}(t)\varphi_{J,k}(x), \tag{2.2}$$

where $J \geq 1$. Substituting (2.2) into (1.1a) and applying the Galerkin discretization scheme, we have

$$\sum_{k=2-N}^{2^J-1} a_{m,k}\frac{du_{J,k}(t)}{dt} = \sum_{k=2-N}^{2^J-1} b_{m,k}u_{J,k}(t), \quad m = 2-N, 3-N, ..., 2^J-1, \tag{2.3}$$

where the coefficients $a_{m,k}$ and $b_{m,k}$ are given by

$$a_{m,k} = \Gamma^0_{k-m}(2^J - m) - \Gamma^0_{k-m}(-m), \tag{2.4}$$
$$b_{m,k} = 2^{2J}[\Gamma^2_{k-m}(2^J - m) - \Gamma^2_{k-m}(-m)]. \tag{2.5}$$

Here $\Gamma^n_\ell(y)$ denotes the connection coefficient which is the integral of the product of the scaling function $\varphi(x)$ and its nth order derivative $\varphi^{(n)}(x-\ell)$, i.e.,

$$\Gamma^n_\ell(y) = \int_0^y \varphi^{(n)}(\xi - \ell)\varphi(\xi)d\xi.$$

For the evaluation of $\Gamma^n_\ell(y)$ and its properties, see [8]. The initial conditions $u_{J,k}(0)$ for the system of differential equations (2.3) are derived from the initial condition $U(x,0)$ of the problem. The initial conditions $u_{J,k}(0)$ can be determined by solving the following system of algebraic equations:

$$\sum_{k=2-N}^{2^J-1} a_{m,k}u_{J,k}(0) \int_0^1 f(x)\varphi_{J,m}(x)dx, \quad m = 2-N, 3-N, ..., 2^J-1. \tag{2.6}$$

For $U(x,t)$ to satisfy the boundary condition (1.1c), the expansion coefficient $u_{J,2^J-1}(t)$ must satisfy the following relation:

$$u_{J,2^J-1}(t) = \frac{1}{\varphi'(1)}\left[2^{-\frac{3J}{2}} \cdot g(t) - \sum_{i=2}^{N-2} u_{J,2^J-i}(t) \cdot \varphi'(i)\right]. \tag{2.7}$$

For $U(x,t)$ to satisfy the condition (1.1d), the expansion coefficients $u_{J,k}(t)$ must satisfy the relation

$$\sum_{k=2-N}^{2^J-1} u_{J,k}(t)[\varphi'(2^J \cdot b - k) - \varphi'(-k)] = 2^{-3J/2} \cdot M_t(t), \tag{2.8}$$

where $M_t(t)$ is the derivative of the function $M(t)$ with respect to t.

To get expansion coefficients $u_{J,k}(t)$ in the approximate solution (2.2), first we have to obtain the initial conditions $u_{J,k}(0)$ from the system (2.6) and then use them with the boundary condition (2.7) and relation (2.8) to solve the first order system of differential equations (2.3). These problems are shown in detail in the next section.

3 Algebraic Problem

Let us define some notation which will be used throughout this section. Define $n = 2^{J-2} + 1$, $J \geq 2$, let $P \in \mathbb{N}$, $\Delta t = \frac{T}{P}, t_i = i\Delta t$, $i = 1, 2, ..., P$, and $u_s^i = u_s(i\triangle t)$, where the subscript s denotes either a double or a single subscript index. Let the integrals in the right-hand side of (2.6) be denoted by $L_{J,m}$, $m = 2 - N, ..., 2^J - 1$.

Let the matrices on the left - and right-hand sides of (2.3) be denoted by A and B, respectively.

Lemma 3.1 *Consider the matrix* $A = [a_{m,k}]; m, k = 2 - N, ..., 2^J - 1$, *given by (2.4). Then* A *has the form*

$$A = \begin{array}{|c|c|c|} \hline I - G & 0 & 0 \\ \hline 0 & \tilde{I} & 0 \\ \hline 0 & 0 & G \\ \hline \end{array} ,$$

where $I - G$ *and* G *are square matrices of size* $(N - 2)$ *and* $\tilde{I}$ *is an identity matrix of size* $(2^J - N + 2)$.

Proof. By definition

$$A = [a_{m,k}] = \tilde{B} - C,$$

where

$$\begin{aligned} \tilde{B} &= [\tilde{b}_{m,k}], \quad \text{with} \quad \tilde{b}_{m,k} = \Gamma_{k-m}(2^J - m), \\ C &= [c_{m,k}], \quad \text{with} \quad c_{m,k} = \Gamma_{k-m}(-m), \end{aligned}$$

and $\Gamma_k(y)$ is given by

$$\Gamma_k(y) = \int_0^y \phi(x)\phi(x - k)dx.$$

It is easy to verify that A is a square matrix of size $2^J + N - 2$ and that $\Gamma_{k-m}(2^J - m) = \Gamma_{m-k}(2^J - k)$ which implies that $\tilde{B}$ is symmetric. Likewise $\Gamma_{k-m}(-m) = \Gamma_{m-k}(-k)$; hence C is symmetric. Moreover, by the orthogonality of $\{\phi(x - k)\}$, we have

$$\Gamma_{k-m}(2^J - m) = \begin{cases} \left.\begin{matrix} 0, \text{ if } k \neq m \\ 1, \text{ if } k = m \end{matrix}\right\} & 2 - N \leq m \leq 2^J - N + 1 \\ \text{other values} & \text{if } m > 2^J - N + 1. \end{cases}$$

This shows that

$$\tilde{B} = \begin{array}{|c|c|} \hline \tilde{\tilde{I}} & 0 \\ \hline 0 & D \\ \hline \end{array}\ ,$$

where $\tilde{\tilde{I}}$ is an identity matrix of size $2^J \times 2^J$ and $D = [d_{m,k}]$ is a square matrix of size $N-2$, with

$$d_{m,k} = a_{1-N+2^J+m,1-N+2^J+k} = \Gamma_{k-m}(N-1-m); \quad k, m = 1, 2, ..., N-2. \tag{3.1}$$

As for the matrix C, we have

$$c_{m,k} = \Gamma_{k-m}(-m) = \int_0^{-m} \phi(x)\phi(x-(k-m))dx.$$

Since the support of ϕ is $[0, N-1]$, it follows that $c_{m,k} = 0$ if $m \geq 0$, and since $m, k = (2-N), ..., 0, 1, .., 2^J - 1$, we have that C is a square matrix of size $2^J + N - 2$ of the form

$$C = \begin{array}{|c|c|} \hline G & O_1 \\ \hline O_2 & O_3 \\ \hline \end{array}\ ,$$

where G is a square matrix of size $(N-2)$, and O_1, O_2, and O_3 are matrices of size $(N-2) \times 2^J$, $2^J \times (N-2)$, and $2^J \times 2^J$, respectively.

Let us denote the elements of G by $g_{m,k}$ with $m, k = 1, 2, ..., N-2$. Then

$$g_{m,k} = c_{1-N+m,1-N+k} = \Gamma_{k-m}(N-1-m). \tag{3.2}$$

Thus, in view of (3.1) and (3.2), we have $D = G$, and the result follows. ■

Corollary 3.2 *The matrix A is a block diagonal matrix, and each block is an $(N-2) \times (N-2)$ matrix if $2^J + N - 2$ is divisible by $N-2$. In particular, each block is a 4×4 matrix if $J \geq 2$ and $N = 6, 10, 18, ...$*

In the sequel, we shall use Daubechies 6, i.e., $N = 6$. The matrix $A = [a_{m,k}], m, k = 2-N, 3-N, ..., 2^J - 1$ is a block diagonal matrix, and each block is a 4×4 matrix. Using relation (2.4) the matrix A has the following structure:

$$A = \begin{bmatrix} A_{11} & 0 & \cdots & 0 \\ 0 & A_{22} & \ddots & \vdots \\ \vdots & \ddots & \ddots & 0 \\ 0 & \cdots & 0 & A_{nn} \end{bmatrix}.$$

Entries of the submatrix A_{nn} are independent of J, i.e., are the same for any level J. These entries can be obtained using (2.4) and the properties of the connection coefficients [8]. The submatrix A_{11} is found to be $I_4 - A_{nn}$ (I_4 is the identity matrix). All other block matrices $A_{ii}, i = 2, ..., n-1$ are identities 4 by 4 each, of course with $J \geq 3$.

The matrix B is a banded matrix with five off diagonal above and five off diagonal below. This matrix can be also partitioned into block submatrices where each block submatrix is 4 by 4. B may have the next structure form:

$$B = \begin{bmatrix} B_{11} & B_{12} & 0 & \cdots & 0 \\ B_{21} & B_{22} & B_{23} & \ddots & \vdots \\ 0 & B_{32} & B_{33} & \ddots & 0 \\ \vdots & \ddots & \ddots & \ddots & B_{n-1,n} \\ 0 & \cdots & 0 & B_{n,n-1} & B_{nn} \end{bmatrix}.$$

Also, entries of submatrix B_{nn} are independent of J and can be obtained using (2.5) and its related properties of the connection coefficients [8]. It can be detected that the submatrices $B_{ii}, i = 2, ..., n-1$ are equal, say, $B_{ii} = \beta, i = 2, ..., n-1$ with $J \geq 3$. Each submatrix $B_{ij}, i \neq j$ is a triangle with $B_{ij} = B_{ji}^T$, also $B_{ij}, i < j$ are equal, say, $B_{ij} = \alpha, i < j$.

Now, we are ready to solve the system (2.6) for the initial conditions $u_{J,k}(0)$. The vector $u_{J,k}(t), k = 2-N, ..., 2^J - 1$ in (2.2) can be subdivided into subvectors where each has 4 elements, say, $u_\ell(t), \ell = 1, 2, ..., n$ for a fixed J. Also the vector $L_{J,m}$ may be subdivided into subvectors where each has 4 elements, say $L_\ell, \ell = 1, 2, ..., n$. Using condition (2.7) with the subsystem $A_{nn}u_n(0) = L_n$, and the relation (2.8) with subsystem $A_{11}u_1(0) = L_1$, the initial conditions $u_i(0), i = 1, ..., n$ are determined, where it is noted that $u_i(0) = l_i, i = 2, ..., n-1$. Using the above partitioning for both matrices A, B, and the vector $u_\ell(t), \ell = 1, 2, ..., n$, the system

(2.3) can be approximated by the following subsystems:

$$A_{11} \cdot \frac{1}{\Delta t}(u_1^i - u_1^{i-1}) = B_{11} \cdot \frac{1}{2}(u_1^i + u_1^{i-1}) + B_{12} \cdot \frac{1}{2}(u_2^i + u_2^{i-1}),$$

$$\frac{1}{\Delta t}(u_\ell^i - u_\ell^{i-1}) = \alpha^T \cdot \frac{1}{2}(u_{\ell-1}^i + u_{\ell-1}^{i-1}) + \beta \cdot \frac{1}{2}(u_\ell^i + u_\ell^{i-1})$$

$$+ \alpha \frac{1}{2}(u_{\ell+1}^i + u_{\ell+1}^{i-1}), \quad \ell = 2, ..., n-1,$$

$$A_{nn} \cdot \frac{1}{\Delta t}(u_n^i - u_n^{i-1}) = \alpha^T \cdot \frac{1}{2}(u_{n-1}^i + u_{n-1}^{i-1}) + B_{nn} \cdot \frac{1}{2}(u_n^i + u_n^{i-1}),$$

$$i = 1, 2, ..., P$$

or

$$\left(A_{11} - \frac{\Delta t}{2} \cdot B_{11}\right) u_1^i = A_{11} u_1^{i-1} + \frac{\Delta t}{2} \cdot B_{11} u_1^{i-1} + \frac{\Delta t}{2} \cdot B_{12}(u_2^i + u_2^{i-1}),$$

$$\left(I_4 - \frac{\Delta t}{2}\beta\right) u_\ell^i = \frac{\Delta t}{2} \cdot \alpha^T (u_{\ell-1}^i + u_{\ell-1}^{i-1}) + \frac{\Delta t}{2} \cdot \alpha(u_{\ell+1}^i + u_{\ell+1}^{i-1})$$

$$+ \left(I_4 + \frac{\Delta t}{2}\beta\right) u_\ell^{i-1}, \ell = 2, ..., n-1, \tag{3.1}$$

$$\left(A_{nn} - \frac{\Delta t}{2} \cdot B_{nn}\right) u_n^i = \frac{\Delta t}{2} \cdot \alpha^T (u_{n-1}^i + u_{n-1}^{i-1}) + \left(A_{nn} + \frac{\Delta t}{2} \cdot B_{nn}\right) u_n^{i-1},$$

$$i = 1, 2, ..., P.$$

Using condition (2.7), the system (3.1) can be solved for $u_\ell(t)$ at any time t by back substitution.

4 Some Numerical Comparisons

For the Crank-Nicolson-Galerkin approximation in [19], the experiments were performed using piecewise linear functions as the basis for the Galerkin method. For each of the first three examples, $\Delta t = \Delta x = 0.0025$ was employed in the Crank-Nicolson-Galerkin method, while $\Delta t = 0.0025$ was employed in the Wavelet-Galerkin method. The Crank-Nicolson-Galerkin method was initialized directly from $f(x)$, while the Wavelet-Galerkin method required utilizing a power series approximation of $f(x)$ with the quadrature rules given in [8]. To evaluate the integral term in $D(t)$ of the Crank-Nicolson-Galerkin method we used the trapezoidal rule. For the evaluation of the theta function $\theta(x,t)$ and $\theta_{xx}(x,t)$, we used the first 21 terms $m = -10$ to 10 in its series representation in terms of the fundamental solution $K(x,t)$ of the heat equation. See (1.5) and (1.6) for the definitions of $K(x,t)$ and $\theta(x,t)$. In the first three cases, we took $b = 0.75$ and $x = 0.25$.

Case 1.

$$f(x) = \frac{1}{2}x^2,$$
$$g(t) = 1.0,$$
$$M(t) = 0.75t + \frac{(0.75)^3}{6}.$$

Exact solution:

$$U(x,t) = \frac{1}{2}x^2 + t.$$

Crank-Nicolson-Galerkin results:

t	$W(x,t)$	$U(x,t)$	$\lvert U-W\rvert/\lvert U\rvert$
0.01	0.041170	0.041250	0.001929
0.025	0.0561994	0.056250	0.000993
0.1	0.131491	0.131250	0.001834

Wavelet-Galerkin ($J = 8$) results:

t	$W(x,t)$	$U(x,t)$	$\lvert U-W\rvert/\lvert U\rvert$
0.01	0.0412499841	0.04125	0.0000003846
0.025	0.0562499791	0.05625	0.0000003720
0.1	0.1312499704	0.13125	0.0000002254

Case 2.

$$f(x) = \cos(x),$$
$$g(t) = -\sin(1.0)e^{-t},$$
$$M(t) = \sin(.75)e^{-t}.$$

Exact solution:

$$U(x,t) = e^{-t}\cos(x).$$

Crank-Nicolson-Galerkin results:

t	$W(x,t)$	$U(x,t)$	$\lvert U-W\rvert/\lvert U\rvert$
0.01	0.959186	0.959272	0.000089
0.025	0.944932	0.944990	0.000060
0.1	0.876955	0.876708	0.000281

Wavelet-Galerkin ($J = 8$) results:

t	$W(x,t)$	$U(x,t)$	$\|U - W\|/\|U\|$
0.01	0.9592705921	0.9592715820	0.0000010319
0.025	0.9449865067	0.9449898888	0.0000035789
0.1	0.8767009222	0.8767082139	0.0000083172

Case 3.

$$f(x) = \sin \pi x,$$
$$g(t) = -\pi e^{-\pi^2 t},$$
$$M(t) = \frac{1}{\pi}\left(\frac{1}{\sqrt{2}} + 1\right) e^{-\pi^2 t}.$$

Exact solution:

$$U(x,t) = e^{-\pi^2 t} \sin \pi x.$$

Crank-Nicolson-Galerkin results:

t	$W(x,t)$	$U(x,t)$	$\|U - W\|/\|U\|$
0.01	0.640649	0.640651	0.000004
0.025	0.552491	0.552493	0.000003
0.1	0.263543	0.263545	0.000007

Wavelet-Galerkin ($J = 8$) results:

$$\triangle t = 0.0025$$

t	$W(x,t)$	$U(x,t)$	$\|U - W\|/\|U\|$
0.01	0.6406490701	0.6406515111	0.0000038101
0.025	0.5524861071	0.5524934503	0.0000132911
0.1	0.2635247151	0.2635442402	0.0000740868

$$\triangle t = 0.001$$

t	$W(x,t)$	$U(x,t)$	$\|U - W\|/\|U\|$
0.01	0.6406511036	0.6406515111	0.0000006361
0.025	0.5524922999	0.5524934503	0.0000020822
0.1	0.2635411446	0.2635442402	0.0000117463

Finally, we present some calculations which demonstrate that, for $\triangle t$ sufficiently small, convergence will occur. In the tables below, the figures represent the relative error $|U - W|/|U|$ in the solution for the displayed data. We set $\triangle x = 1/N$. Thus, the table displays the relative error at $x = 0.25$ and $t = 0.1$ as a function of $\triangle x$ and $\triangle t$ for the Crank-Nicolson-Galerkin method. With respect to the Wavelet-Galerkin method, the scale index J plays the role of $\triangle x$ in the corresponding table.

Case 4.

$$\begin{aligned} f(x) &= \sin \pi x, \\ g(t) &= -\pi e^{-\pi^2 t}, \\ M(t) &= \frac{1}{\pi}\left(\frac{1}{\sqrt{2}} + 1\right) e^{-\pi^2 t}. \end{aligned}$$

Exact solution:

$$U(x,t) = e^{-\pi^2 t} \sin \pi x.$$

Remark 4.1 From the wavelet literature one can find results such as the following: for certain conditions on the scaling function φ and the mother wavelet ψ, if f belongs to the Sobolev space $W_p^{N+1}(\mathbb{R})$, then

$$\|K_j f - f\|_p = O(2^{-j(N+1)}), \quad \text{as} \quad j \to \infty. \tag{4.1}$$

For any $p \in [1, \infty]$, where K_j is the wavelet projection kernel on V_j,

$$K_j(x,y) = \sum_k 2^j \varphi(2^j x - k)\, \overline{\varphi(2^j y - k)}\,. \tag{4.2}$$

When we are able to closely approximate the initial conditions and the boundary conditions, the smoothness of the solution to the heat equation and (4.1) we expect to see good agreement between the solution and the approximation. Cases 1 and 2 above illustrate this expectation as the series for $\cos x$ converges rapidly on $0 \le x \le 1$ and can be replaced by a few terms in the expansion. The initial condition of $\sin \pi x$ in Cases 3 and 4 requires a lot more work to get a good approximation over $0 \le x \le 1$. This can be seen more clearly in Case 4. Overall, the Wavelet-Galerkin method competes very well with the Crank-Nicolson-Galerkin method with respect to accuracy and ease of implementation.

Crank-Nicolson-Galerkin results:

$\triangle x\backslash\triangle t$	0.05	0.01	0.005	0.0025	0.001
0.25	0.0621336863	0.0733938811	0.0711146146	0.0715135139	0.0714996596
0.05	0.0131523568	0.0027760947	0.0028219241	0.0028334395	0.0028347049
0.01	0.0161660572	0.0000239519	0.0001167389	0.0001136604	0.0001145364
0.005	0.0162602354	0.0001114245	0.0000327017	0.0000286646	0.0000295681
0.0025	0.0162836944	0.0001332437	0.0000113031	0.0000072892	0.0000081151
0.001	0.0162919056	0.0001393060	0.0000050953	0.0000010501	0.0000000351

Wavelet-Galerkin results:

$J\backslash\triangle$ t	0.05	0.01	0.005	0.0025	0.001
4	0.0209785202	0.0013310794	0.0007210632	0.0005748999	0.0005342058
5	0.0273778068	0.0011338764	0.0003405053	0.0001423377	0.0000868577
6	0.0292679027	0.0011531885	0.0002943706	0.0000798350	0.0000197767
7	0.0299744908	0.0011777341	0.0002861752	0.0000740868	0.0000123248
8	0.0302739765	0.0011907001	0.0002974173	0.0000742293	0.0000117463

References

[1] E. Hernands and G. Weiss, *A First Course on Wavelets*, CRC Press, Boca Raton, FL (1996).

[2] I. Daubechies, Orthonormal bases of compactly supported wavelets, *Commun. Pure Appl. Math.*, **41** (1988), 909–996.

[3] J. R. Williams and K. Amaratunga, Introduction to wavelets in engineering, *Int. J. Numer. Methods Eng.*, **37** (1994), 2365–2388.

[4] G. Beylkin, On the representation of operators in bases of compactly supported wavelets, *SIAM J. Numer. Anal.*, **29** (1993), 507–537.

[5] A. Latto, H. L. Renikoff, and E. Tenenbaum, The evaluation of connection coefficients of compactly supported wavelets, *Proc. French-USA Workshop on Wavelets and Turbulence*, Princeton University, June 1991, New York, Springer-Verlag, Heidelberg (1994).

[6] W. Dahmen and C. A. Micchelli, Using the refinement equation for evaluating integrals, *SIAM J. Numer. Anal.*, **30** (1993), 507–537.

[7] K. Amaratunga and J. R. Williams, Wavelet-Galerkin solution for one-dimensional partial differential equations, *Int. J. Numer. Methods Eng.*, **37** (1994), 2703–2716.

[8] M. Q. Chen, C. Hwang, and Y. P. Shif, The computation of Wavelet-Galerkin approximation on bounded interval, *Int. J. Numer. Methods Eng.*, **39** (1996), 2921–2944.

References for Diffusion Subject to Specification of Mass

[9] K. T. Andrews, P. Shi, M. Shillor, and S. Wright, Thermoelastic contract with Barber's heat exchange condition, *Appl. Math. Optim.*, **28** (1993), 11–48.

[10] I. Barradas and S. Perez-Esteva, Solution to a parabolic equation with integral type boundary conditions, to appear.

[11] B. Cahlon, D. M. Kulkarni, and P. Shi, Stability of a finite-difference scheme for the heat equation with a nonlocal constraint, to appear.

[12] J. R. Cannon, The one-dimensional heat equation, *The Encyclopedia of Mathematics*, Vol. **23**, Addison-Wesley, Reading, MA (1984).

[13] J. R. Cannon, The solution of the heat equation subject to the specification of energy, *Quart. Appl. Math.*, **21** (1963), 155–160.

[14] J. R. Cannon and J. van der Hoek, The existence of and a continuous dependence result for the solution of the heat equation subject to the specification of energy, *Boll. Un. Mat. Ital. Suppl.*, **1** (1981), 253–282.

[15] J. R. Cannon and J. van der Hoek, The one phase Stefan problem subject to the specification of energy, *J. Math. Anal. Appl.*, **86** (1982), 281–291.

[16] J. R. Cannon and J. van der Hoek, The classical solution of the one-dimensional two phase Stefan problem with energy specification, *Annali Mat. Pura ed Appl.* (IV), **CXXX**, 385–398.

[17] J. R. Cannon and J. van der Hoek, An implicit finite difference scheme for the diffusion equation subject to the specification of mass in a portion of the domain, *Numerical Solutions of Partial Differential Equations*, Editor J. Noye, North-Holland, Amsterdam (1982), 527–539.

[18] J. R. Cannon and J. van der Hoek, Diffusion subject to specification of mass, *J. Math. Anal. Appl.*, **115** (1986), 517–529.

[19] J. R. Cannon, S. Perez-Esteva, and J. van der Hoek, A Galerkin procedure for the diffusion equation subject to the specification of mass, *SIAM J. Numer. Anal.*, **24** (1987), 499–515.

[20] J. R. Cannon and Y. Lin, Determination of a parameter $p(t)$ in some quasi-linear parabolic differential equations, *Inverse Problems*, **4** (1988), 35–45.

[21] J. R. Cannon and Y. Lin, Determination of a parameter $p(t)$ in Holder classes for semi-linear parabolic equations, *Inverse Problems*, **4** (1988), 595–606.

[22] J. R. Cannon and H. M. Yin, A class of non-linear non-classical parabolic equations, *J. Diff. Equ.*, **79** (1989), 266–288.

[23] J. R. Cannon and Y. Lin, An inverse problem of finding a parameter in a semi-linear heat equation, *J. Math. Anal. Appl.*, **145** (1990), 470–484.

[24] J. R. Cannon and Y. Lin, A Galerkin procedure for diffusion equations with boundary integral conditions, *Int. J. Eng. Sci.*, **28** (1990), 579–587.

[25] J. R. Cannon, Y. Lin, and S. Wang, An implicit finite difference scheme for the diffusion equation subject to mass specification, *Int. J. Eng. Sci.*, **28** (1990), 573–578.

[26] J. R. Cannon, Y. Lin, and J. van der Hoek, A quasi-linear parabolic equation with nonlocal boundary conditions, *Rendiconti di Matematica, Serie*, VII **9** (1989), 239–264.

[27] J. R. Cannon, Y. Lin, and J. van der Hoek, Semi-linear heat equation subject to the specification of energy, modeling and analysis of diffusive and advective processes in geosciences, Proceedings of the SIAM Conference on Mathematical and Computational Issues, Houston, September 1989, in *Geophysical Fluid and Solid Mechanics*, Edited by W. E. Fitzgibbon and M. F. Wheeler, SIAM, Philadelphia, (1992).

[28] J. R. Cannon, Y. Lin, and S. Wang, Determination of a control parameter in a partial differential equation, *J. Austral. Math. Soc. Ser. B*, **33** (1991), 149–163.

[29] J. R. Cannon, Y. Lin, and S. Wang, Determination of source parameter in parabolic equations, *Meccanica*, **27** (1992), 84–95.

[30] J. R. Cannon and A. Matheson, On the numerical solution of diffusion subject to the specification of mass, *Int. J. Eng. Sci.*, **131** (1993), 347–355.

[31] J. R. Cannon, Y. Lin, and A. Matheson, The solution of the diffusion equation in two space variables subject to the specification of mass, *Appl. Anal.*, **104** (1993), 325–331.

[32] J. R. Cannon and S. Perez-Esteva, Determination of the coefficient of u_x in a linear parabolic equation, *Inverse Problems*, **10** (1994), 521–531.

[33] J. R. Cannon and C. Denson Hill, On the movement of a chemical reaction interface, *Indiana University Math. J.*, **20** (1970), 429–454.

[34] J. R. Cannon and A. Fasano, Boundary value multidimensional problems in fast chemical reactions, *Arch. Rat. Mech. Anal.*, **53** (1973), 1–13.

[35] J. R. Cannon and E. DiBenedetto, On the existence of solutions of boundary-value problems in fast chemical reactions, *Bull. della Unione Mat. Italiana*, **15** (1978), 835–843.

[36] J. M. Chadam and H.-M. Yin, An iteration procedure for a class of integro-differential equations of parabolic type, *J. Integral Equ. Appl.*, **2** (1989), 31–47.

[37] Y. S. Choi and K. Y. Chan, A parabolic equation with nonlocal boundary conditions arising from electrochemistry, *Nonlinear Analysis, TMA,* **18** (1992), 317–331.

[38] P. Colli, On the Stefan problem with energy specification, *Accad. Naz. Lincei Rend. Cl. Sci. Fis. Mat. Natur.*, **75** (1983), 303–312.

[39] E. Comparini and D. A. Tarzia, A Stefan problem for the heat equation subject to an integral condition, *Rend. Sem. Mat. Univ. Padova,* **73** (1985), 119–136.

[40] W. A. Day, Extensions of a property of the heat equation to linear thermoelasticity and other theories, **40** (1982), 319–330.

[41] W. A. Day, A decreasing property of solutions of parabolic equations with applications to thermoelasticity, *Quart. of Appl. Math.*, **41** (1983), 468–475.

[42] J. Douglas, Jr., *Survey of Numerical Methods for Parabolic Differential Equations, Advances in Computers*, Vol. 2, Academic Press, New York, (1961), 1–54.

[43] J. Douglas, Jr. and T. Dupont, Galerkin procedures for parabolic equations, *SIAM J. Numer. Anal.*, **7** (1970), 575–626.

[44] M. T. Dzhenaliev, A boundary value problem for a linear loaded parabolic equation with nonlocal boundary conditions, *Differecial'nye Uravneniya,* **27** (1991), 1825–1827, 1839.

[45] G. Ekolin, Finite difference methods for a nonlocal boundary value problem for the heat equation, *BIT, Computer Sc. Num. Math.*, **31** (1991), 245–261.

[46] G. Fairweather and R. D. Saylor, The reformulation and numerical solution of certain non-classical initial-boundary value problems, *SIAM J. Sci. Stat. Comp.*, **12** (1991), 127–144.

[47] G. Fairweather and J. C. Lopez Marcos, An explicit extrapolated box scheme for the Curtin-MacCamy equation, *Comput. Math. Appl.*, **27** (1994), 41–53.

[48] G. Fairweather and R. D. Saylor, Keller's box scheme for the diffusion equation subject to the specification of mass, preprint/private communication.

[49] A. Friedman, Monotonic decay of solutions of parabolic equations with nonlocal boundary conditions, *Quart. Appl. Math.*, **44** (1986), 401–407.

[50] A. Friedman, Parameter identification in a reaction diffusion model, *Mathematics in Industrial Problems Part* 4, Springer-Verlag, New York, **Chapter 15**, Springer-Verlag, New York, 146–152, 1991.

[51] A. Friedman and F. Reitich, Parameter identification in reaction diffusion models, *Inverse Problems,* **8** (1992), 187–192.

[52] N. I. Ionkin, Solution of a boundary problem in heat conduction with a non-classical boundary condition, *Differencial'nye Uravenija,* **13** (1977), 294–304 (English translation: *Differential Equations*, **13** (1977), 204–211).

[53] N. I. Ionkin, Stability of a problem in heat transfer theory with a non-classical boundary condition, *Differencial'nye Uravenija*, **15** (1979), 1279–1283 (English translation: *Differential Equations*, **15** (1980), 911–914).

[54] N. I. Ionkin and E. I. Moiseev, Problem for the heat-transfer equation with two point boundary conditions, *Differencial'nye Uravenija*, **15** (1979), 1284–1295 (English translation: *Differential Equations*, **15** (1980), 915–923).

[55] N. I. Ionkin and D.G. Furletov, Uniform stability of difference schemes for a nonlocal non-self-adjoint boundary value problem with variable coefficients, *Differential Equations*, **27** (1991).

[56] D. Jackson, Existence and uniqueness of solutions to semi-linear nonlocal parabolic equations, *J. Math. Anal. Appl.*, **172** (1993), 256–265.

[57] L. I. Kamynin, A boundary value problem in the theory of heat conduction with a non-classical boundary condition, *Zh. Vycial. Mat. i Mat. Fiz.*, **4** (1964), 1006–1024 (English translation: *USSR Comp. Math. and Math. Physics* (1964), 33–59).

[58] B. Kawohl, Remarks on a paper by W. A. Day on a maximum principle under nonlocal boundary conditions, *Quart. Appl. Math.*, **44** (1987), 751–752.

[59] H. B. Keller, A new difference scheme for parabolic problems, *Numerical Solution of Partial Differential Equations - II*, Editor, B. Hubbard, Academic Press, New York (1971), 327–350.

[60] M. Lees, *A. priori* estimates for the solutions of difference approximation to parabolic partial differential equations, *Duke J. Math.*, **27** (1960), 297–311.

[61] Z. Y. Liang, The two-phase Stefan problem in the heat equation with energy boundary condition, *J. Harbin Inst. Tech.*, (1987), 1–11.

[62] Z. Y. Liang and H. Cheng, The one phase Stefan problem of energy integral type, *J. Harbin Inst. Tech.*, **23** (1991), 15–22.

[63] Z. Y. Liang, The existence of classical solution for quasilinear heat equation with phases free boundary problem subject to the specification of energy integral form, to appear.

[64] Y. Lin, Parabolic partial differential equations subject to non-local boundary conditions, Ph.D. Thesis, Washington State University, Pullman, Washington, 1988.

[65] Y. Lin and S. Wang, A numerical method for diffusion equation with non-local boundary specifications, *Int. J. of Eng. Sci.*, **28** (1990), 543–546.

[66] V. L. Markov and D. T. Kulyev, Solution of a boundary-value problem for a quasilinear parabolic equation with a nonclassical boundary condition, *Differencial'nye Uravneniya*, **21** (1985), 296–305 (English translation: *Differential Equations*, **21** (1985), 216–223).

[67] V. L. Makarov, T. Arazmyradov, and D. T. Kulyev, Solution of nonlocal boundary value problem in classes of generalized function, *Izv. Akad. Nauk Turk m. Ser. Fiz.-Mat. Tekhn. Khim, Geol. Nauk*, (1992), 3–8.

[68] A. K. Pani, A finite element method for a diffusion equation with constrained energy and nonlinear boundary conditions, *J. Austral. Math. Soc. Ser. B*, **35** (1993), 87–102.

[69] P. Shi and M. Shillor, On design of contract patterns in one dimensional thermoelasticity, *Theoretical Aspects of Industrial Design,* SIAM, Philadelphia, 1992.

[70] P. Shi, Weak solution to an evolution problem with a nonlocal constraint, *SIAM J. Math. Anal.*, **24** (1993), 46–58.

[71] R. Spigler and C. Sartori, The Huber polygonal method for the one-phase Stefan problem with the specification of energy, *Z Agnew Mat. Mech.*, **69** (1989), 447–456.

[72] N. I. Yurchuk, Mixed problem with an integral condition for certain parabolic equations, *Differencial'nye Uravneniya*, **22** (1986), 2117–2126 (English translation: *Differential Equations*, **22** (1986), 1457–1463).

Chapter 6

TRENDS IN WAVELET APPLICATIONS

K. M. Furati

King Fahd University of Petroleum & Minerals

P. Manchanda

Gurunanak Dev University

M. K. Ahmad

Aligarh Muslim University

A. H. Siddiqi

King Fahd University of Petroleum & Minerals

Abstract

The study of wavelet analysis which was formally developed in the late 1980s has progressed very rapidly. There exists a vast literature on its applications to image processing and partial differential equations. However, Black-Scholes equation of pricing, Maxwell's equations, variational inequalities, and complex dynamic optimization problems related to real-world phenomena are some of the areas where applications of wavelet methods have not been fully explored. Wavelet packet analysis which includes wavelet analysis as a special case has wide scope for further research from both theoretical and applied viewpoints. As we know, wavelet methods are refinements of Fourier analysis, finite element, and boundary element methods. Compared to classical methods, wavelet methods yield better results. We briefly introduce applications of wavelets to image processing and partial differential equations. Important publications have been cited, in particular those of Cohen, Dahmen, DeVore, and Meyer, and some of their contributions have been discussed. The other important topics such as Black-Scholes equation of option pricing, Maxwell's equation, variational inequalities, modeling real-world problems, and complex dynamic optimization problems have been discussed, and some open problems are mentioned. The main aim of this chapter is to present an overview of certain applications of wavelets which may provide motivation for further research. We have briefly mentioned the properties of wavelet packets and have given updated references for further study.

1 Introduction

The concept of wavelets is viewed as a synthesis of ideas which have originated in the last three decades from engineering, namely, subband coding; physics, especially coherent states and renormalization group; and pure mathematics, related to Caldéron-Zygmund operators. This multitude of origins aroused great interest in many areas of science and technology. The current research landscape of wavelets is very wide, such as construction of wavelets and their generalizations, wavelets as a modeling tool, wavelets as an analysis tool, and multiscale geometric representation. Wavelet methods are applied to diverse fields such as signal analysis and image processing, numerical treatment of partial differential equations, dynamic optimization, Maxwell's equations, Black-Scholes model of option pricing, time series analysis and turbulence.

There exists a vast literature on applications of wavelet methods to image processing and partial differential equations, including books and excellent review articles. However, some other areas like applications to dynamic optimization, Maxwell equations, and option pricing are not fully explored and require more attention. One of the goals of this chapter is to present an overview of the current trends in applications of wavelet methods to numerical analysis of partial differential equations and image processing reflected in the latest work of Cohen, Dahmen, DeVore, and Meyer.

In Section 2, essential basic concepts, properties, and advantages of wavelets are presented. Wavelet packets are also introduced; of course, for comprehensive study we give updated references. Section 3 is devoted to a brief introduction of applications of wavelets to signal and image processing. However, for an updated account and outstanding problems, we refer to Meyer's series of lectures at the Abdus Salam International Centre of Theoretical Physics, Italy, September 2000 [97]. We introduce wavelet methods for the numerical solution of partial differential equations in Section 4. The presentation is based on Cohen [27], Cohen, Dahmen, and DeVore [29], Dahlke [36], Dahmen [40], Jaffard and Laurencot [78], and Liandrat, Perrier, and Tchamitchian [79]. Essentially, we focus on the main points of [29, 40] and advise interested readers on this aspect of wavelets to go through these to have a clear picture.

Section 5 deals with dynamic optimization problems of chemical engineering that are essentially based on Binder, et al. [13] and Blank [16]. We present a wavelet method for the Black-Scholes model of option pricing [15] In Section 6. Wavelets for financial time-series analysis are given in Section 7. We discuss the application of wavelets to Maxwell's equations in Section 8. In Section 9 we indicate the application of wavelet transform to turbulence analysis. In Section 10 we indicate some limitations of wavelet theory and introduce ridgelets and curvelets. We also mention a few open problems in Section 11.

2 Basic Tools

The idea of wavelets as a family of functions constructed from translation and dilation of a single function came from Morlet et al. [101, 102]. They are defined by

$$\psi_{a,b}(t) = |a|^{-\frac{1}{2}}\psi\left(\frac{t-b}{a}\right), a, b \in R, a \neq 0, \tag{2.1}$$

where a and b are scaling and translation parameters, respectively. In order for an analog of the Fourier transform inversion formula to be valid, it is assumed that ψ satisfies condition (2.6). This condition implies that $\int_{-\infty}^{\infty} \psi(t)dt = 0$. According to Mallat [90] a wavelet is a function $\psi \in L_2(R)$ such that $\int_{-\infty}^{\infty} \psi(t)dt = 0$.

Grossman and Morlet [76] recognized the importance of the Morlet wavelets which are somewhat similar to the formalism of coherent states in quantum mechanics, and they developed an inversion formula. They are often called affine coherent states because they are associated with an affine group "$ax+b$". Thus, the wavelets $\psi_{a,b}$ are, in fact, the result of the action of the operators $U(a,b)$ on the function ψ so that

$$[U(a,b)\psi](x) = |a|^{-\frac{1}{2}}\psi\left(\frac{t-b}{a}\right). \tag{2.2}$$

These operators are unitary on the Hilbert space $L_2(\mathbb{R})$ and constitute a representation of the "$ax+b$" group:

$$U(a,b)U(c,d) = U(ac, b+ad). \tag{2.3}$$

For details we refer to Torresani [133].

The *wavelet transform* of $f \in L_2(\mathbb{R})$ is defined by

$$W_\psi[f](a,b) = \langle f, \psi_{a,b}\rangle = |a|^{-\frac{1}{2}} \int_{-\infty}^{\infty} f(t)\overline{\psi\left(\frac{t-b}{a}\right)}dt. \tag{2.4}$$

Grossman and Morlet [76] proved that a function f can be constructed from its wavelet transform by the formula

$$f(t) = C_\psi^{-1}\int_{-\infty}^{\infty}\int_{-\infty}^{\infty}\{W_\psi[f](a,b)\}\psi_{a,b}(t)\cdot a^{-2}dadb, \tag{2.5}$$

provided ψ satisfies the so-called *admissibility condition*

$$C_\psi = \int_{-\infty}^{\infty}\frac{|\widehat{\psi}(w)|^2}{|\psi|}dw < \infty \tag{2.6}$$

where $\widehat{\psi}(w)$ is the Fourier transform of the wavelet $\psi(t)$.

In fact, this type of formula was already proved by A.P. Caldéron in 1964, which was rediscovered by Morlet and Grossmann. Therefore, this formula is often known by the name of Caldéron, Grossmann, Morlet. In practical applications involving fast numerical algorithms, the continuous wavelet can be computed at discrete grid points by replacing a with $a_0^m (a_0 \neq 0, 1)$, and b with $nb_0a_0^m (b_0 \neq 0)$, where m and n are integers.

Definition 2.1 *(i) A sequence $\{\varphi_n\}$ of $L_2(R)$ is called a frame if there exists positive constants A and B such that*

$$A\|f\|^2 \leq \sum_{n=1}^{\infty}|\langle f, \varphi_n\rangle|^2 \leq B\|f\|^2 \quad \textit{for all } f \in H,$$

where $\langle f, \varphi_n \rangle = \int_R f(x)\overline{\varphi_n(x)}dx$. *Constants* A *and* B *are called frame bounds. If* $A = B = 1$, *the frame is called a tight exact frame.*

(ii) A sequence $\{\varphi_n\}$ *in* $L_2(R)$ *is called an orthonormal system (sequence) if*

$$\begin{aligned}\langle \varphi_n, \varphi_m \rangle &= 0 \text{ if } n \neq m \\ &= 1 \text{ if } n = m.\end{aligned}$$

(iii) A sequence $\{\varphi_n\}$ *of* $L_2(R)$ *is called an orthonormal basis if* $\{\varphi_n\}$ *is an orthonormal seuence and for every*

$$f \in L_2(R),\ f = \sum_{n\in Z} \langle f, \varphi_n \rangle \varphi_n.$$

It is clear that an orthonormal basis is a tight exact frame.

(iv) A sequence of functions $\{\varphi_n(x)\}$ *of* $L_2(R)$ *is called a Riesz sequence if there exist constants* $0 < A \leq B$ *such that*

$$A\|\alpha\| \leq \|\sum_{n\in Z} \alpha_n \varphi_n(x)\| \leq B\|\alpha\|,$$

where $\alpha = (\alpha_1, \alpha_2, \ldots, \alpha_n, \ldots)$ *is a sequence of arbitrary scalars and* $\|\alpha\| = \left(\sum_{n\in Z} |\alpha_n|^2\right)^{1/2}$. *A Riesz sequence is called a Riesz basis if* $span\{\varphi_n(x)\}_{n\in Z} = L_2(R)$.

(v) A wavelet $\psi \in L_2(R)$ *is called orthonormal if the family of functions* $\psi_{m,n}(t) = 2^{m/2}\psi(2^m t - n)$, *generated by translation by* n *and dilation by* 2^m, *is an orthonormal system, that is*

$$\begin{aligned}\langle \psi_{m,n}(t), \psi_{m',n'}(t) \rangle &= 0 \text{ if } m \neq m' \text{ and } n \neq n' \\ &= 1 \text{ if } m = m' \text{ and } n = n'.\end{aligned}$$

(vi) A wavelet ψ *is called an orthonormal basis of* $L_2(R)$ *if it is orthonormal and for every* $f \in L_2(R)$, *we halve*

$$f = \sum_{m\in Z} \sum_{n\in Z} \langle f, \psi_{n,m}(t) \rangle \psi_{m,n}(t).$$

Usually, an arbitrary c_n is chosen in place of $\langle f, \psi_{n,m}(t) \rangle$.

$$d_{j,k} = W[f](m,n) = \langle f, \psi_{n,m} \rangle = \int_{-\infty}^{\infty} f(t)\psi_{m,n}(t)dt,$$

where $\psi(t)$ is real, are called wavelet coefficients. The series

$$\sum_{m\in Z}\sum_{n\in Z}\langle f,\psi_{m,n}(t)\rangle\,\psi_{m,n}(t)$$

is called the wavelet series associated with a given function $f\in L_2(R)$.

In practical applications, especially those involving fast algorithms, the continuous wavelet transform is computed on a discrete grid of points (a_n,b_n), $n\in Z$. The important issue is the choice of this sampling so that it contains all the information on the function f. Daubechies [44] proved in 1988 that the sampling

$$\psi_{m,n}(x)=a_0^{m/2}\psi(a_0^n x-b_0 n),\quad m,n\in Z$$

generates a frame of $L_2(R)$ if $a_0>1$ and $b_0>0$ are chosen small enough. Subsequently, Meyer proved the existence of an orthonormal basis.

It may be observed that $\psi_{m,n}(t)$ is more suited for representing finer details of a signal as it oscillates rapidly. The wavelet coefficients $d_{m,n}$ measure the amount of fluctuation about the point $t=2^{-m}n$ with a frequency determined by the dilation index m. It is interesting to note that

$$d_{m,n}=T_\psi f(2^{-m},n2^{-m})=\text{ wavelet transform of } f$$

with respect to wavelet ψ at the point $(2^{-n},n2^{-m})$.

2.1 Multiresolution Analysis (MRA)

A natural framework for wavelet theory is multiresolution analysis (MRA), which is a mathematical construction that characterizes wavelets in a general way. The MRA yields fundamental insights into wavelet theory and leads to important algorithms as well. The idea of a multiresolution is to write the L_2-function f as a limit of successive approximations. In general, each approximation is a smoothed version of f. The successive approximations thus use a different resolution. For a complete description of the theory we refer to [26, 31, 44, 45, 89].

Definition 2.2 [Mallat, 1989]. *A multiresolution analysis is a sequence $\{V_j\}$ of subspaces of $L_2(R)$ such that*

(i) $\cdots\subset V_{-1}\subset V_0\subset V_1\subset\cdots,$

(ii) Span $\bigcup_{j\in Z}V_j=L_2(R),$

(iii) $\bigcap_{j \in Z} V_j = \{0\}$,

(iv) $f(x) \in V_j$ *if and only if* $f(2^{-j}x) \in V_0$,

(v) $f(x) \in V_0$ *if and only if* $f(x - m) \in V_0$ *for all* $m \in Z$, *and*

(vi) there exists a function $\varphi \in V_0$ *called the scaling function such that the system* $\{\varphi(t - m)\}_{m \in Z}$ *orthonormal basis in* V_0.

Remark 2.3 (a) Conditions (i) to (iii) mean that every function in $L_2(R)$ can be approximated by elements of the subspaces V_j, and as j approaches ∞, the precision of approximation increases.

(b) Conditions (iv) and (v) express the invariance of the system of subspaces $\{V_j\}$ with respect to the translation and dilation operators.

(c) Condition (v) follows from (vi).

(d) Condition (vi) can be rephrased for each $j \in Z$ that the system $\{2^{j/2}\varphi(2^j x - k)\}_{k \in Z}$ is an orthonormal basis of V_j.

(e) For a given MRA$\{V_j\}$ in $L_2(R)$ with the scaling function φ, a wavelet is obtained in the following manner. Let the subspace W_j of $L_2(R)$ be defined by the condition

$$\begin{aligned} V_j \oplus W_j &= V_{j+1},\ V_j \perp W_j\ \forall\, j \\ V_{j+1} &= J_j(V_0 \oplus W_0) = J_j(V_0) \oplus J_j(W_0) = V_j \oplus J_j(W_0), \end{aligned}$$

where J_j (for an integer j, J_j is defined as $J_j(f(x)) = f(2^j x)\ \forall\, f \in L_2(R)$) is an isometry, $J_j(V_1) = V_{j+1}$.

$$V_m = \bigoplus_{j \geq m+1} W_j.$$

This gives

$$W_j = J_j(W_0) \text{ for all } j \in Z.$$

From conditions (i) to (iii), we obtain an orthogonal decomposition

$$\begin{aligned} L_2(R) &= \oplus \sum_{j \in Z} W_j = W_1 \oplus W_2 \oplus W_2 \oplus \\ &= \bigoplus_{j \in Z} W_j. \end{aligned}$$

Let $\psi \in W_0$ be such that $\{\psi(t - m)\}_{m \in Z}$ is an orthonormal basis in W_0. This function is a wavelet. Let $\varphi(x) = \sum_{n \in Z} c_n \varphi(2x - n)$, where c_n is an appropriate constant, and then $\psi(x) = \sum_{n \in Z} (-1) c_{n+1} \varphi(2x + n)$.

(f) It may be noted that the convention of increasing subspaces $\{V_j\}$ is not universal. Very often decreasing sequences of subspaces $\{V_j\}$ are used in the definition. However, one gets similar results.

The following theorems provide a relationship between scaling functions, MRA, and wavelets.

Theorem 2.4 *Let $\varphi \in L_2(R)$ satisfy*

(i) $\{\varphi(t-m)\}$ is a Riesz sequence of $L_2(R)$,

(ii) $\varphi(x/2) = \sum_{k \in Z} a_k \varphi(x-k)$ converges on $L_2(R)$, and

(ii) $\hat{\varphi}(\xi)$ is continuous at 0 and $\hat{\varphi}(0) \neq 0$, where $\hat{\varphi}$ denotes the Fourier transform of φ.

Then the spaces $V_j = Span\{\varphi(2^j x - k)\}_{k \in Z}$ with $j \in Z$ form an MRA.

Theorem 2.5 *Let $\{V_j\}$ be an MRA with a scaling function $\varphi \in V_0$. The function $\psi \in W_0 = V_1 \ominus V_0 (W_0 \oplus V_0 = V_1)$ is a wavelet if and only if*

$$\hat{\psi}(\xi) = e^{i\xi/2} v(\xi) \overline{m\varphi(\xi/2+\pi)} \hat{\varphi}(\xi/2)$$

for some 2π-periodic function $v(\xi)$ such that $|v(\xi)| = 1$ almost everywhere, where $m_\varphi(\xi) = \frac{1}{2} \sum_{n \in Z} a_n e^{-n\xi}$. Each such wavelet ψ has the property that $Span\{\psi_j\}_{k \in Z, j<s} = V_s$ for every $s \in Z$.

2.2 Wavelet Decomposition

We now describe the wavelet transform of Mallat [89] in two dimensions using the concept of tensor product.

Proposition 2.6 *Let V_j, $j \in Z$, be an MRA of $L_2(R)$. Then $\widetilde{V}_j = V_j \otimes V_j$ is a multiresolution analysis of $L_2(R^2)$.*

Proof. See Daubechies [44]. ■

$\widetilde{W}_j$ is defined as the orthogonal complement of $\widetilde{V}_j$ into $\widetilde{V}_{j+1}$. Thus, we have the following orthonormal bases:

- basis for $\widetilde{V}_j$:
  $(\phi^j_{m,n}) = (\phi_{j,m}(x)\phi_{j,n}(y))_{m,n}$, $m, n \in Z$,

- basis for $\widetilde{W}_j$:
  $(\psi^{j,1}_{m,n}) = (\phi_{j,m}(x)\psi_{j,n}(y))_{m,n}$, $(\psi^{j,2}_{m,n}) = (\psi_{j,m}(x)\phi_{j,n}(y))_{m,n}$,
  $(\psi^{j,3}_{m,n}) = (\psi_{j,m}(x)\psi_{j,n}(y))_{m,n}$, $m, n \in Z$.

2.3 Wavelet Packets

In computing the discrete wavelet transformation (DWT) of a signal $c_j(k)$, we split it into a pair of sequences $c_{j+1}(k)$ and $d_{j+1}(k)$ by the action of filters H and G (the low- and high- pass filters, respectively), that is,

$$c_{j+1} = Hc_j \text{ and } d_{j+1} = Gc_j.$$

In wavelet packets, we apply the operators not only on the sequences c_j's but also on the sequences d_j's. Thus, we can obtain various bases, e.g., the wavelet basis, the Walsh basis, the subband basis etc.

We take $w^0(x) = \phi(x)$ and $w^1(x) = \psi(x)$ in the subsequent definition.

Definition 2.7 *Let $w^0(x)$ be an orthonormal scaling function with corresponding scaling filter $h(n)$. Define the sequence $\{w^m(x)\}_{m\in Z^+}$ of wavelet packets by*

$$w^{2n}(x) = \sum_k h(k)w^n_{1,k}(x)$$

$$w^{2n+1}(x) = \sum_k g(k)w^n_{1,k}(x),$$

where $H = \{h(k)\}$ is the filter satisfying $\sum_{n\in Z} h(n-2k)h(n-2l) = \delta_{k,l}$, $\sum_{n\in Z} h(n) = \sqrt{2}$, and $g(k) = (-1)^k h(1-k)$.

Theorem 2.8 *The collection $\{w^n_{0,k}(x)\}_{k\in Z, n\in Z^+}$ is an orthonormal basis on R.*

Proof. See [137], pp. 347–348. ■

In case of wavelet packets with mixed scales, we have the following theorem.

Theorem 2.9 *Suppose $\mathcal{P}$ is a collection of intervals of the form $A_{j,n}$ $\left(A_{j,n} = [2^{j-1}k(n), 2^{j-1}(k(n)+1))\right)$ for $j, n \in Z^+$, that forms a disjoint partition of $[0,\infty)$, that is,*

(i) if $I, J \in \mathcal{P}$ with $I \neq J$, then $I \cap J = \emptyset$, and

(ii) $\bigcup_{I\in\mathcal{P}} I = [0,\infty)$.

Then the collection $\{w^n_{j,k}(x) \; : \; k \in Z, A_{j,n} \in \mathcal{P}\}$ is an orthonormal basis of R.

Proof. See [137], pp. 352–353. ■

2.4 Best Basis

Wavelet packets provide a family of orthonormal bases for $L_2(R)$. We have many choices to represent the data as the direct sum of orthonormal basis subsets. The optimal representation of the data within the library of wavelet packets is obtained by using the so-called best basis.

Definition 2.10 *A function $\mathcal{M}$ is an additive cost functional if there is a non-negative function $f(t)$ on R such that for all vectors $c \in R^M$ and orthonormal systems $B = \{b_j\} \subseteq R^M$,*

$$\mathcal{M}(c, B) = \sum_j f(|\langle b_j, c\rangle|).$$

Definition 2.11 *Given a vector $c \in R^M$, an additive cost functional $\mathcal{M}$, and a finite collection $\mathcal{B}$, of orthonormal systems in R^M, a best basis relative to $\mathcal{M}$ for c is a system $B \in \mathcal{B}$ for which $\mathcal{M}(c, B)$ is minimized.*

Here, for a given threshold value $0 < \lambda$, we define $\mathcal{M}$ by

$$\mathcal{M}(c, \{b_j\}) = |\{n \ : \ |\langle c, b_j\rangle| \geq \lambda\}|.$$

In the context of signal or image processing, $\mathcal{M}$ measures how many coefficients are "negligible" (i.e., below threshold) in the transformed signal or image and how many are "important".

Nielson [105] has introduced a new class of basic wavelet packets called highly nonstationary wavelet packets. He has considered the representation of the differential operator in such periodic wavelet packets. For a comprehensive discussion on various properties of wavelet packets, we refer to Wickerhauser [139], Nielson [103, 104, 106, 107], Nielson and Zhou [108], Zarowski [145, 146], Manchanda and Siddiqi [93], and references therein. For a comprehensive updated account of wavelet packets, see Wickerhauser [140].

3 Image Processing

An analog image on a domain Ω can be viewed as a function $f(x_1, x_2) = f(x)$ belonging to the Hilbert space $L_2(\Omega)$. The energy of such an image is defined as $\int_\Omega |f(x)|^2 dx$.

In order to sample the analog image into a digital image, we need to fix a grid defined as $N^{-1}Z \times N^{-1}Z$ for some large N. A fine grid is a grid where $N = 2^j$. It is clear that if such grids are denoted by Γ_j, then $\Gamma_j \subseteq \Gamma_{j+1}$. A digital image f_j is a matrix indexed by points in Γ_j. If Ω is a unit square $[0,1] \times [0,1]$, this digital image $f_j \in l_2(\Gamma_j)$ is now a huge matrix $c_{k,l} = f_j(x_1, x_2)$, where $x_1 = k2^{-j}, x_2 = l2^{-j}$, and k and l range from 0 to $2^j - 1$. These entries $c_{k,l}$ are called pixels, and each $c_{k,l}$ measures the gray level of the given image at $(k2^{-j}, l2^{-j})$. This is the case for a black and white image, and a color image has a similar definition with the difference that $c_{k,l}$ is now vector valued. A digital image can be viewed as a vector inside a 4^j-dimensional vector space. The gray levels $c_{k,l}$ are finally quantized with an 8-bit precision which provides 256 gray levels. This discrete representation of an image needs to be compressed for efficient storage or transmission. An image is usually given in a pixel-valued basis, and a wavelet representation is equivalent to a basis change.

For an image $f \in L_2(R^2)$, its approximation at scale 2^j is given by its orthogonal projection onto $\widetilde{V}_j$, i.e.,

$$f_{\widetilde{V}_j}(x,y) = \sum_{m,n \in Z} \left\langle f, \phi^j_{m,n} \right\rangle \phi^j_{m,n}(x,y)\,.$$

This approximation is thus characterized by the sequence $S^j_{m,n} = 2^j \langle f, \phi^j_{m,n} \rangle$. The sequence $S_j = \left(S^j_{m,n}\right)_{m,n \in Z}$ is called discrete approximation of f at the resolution 2^j. The additional details from the scale 2^j to 2^{j+1} are given by its orthogonal projection onto $\widetilde{W}_j$.

$$f_{\widetilde{W}_j}(x,y) = \sum_{\substack{(m,n) \in Z^2 \\ d=1,2,3}} \left\langle f, \psi^{j,d}_{m,n} \right\rangle \psi^{j,d}_{m,n}(x,y)\,.$$

This component is thus characterized by the sequences

- $D^{j,1}_{m,n} = 2^j \left\langle f, \phi_{j,m} \psi_{j,n} \right\rangle$,
- $D^{j,2}_{m,n} = 2^j \left\langle f, \psi_{j,m} \phi_{j,n} \right\rangle$, and
- $D^{j,3}_{m,n} = 2^j \left\langle f, \psi_{j,m} \psi_{j,n} \right\rangle$.

These sequences

$$D^d_j = \left(D^{j,d}_{m,n}\right)_{m,n \in Z},\ d = 1,2,3$$

are called details of f at the resolution 2^{j+1}.

3.1 Image Compression

In the image compression domain, wavelets have been successful in providing a high rate of compression while maintaining good recognizability. Because wavelet coefficients only indicate changes, areas with no change (or very small change) give small or zero coefficients. These small coefficients can be ignored, reducing the number of coefficients that have to be kept to encode the information. The reduced coefficients are then quantized, making possible even the higher compression rates.

In image compression we get a discretized image at resolution, say, 2^j, and the goal is to decompose it into lower resolutions. According to Mallat the algorithm, S_j could be computed from S_{j+1} with the action of a low-pass filter $\widetilde{h}(n)$ followed by a decimation.

$$S^j_{m,n} = \sum_{k,l \in Z} S^{j+1}_{k,l} \widetilde{h}(2m-k, 2n-l).$$

The bi-dimensional filter $\widetilde{h}$ is the tensor product of the same 1-dimensional (1-D) low-pass filter h defined as $\widetilde{h}(m,n) = h(-m)h(-n)$ with $h(n) = 2^{-\frac{1}{2}} \langle \phi_{-1,0}, \phi_{0,n} \rangle$.

Similarly, D^d_j, $d = 1, 2, 3$, can be computed from S_{j+1} by the action of a high-pass filter $\widetilde{g}_d$ ($\widetilde{g}_d$ is the tensor product of 1-D high- and low-pass filters g and h, respectively) followed by the same decimation.

$$D^{j,d}_{m,n} = \sum_{k,l \in Z} S^{j+1}_{k,l} \widetilde{g}_d(2m-k, 2n-l),$$

where $\widetilde{g}_1(m,n) = h(-m)g(-n)$, $\widetilde{g}_2(m,n) = g(-m)h(-n)$, $\widetilde{g}_3(m,n) = g(-m)g(-n)$, and $g(n) = 2^{-\frac{1}{2}} \langle \psi_{-1,0}, \psi_{0,n} \rangle$.

The filters $H = \{h(n) : n \in Z\}$ and $G = \{g(n) : n \in Z\}$ are called quadrature mirror filters (QMF). H and G correspond to low- and high-pass filters, respectively. The rows of the image are filtered by computing their correlation with the low- and high-pass filters H and G followed by $2:1$ decimation. The same procedure is then applied to columns. Thus, we get a four-channel orthogonal subband decomposition using separable QMF. That is, from the discretized image S_j we get the four-channel decomposition $S_{j-1}, D^1_{j-1}, D^2_{j-1}$, and D^3_{j-1}. The same is repeated on channel S_{j-1} and so on. The transformed wavelet coefficients are then quantized [4] and followed by entropy encoding.

Figure 1 represents the general scheme of wavelet decomposition.

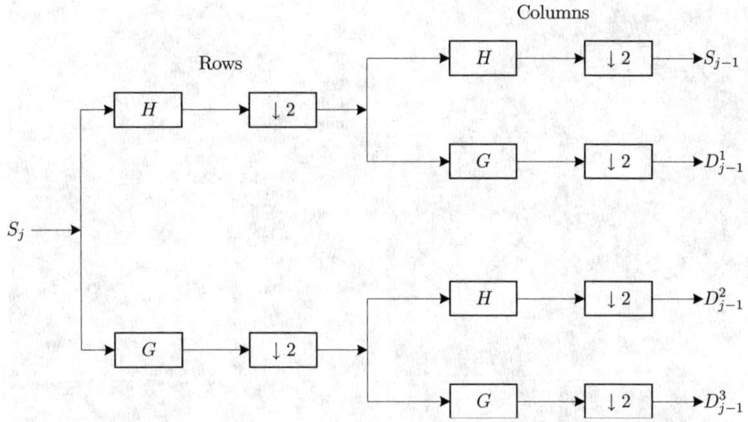

Figure 1: One stage in wavelet decomposition

Compression Results

Ahmad and Siddiqi [1] defined the Information function which measures oscillatory behavior of images. Based on the Information function they classified images and discussed its application in choosing a suitable compression technique. It was found that wavelet image compression (WIC) is the best algorithm for higher compression ratios (more than 16:1) for a certain class of images if the fine details are of interest. A comparison of performance in terms of compression ratio and visual appearance for several kinds of images through fractal, wavelet, and JPEG techniques is also given in Siddiqi, Ahmad, and Mukheimer [123]. Again, it was found that the wavelet technique is better in terms of visual quality of images after compression followed by fractal and JPEG techniques.

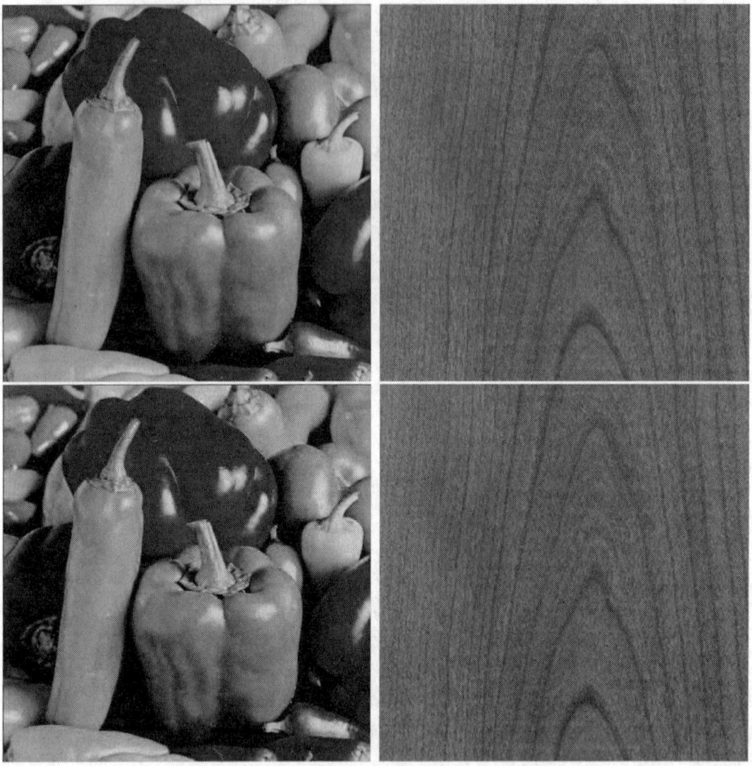

Figure 2: Top: Original images of peppers and Holz8. Bottom: Compressed images with wavelets, CR 32:1 and 20:1, respectively.

3.2 Denoising

The idea of using wavelets for denoising was successfully initiated by Donoho and Johnstone [54], which was followed by a number of articles (see [33, 51]). Consider a given signal to be denoised. Take the discrete wavelet transform of the signal and a suitably chosen positive threshold λ. Define a new wavelet coefficient at a given location and level in terms of the old one as follows:

$$y(t)_{new} = \begin{cases} y(t)_{old} - \lambda & \text{if } y(t)_{old} > \lambda \\ 0 & \text{if } |y(t)_{old}| \leq \lambda \\ y(t)_{old} + \lambda & \text{if } y(t)_{old} < -\lambda. \end{cases}$$

Then apply the inverse wavelet transform and reconstruct the signal.

These ideas have been used in a number of areas including medical images and classical music by R. Coifman and his team at Yale University, as well as in applications to synthetic aperture radar images in collaboration between the Computational Mathematics Laboratory at Rice and M.I.T. Lincoln Laboratory.

Wavelets have also been used as a feature extraction method for discrimination, recognition of handprinted characters, video compression, etc. and have been proven to be very effective.

A systematic study of applications of wavelet methods to various themes is in the references: Siddiqi, Ahmad, and Mukheimer et al. [123], Aldroubi and Unser [3], Brislawn [19], Meyer [97], Chambolle [25] et al. Dhalke, Maass, and Teschke [38], Efi-Foufoula and Kumar [55], Gencay and Seluk [69], Mallat [90], Manchanda and Siddiqi [93], Neunzert and Siddiqi [109], Shumway and Stoffer [121], Strang and Nguyen [128], Teolis [129], and Vetterli and Kovačevic [135].

3.3 Miscellaneous Issues – Quantization and Blowup of Solutions

We briefly discuss quantization and blowup of solutions of well-known equations such as nonlinear heat, Navier-Stokes, and Schrödinger equations. It may be observed that the interaction between image processing and wavelet analysis, in particular, and functional analysis, in general, has also benefitted partial differential equations. In fact, new estimates on wavelet coefficients of functions with bounded variation contribute to a better understanding of blowup phenomena for solutions of some nonlinear evolution equations.

Meyer [97] delivered a series of lectures at the Abdus Salam Centre for Theoretical Physics, Trieste, Italy, where he explained the performance of JPEG-2000 through several modeling of images and talked about "wavelets and functions with bounded variation from image processing to pure mathematics" and the role of oscillations in some nonlinear problems. Among many models, he based his discussion on the Stan Osher and Leonid Rudin model and the related studies concerning nonlinear approximations and the space of functions of bounded variation on R^2 by Cohen, DeVore, Petrochev, and Xu. In this model, the simplified image is assumed to be a function with bounded variation. The efficiency of wavelet-based algorithms will be related to the remarkable properties of wavelet expansions of functions with bounded variation. Usually, the class of such functions is denoted by BV.

In practical computation, the actual coefficients arising in some expansion will be replaced by approximations to a given precision. What happens to the expansion after a quantization is performed is a problem to be addressed. Wavelet expansions have the advantage that the effect of small changes over the coefficients will have only a local effect. This fact is related to the following observation. If $\{\psi_{j,k}(t)\}$ is an orthonormal wavelet basis, and if $m_\lambda, \lambda = (i,j)$ is a multiplier sequence which is used to shrink the wavelet coefficients $d_{i,j} = \langle f, \psi_{c,j} \rangle$, then the corresponding multiplier operator M defined by $M(\psi_{i,j}) = m_{i,j}\psi_{i,j}$ is a Caldéron-Zygmund operator. In contrast, when a trigonometric system is used, any change on any coefficient will affect the resulting function globally. More precisely, let us consider a nonlinear operator $Q_\epsilon(t)$ (quantization) defined by a 2π-periodic function by the following algorithm. One starts with the Fourier series expansion of a 2π-periodic function $f(x)$ and replaces by 0 all coefficients whose absolute value is less than ϵ. We then obtain f_ϵ and write $f_\epsilon = Q_\epsilon(f)$. The following theorem describes the behavior of the operator, $Q_\epsilon(f)$, as ϵ tends to 0.

Theorem 3.1 [97] *For each exponent α less than 1/2 there exists a 2π-periodic function $f(x) = f^\alpha(x)$ belonging to the Hölder space C^α such that L^∞ norm of $Q_\epsilon(f) = f_\epsilon$ tends to infinity as ϵ tends to 0. More precisely,*

$$\|f_\epsilon\|_\infty > C\epsilon^{-\beta},$$

where $C = C(f)$ is a constant and β which is defined by $\beta = \dfrac{1-2\alpha}{1+2\alpha}$ is also positive.

The blowup described by this theorem cannot occur with wavelet expansion. In fact, Hölder space C^α is characterized by size conditions on the wavelet coefficients [45].

The following theorems are very useful in the study of Osher-Rudin and Cohen et al. [28] methods for image processing problems, particularly compression, noise removal, and feature extraction. One can find a detailed discussion and relevant references in [97] problems including compression, noise removal, and feature extraction.

Theorem 3.2 *Let $f(x) \in BV(R^2)$, and let $\psi_{j,k}(t) = 2^{j/2}\psi(2^j t - k)$. $j \in Z$, $k \in Z^2$, is an orthonormal wavelet basis of $L_2(R^2)$ where ψ is a smooth* *wavelet. Then the corresponding wavelet coefficients* $d_{j,k} = \langle f, \psi_{j,k}\rangle$ *satisfy*

$$\sum_j \left\{ \sum_k |d_{j,k}|^p \right\}^{1/p} \le C/(p-1)\|f\|_{BV}, \qquad 1 < p < 2.$$

Theorem 3.3 *Let* $\psi_\Lambda, \lambda \in \Lambda = \{(j,k), \quad j \in Z, k \in Z^2\}$ *be a two-dimensional orthonormal wavelet base described in Theorem 3.2. Let* $d_{j,k}$ *be the corresponding wavelet coefficients. Then for every* $f \in BV(R^2)$, *if* $d_{j,k}$ *is such that* $|d_{j,k}| > \lambda, \ \lambda \in \Lambda$ *are arranged in decreasing sequence* c_n^* *satisfying* $c_n^* \le C/n$ *for* $1 \le n$.

Theorem 3.4 $w(\mu) = inf\{J(u) = \|u\|_{BV} + \mu\|v\|_{L_2}, f = u + v\} = 0(\mu^\gamma)$, $\mu \to \infty$ *if and only if the arranged wavelet coefficients of* $f(x)$ *as in Theorem 3.3 satisfy* $c_n^* = 0(n^{-\alpha})$ *where* $\alpha = 1 - \gamma/2$.

Theorem 3.4 is of vital importance for the Cohen et al. approach where $L_2(R^2)$ is decomposed in two parts and one wants to solve the variational problem

$$w(\mu) = \inf\{J(u) = \|u\|_{BV} + \|v\|_{L_2}, f = u + v\}.$$

The tuning given by the large factor $\lambda = \epsilon^{-1}$ implies that the L_2 norm of v should be of the order of magnitude of ϵ. An interesting mathematical question is to relate growth of $w(\mu)$ as λ tends to infinity to some properties of the given function in L_2. Theorem 3.4 provides an answer to this question.

Meyer has also discussed applications of theorems mentioned above to a better understanding of blowup phenomena for solutions of some nonlinear evolution equations such as nonlinear heat equations, the Navier-Stokes equations, and the nonlinear Schrödinger equations.

4 Partial Differential Equations

Wavelets have been used by a number of authors for solving various problems in differential and integral equations, e.g., one-dimensional nonlinear wave equation (Burger's Equation) by Glowinski et al. [70]; solution of integral equations by Beylkin, Coifman, and Rokhlin [11]; a new multiscale approach to the one-dimensional problem by Bacry, Mallat, and Papanicolaou [5]; work on turbulence problems involving stability issues [114, 138]; a series of papers by Dahmen et al. [39, 40, 42, 43], involving various aspects of using wavelet analysis for solving partial differential equations, and some interesting work on the solution of integral equations on the boundary of elliptic boundary value problems [113].

Wavelets provide efficient algorithms to solve partial differential equations in the following sense.

(a) The quality of approximation, that is, the set of approximation ε to which belongs the computed solution, must be close to the exact solution u, that is, $\inf_{v \in \varepsilon} d(u, v) \ll 1$ and ε need to be small enough to allow the computation of the numerical solution.

(b) The algorithm needs to be fast (less time consuming).

(c) In order to reduce the time of computation, the algorithm needs to select the minimal set of approximation at each step so that the computed solution remains close to the exact solution. This point is called adaptivity. It ensures that no unnecessary quantity is computed.

Numerical algorithms may meet the above requirements. For example, if the solution of partial differential equations we wish to compute is smooth in some regions, only a few wavelet coefficients will be needed to get a good approximation of the solution in those regions. Practically, only the wavelet coefficients of the low frequencies, whose supports are in these regions, will be needed. On the other hand, the greatest coefficients (in absolute value) will be localized near the singularities; this allows us to define and implement the criteria of adaptivity through time evolution. These issues are discussed in detail in [7, 27, 29, 40, 71, 84, 109, 117, 125]. Very often partial differential equations and their boundary conditions are all converted to the wavelet domain, thus reducing the problem to find a solution of algebraic-equations. The resulting algebraic equations are then solved by the well-known numerical methods such as direct methods, e.g., LU and Cholesky factorization, or iterative methods, e.g., conjugate gradient and multigrid.

4.1 Error Estimation by the Coifman Orthogonal System

Coifman wavelets or Coiflets have been found useful in the numerical simulation of the partial differential equations. An orthonormal wavelet system with compact support is called a Coifman wavelet system of degree K if the moments of associated scaling function φ and wavelet ψ satisfy the conditions

$$\begin{aligned}
\mathrm{Mom}_\ell(\varphi) &= \int_R x^\ell \varphi(x)dx = 1 \text{ if } \ell = 0 \\
\mathrm{Mom}_\ell(\varphi) &= \int_R x^\ell \varphi(x)dx = 0 \text{ if } \ell = 1, 2, 3, \ldots, K \\
\mathrm{Mom}_\ell(\psi) &= \int_R x^\ell \psi(x)dx = 0 \text{ if } \ell = 0, 1, 2, 3, \ldots, K.
\end{aligned}$$

To estimate approximations to solutions of partial differential equations, we need a suitable function space in which to measure such approximations. The variational form (Galerkin's method) of the differential equation leads naturally to the use of Sobolev spaces in which to make our estimates. We use the Sobolev spaces $H^n(\Omega)$, for Ω open in $\mathbb{R}^n$, with the norm

$$\|f\|^2_{H^n(\Omega)} = \sum_{|\alpha| \le n} \|D^\alpha f\|_{L_2(\Omega)}.$$

Essentially, $H^n(\Omega)$ is the closure of the space of infinitely differentiable functions with compact support denoted by $C_0^\infty(\Omega)$ with respect to the norm defined here. For noninteger n, one needs to use a Fourier transform type definition. For details, we refer to ([71]), p. 287] where one can find relevant details of the following theorem (one can also see Siddiqi's applied functional analysis book cited in the later part of the chapter).

Theorem 4.1 *Let an orthonormal Coifman wavelet system of degree K with scaling function $\varphi(x)$ be given, and let $\varphi_{j,p}(x)$ denote the dilation and translation of $\varphi(x)$, that is, $\varphi_{j,p}(x) = 2^{j/2}\varphi(2^j x - p)$. Let $f \in C^2(\overline{\Omega})$, where Ω is open and bounded in $\mathbb{R}^n$. If*

$$S^j(f)(x) := \sum_{\{p,q\} \in \Lambda} f(x, y)\phi_{j,p}(x)\phi_{j,q}(y), \ (x, y) \in \Omega,$$

then

$$\|f - S^j(f)\|_{H^1(\Omega)} \le \lambda 2^{-j(K-1)}$$

for some constant λ and degree $K \ge 2$.

By applying Theorem 4.1, we get a better estimation for a solution of an elliptic equation with the Neumann boundary condition ([71], Theorem 12.2).

Theorem 4.2 *Let $\varphi(x)$ be a scaling function, $\{V_j\}$ be an MRA, and G_j be a finite-dimensional space of $V_j \otimes V_j$ and $G_j \subset H^1(\Omega)$. Let u be a solution to the equation*

$$-\Delta u + u = f \text{ in } \Omega \subset R^2$$
$$\frac{\partial u}{\partial n} = g \text{ on } \partial\Omega,$$

and let u^j be a solution to the equation

$$\int \Delta u \Delta h dxdy + \int_\Omega uhdxdy = \int_\Omega fhdxdy + \int_{\partial\Omega} ghds,$$

where $u^j \in G_j, h \in G_j$. Then $\|u - u^j\|_{H^1(\Omega)} \leq C2^{-j(K-1)}$, where C is a positive constant depending on Ω.

The estimates in Theorem 4.2 can also be proved for Daubechies wavelet system.

4.2 Preconditioning Based on Wavelet Properties

Large classes of boundary value problems can be expressed in the form of the operator equation

$$\mathcal{A}u = f, \tag{4.1}$$

where $\mathcal{A}$ is an appropriate operator defined on a Hilbert space H and $f \in H^*$ (dual space of H). Usually, H are function spaces like $L_2(R)$, $H^1(\Omega)$, $H_0^1(\Omega)$ and $H^2(\Omega)$, etc. Using wavelets, one can rewrite the operator as an infinite system of linear equations. The two most important issues related to the numerical solution of the operator equation are preconditioning and adaptive (efficient) schemes. Dahmen [39] has discussed in detail results concerning preconditioning obtained till 1997. In two recent article, one in 2001 [40] and the other jointly with Cohen and DeVore also in the same year [29], an excellent exposition of these two concepts of vital importance, preconditioning and adaptive algorithms can be found. We describe here a few results of these articles. However, interested readers may find a lot of other results in this area in the cited papers. Preconditioning of the Helmholtz problem is also discussed in this section.

For a proper understanding of the presentation, we would like to recall some basic notions and concepts. Let $A \lesssim B$ mean that there exists a positive constant c such that $A \leq cB$, and $A \sim B$ mean that $A \lesssim B$ and let $B \lesssim A$. Let

$$\Psi = \{\psi_\lambda / \lambda \in \mathcal{J}\}, \tag{4.2}$$

where $\mathcal{J}$ denotes some infinite set of indices. Each index $\lambda \in \mathcal{J}$ takes the form $\lambda = (j, k)$, where

$$|\lambda| = j \in Z \quad \text{(set of integers)} \tag{4.3}$$

denotes the scale or level and k represents the location of ψ_λ in space as well as the type of elements. Moreover, we assume that there exist infinitely many nonempty subsets $\mathcal{J}_j = \{\lambda \in \mathcal{J}/|\lambda| = j\}$ so that Ψ is a multiscale system.

A family Ψ is called a generalized wavelet basis for H, if

(a) Ψ is a generalized Riesz basis, that is, it spans H

$$H = \operatorname{clos}_{\|\cdot\|_H} \operatorname{span}\Psi \tag{4.4}$$

and there exists an invertible operator T on $\ell_2(\mathcal{J})$ into itself such that

$$\|\mathbf{d}^t\Psi\|_H = \|\sum_{\lambda\in\mathcal{J}} d_\lambda\psi_\lambda\|_H \sim \|T\mathbf{d}\|_{\ell_2(\mathcal{J})} = \left(\sum_{\lambda\in\mathcal{J}} |(Td)_\lambda|^2\right)^{1/2} \tag{4.5}$$

for $\mathbf{d} = \{d_\lambda\}_{\lambda\in\mathcal{J}}$, where $\mathbf{d}^t$ denote transpose of $\mathbf{d}$.

(b) The functions are local in the sense that the support of ψ_λ scales like

$$\operatorname{diam}(\operatorname{supp}\psi_\lambda) \sim 2^{-|\lambda|}, \quad \lambda \in \mathcal{J}. \tag{4.6}$$

If the operator T in (4.5) is diagonal, then Ψ is called a wavelet basis of H. We consider here compactly supported functions. For $H = L_2(\Omega)$ and $T = I$, the identity operator generalized Riesz basis is a nothing but a Riesz basis. Usually, T is taken as a diagonal matrix. In this case (4.5) implies that a properly scaled system forms a Riesz basis for H. If T is not diagonal, it represents a change of base in the sense that (4.5) implies that $T^{-t}\Psi$ (T^{-t} denotes the usual conjugate transpose) is a Riesz basis for H. Generalized Riesz bases are more suited for the numerical solution of Maxwell equations.

By the Riesz representation theorem, there exists a dual wavelet basis

$$\widetilde{\Psi} = \{\widetilde{\psi}_\lambda : \lambda \in \mathcal{J}\} \subset L_2(\Omega),$$

that is,

$$\langle \psi_\lambda, \widetilde{\psi}_{\lambda'} \rangle_{L_2(\Omega)} = \delta_{\lambda,\lambda'}, \quad \lambda, \lambda' \in \mathcal{J}.$$

It can be shown that $\widetilde{\Psi}$ is a Riesz basis for $L_2(\Omega)$, that is,

$$\left\| \sum_{\lambda \in \mathcal{J}} \widetilde{d}_\lambda \widetilde{\psi}_\lambda \right\| \sim \|\widetilde{\mathbf{d}}\|_{\ell_2(\mathcal{J})},$$

$$\widetilde{\mathbf{d}} = \{\widetilde{d}_\lambda\}_{\lambda \in \mathcal{J}},$$

and $\{\Psi, \widetilde{\Psi}\}$ is referred to as the biorthogonal system. An MRA associated with $\widetilde{\Psi}$ is called the dual MRA.

It may be observed that the Riesz basis may be seen to determine the performance of preconditioners for discretized elliptic problems over the energy space.

Given a wavelet basis, seeking solution u of (4.1) is equivalent to finding the expansion sequence $\mathbf{d}$ of $u = \mathbf{d}^t\Psi$. Inserting this into (4.1) yields $(\mathcal{A}\Psi)^t\mathbf{d} = f$. Now, letting $\Phi = \{\theta_\lambda/\lambda \in \mathcal{J}\}$ be another basis satisfying $\langle v, \Phi\rangle = 0$ implies $v = 0$ for $v \in H^*$. The $(\mathcal{A}\Psi)^t\mathbf{d} = f$ becomes the infinite system

$$\langle \mathcal{A}\Psi, \Phi\rangle^t \mathbf{d} = \langle f, \Phi\rangle^t. \tag{4.7}$$

The objective is to find collections Ψ and Φ for which (4.7) is efficiently solvable. In particular, one can choose either the set of all scaling functions $\widetilde{\Psi}$ or Ψ itself. We choose $\Phi = \Psi$, then a simple diagonal scaling transform $\langle \mathcal{A}\Psi, \Psi\rangle$ into I. Thus, one could search those bases Ψ such that for a suitable diagonal matrix T,

$$\mathcal{B} = T\langle \mathcal{A}\Psi, \Psi\rangle^t T \asymp I \tag{4.8}$$

is spectrally equivalent to the identity, in the sense that $\mathcal{B}$ and its inverse $\mathcal{B}^{-1}$ are bound in the norm $\|\mathbf{d}\|_{\ell_2}^2 = \mathbf{d}^*\mathbf{d}$, where $\mathbf{d}^* = \overline{\mathbf{d}}^t$ is the usual complex conjugate transpose. The principal matrix of the infinite matrix $\mathcal{B}$ corresponds to the stiffness matrices arising from a Galerkin scheme applied to (4.1) based on trial spaces spanned by subsets of Ψ. Relation (4.8) means that these linear systems are uniformly well conditioned. A wide class of operators wavelet bases have that property. The precise choice T depends on $\mathcal{A}$ or on the underlying space. One may observe that in place of H and H^*, two different Hilbert spaces H_1 and H_2 can be chosen, of course, satisfying certain relationships. It may be emphasized [39] that Sobolev spaces play a central role, and the question of preconditioning is intimately connected with the characterization of Sobolev spaces in terms of certain discrete norms induced by wavelet expansions.

It may be recalled that if $\mathcal{A}$ is self-adjoint in the sense that

$$(\mathcal{A}u, v) = a(u, v) \tag{4.9}$$

is a symmetric bilinear form and $a(\cdot,\cdot)$ is elliptic, that is,

$$\|\cdot\|^2 = a(\cdot,\cdot)^2 \sim \|\cdot\|_{H^n(\Omega)}, \tag{4.10}$$

then the Galerkin scheme is trivially stable.

If the trial subspaces have large dimension, then direct solvers based on factorization techniques are very expensive in storage and computing time and consequently are prohibited to be applied. For the symmetric case (4.10) the speed of coverage of iterative methods is known to be governed by the condition numbers

$$k_2(\mathcal{A}_\Lambda) = \lambda_{\max}(\mathcal{A}_\Lambda)/\lambda_{\min}(\mathcal{A}_\Lambda), \tag{4.11}$$

where

$$\lambda_{\max}(\mathcal{A}_\Lambda) = \sup_{v \in S(\Psi_\Lambda)} \frac{\langle Av, v\rangle}{\langle v, v\rangle} \tag{4.12}$$

$$\lambda_{\min}(\mathcal{A}_\Lambda) = \inf_{v \in S(\Psi_\Lambda)} \frac{\langle Av, v\rangle}{\langle v, v\rangle}, \tag{4.13}$$

and

$$S(\psi_\Lambda) = \operatorname{clos}_{H^n(\Omega)}(\operatorname{span}\{\psi_\Lambda\}). \tag{4.14}$$

In view of the norm equivalence we obtain

$$\lambda_{\min}(\mathcal{A}_\Lambda) \le \langle \mathcal{A}\psi_\lambda, \psi_\lambda\rangle / \|\psi_\lambda\|_{L_2} \sim 2^{2n|\lambda|}, \tag{4.15}$$

while

$$\lambda_{\max}(\mathcal{A}_\Lambda) \ge \langle \mathcal{A}\psi_\lambda, \psi_\lambda\rangle / \|\psi_\lambda\|_{L_2} \sim 2^{2n|\lambda|} \tag{4.16}$$

$$k_2(\mathcal{A}_\Lambda) \gtrsim 2^{2n|\Lambda|}, \tag{4.17}$$

where

$$|\Lambda| = \max\{|\lambda| - |\lambda'|/\lambda, \lambda' \in \Lambda \subset \mathcal{J}\}. \tag{4.18}$$

Therefore, in such cases the objective is to find a symmetric definite operator $\mathbf{C}_\Lambda$ such that $k_2(\mathbf{C}_\Lambda \mathcal{A}_A)$ remains possibly uniformly bounded, so that algorithms such as

$$u_\Lambda^{\ell+1} = u_\Lambda^\ell + \mathbf{C}_\Lambda(\mathcal{A}_\Lambda u_\Lambda^\ell - f), \tag{4.19}$$

or, better, correspondingly preconditioned conjugate gradient iterations would converge rapidly.

Let

$$\|T\mathbf{d}\|_{\ell_2(I)} \sim \|\mathbf{d}^{-t}\Psi\|_H, \tag{4.20}$$

where T is a fixed positive diagonal matrix. By combining this with the ellipticity, there exists $c_1, c_2 > 0$ such that

$$c_1\|v\|_{\ell_2(\mathcal{J})} \leq \|\mathcal{A}v\|_{L_2(\mathcal{J})} \leq c_2\|v\|^2_{\ell_2(\mathcal{J})}, \tag{4.21}$$

and

$$c_2^{-1}\|v\|_{\ell_2(\mathcal{J})} \leq \|\mathcal{A}^{-1}v\|_{\ell_2(\mathcal{J})} \leq c_1^{-1}\|v\|_{\ell_2(\mathcal{J})}. \tag{4.22}$$

In particular, the condition number $k = \|\mathcal{A}\| \, \|\mathcal{A}^{-1}\|$ of $\mathcal{A}$ satisfies $k \leq c_2 c_1^{-1}$. Then $\mathcal{B} = T\langle \mathcal{A}\Psi, \Psi\rangle^t T$ is an isomorphism on $\ell_2(\mathcal{J})$. Denoting by $b_{\lambda,\lambda'}$ the entries of $\mathcal{B}$ and by $\mathcal{B}_\Lambda = (b_{\lambda,\lambda'})_{\lambda,\lambda'\in\Lambda}$ the section $\mathcal{B}$ restricted to the set Λ, we obtain from the positive definiteness of $\mathcal{B}$ that

$$\|\mathcal{B}_\Lambda\| \leq \|\mathcal{B}\|, \|\mathcal{B}_\Lambda^{-1}\| \leq \|\mathcal{B}^{-1}\|. \tag{4.23}$$

and that the condition of the submatrices remains uniformly bounded for any subset Λ of $\mathcal{J}$, that is,

$$k(\mathcal{B}_\Lambda) = \|\mathcal{B}_\Lambda\| \, \|\mathcal{B}_\Lambda^{-1}\| \leq k. \tag{4.24}$$

A typical example of the above setting involves $H = H^1(\Omega)$, $\mathcal{A} = -\Delta$, and $T_{\lambda;\lambda'} = 2^{-|\lambda|}\delta_{\lambda,\lambda'}$. For more results, see ([2] pp. 36-37).

Theorem 4.3 *Let $A : H \to H^*$ be $H-$elliptic, and suppose $T^{-t}\Psi$ and $T^t\widetilde{\Psi}$ are generalized wavelet bases for H and H^*, i.e.,*

$$\|T\mathbf{d}\|_{\ell_2} \sim \|\mathbf{d}^t\Psi\|_H, \quad \|T^{-t}\mathbf{d}\|_{\ell_2} \sim \|\mathbf{d}^t\widetilde{\Psi}\|_{H^*}, \tag{4.25}$$

for all $\mathbf{d} \in \ell_2$ and an invertible operator $T \in GL(\ell_2, \ell_2)$. Then

1. *the function $u = \mathbf{d}^T\Psi \in H$ solves the original operator equation (4.9) if and only if the sequence*

$$u := T\mathbf{d}$$

solves the matrix equation

$$\mathcal{B}u = \mathbf{f},$$

where $\mathbf{f} := T^{-t}\langle\Psi, f\rangle$.

2. *for* $\mathcal{B} = (b_{\lambda,\mu})_{\lambda,\mu\in\mathcal{J}}$, *the operators*

$$\mathcal{B}_\Lambda := (b_{\lambda,\mu})_{\lambda,\mu\in A}, \quad \Lambda \subset \mathcal{J},$$

have uniformly bounded spectral condition numbers $k_2(\mathcal{A}_\Lambda) \leq k, \;\; \Lambda \subset \mathcal{J}$.

Preconditioning of Helmholtz Problem

Let

$$\begin{aligned} -\epsilon\Delta u + u &= f \text{ in } \Omega, \epsilon \ll 1 \\ u &= 0 \text{ on } \Gamma = \partial\Omega. \end{aligned} \tag{4.26}$$

If we consider the corresponding differential operator $\mathcal{A}$ as a mapping from $H_0^1(\Omega)$ to its dual $H^{-1}(\Omega)$, $\mathcal{A}$ is elliptic. The constants in (4.26) depend on ϵ so that $k(\mathcal{A}) = O(\delta^{-1})$, which is not satisfactory. In order to obtain a bounded invertible ℓ_2-problem with ellipticity constant independent on ϵ, that is, robust, we have to consider $\mathcal{A}$ on the energy space $H_0^1(\Omega)$-functions equipped with the norm

$$\|u_\epsilon\|^2 = a_\epsilon(u,u), a_\epsilon(u,v) = \epsilon\langle \text{grad } u, \text{grad } v\rangle_{0,\Omega} + \langle u,v\rangle_{0,\Omega}. \tag{4.27}$$

Denoting the induced Hilbert space by H_ϵ, one can, in fact, easily show that $\mathcal{A}$ is H_ϵ-elliptic independent on ϵ. In order to obtain the desired preconditioner, we have to check the equivalence of norms, that is, (4.25). Assuming that Ψ characterizes both $L_2(\Omega)$ and $H_0^1(\Omega)$ in the sense of

$$\left\| \sum_{\lambda\in\mathcal{J}} d_\lambda \psi_\Lambda \right\|_{m,\Omega} \sim \left(\sum_{\lambda\in\mathcal{J}} 2^{2m|\lambda|} |d_\lambda|^2 \right) \tag{4.28}$$

for $m = 0, 1$, we have

$$a_\epsilon(u,v) = \epsilon\| \text{ grad } u\|_{0,\Omega}^2 + \|u\|_{0,\Omega}^2 \sim \sum_\lambda (\epsilon 2^{2|\lambda|} + 1) d_\lambda^2,$$

for $u = \mathbf{d}^t\Psi$, so that we get

$$T = \text{diag } (\epsilon 2^{2|\lambda|} + 1)^{1.2}.$$

For further details see Cohen and Masson [32] and Cohen et al. [29, 30].

4.3 Illustration of Adaptive Wavelet Methods for Elliptic Operators

For many operators, matrices $\mathcal{B}$ in (4.8), as well as their measures, are nearly sparse. Then it means that replacing entries below a given threshold by zero yields a sparse matrix. When $\mathcal{A}$ is a differentiable operator and the wavelets have compact support, then this statement holds true, even it remains true for certain integral operators [39]. Quantifying this sparcification will depend on $\mathcal{A}$ and on certain properties of the wavelet bases. Once we are in a position to track these wavelets in Ψ needed to represent the solution u of (4.1) accurately, we can, in principle, restrict the computations to the corresponding subspaces. According to Dahmen [39, 42], combining this with the sparse representation of operators is one of the most promising prospectives of wavelet concepts. One may find a lucid presentation of current developments of this theme in the papers of Dahman [39], Dahmen [40], and Cohen, Dahmen, DeVore [29]. All previous relevant references can be seen there. We introduce here this concept based on the presentation in [29, 40].

Developing adaptive solvers is one of the areas of vital importance where wavelet methods can greatly contribute to scientific large-scale computations. Adaptivity has been the subject of numerous studies from different perspectives. Till recently, practically nothing was explored. Keeping the importance of this theme, Dahmen along with his coworkers has vigorously pursued, it and the main results are reported in [29].

Let $\{\psi_\lambda\}_{\lambda,\in\mathcal{J}}$ be a wavelet basis to be used for numerical simulation of the elliptic equation (4.9). The adaptive scheme (algorithm) produces finite sets $\Lambda_j \subset \mathcal{J}$, $j = 1, 2, \ldots$, and the Galerkin approximation u_{Λ_j} to u from the space $S_{\Lambda_j} = \text{span}(\{\psi_\lambda\}_{\lambda\in\Lambda_j})$. The function u_{Λ_j} is a linear combination of $N_j = \#\Lambda_j$ wavelets. Thus, the adaptive method can be viewed as a particular form of nonlinear N-term wavelet approximation, and a benchmark for the performance of such an adaptive method is provided by comparison with best N-term approximation with respect to the energy norm when full knowledge of u is available.

An important feature of N-term approximation is that a near best approximation is produced by thresholding, that is, simply keeping the N largest contributions (measured in the same metric as the approximation error) of the wavelet expansion of v. Ideally, an optimal adaptive wavelet algorithm should produce a result similar to thresholding the exact solution. More precisely, this means that whenever the solution u is in Besov space B^s, the approximate u_{Λ_i} should satisfy

$$\|u - u_{\Lambda_j}\| \leq C\|u\|_{B^s} N_j^{-s}, \quad N_j = \#\Lambda_j, \tag{4.29}$$

where $\|\cdot\|$ is the energy norm, **C** is the positive constant, and $\|u\|_{B^s}$ is the norm B^s (see [39], or [40] for Besov spaces). It is a larger class than the corresponding Sobolev space H^s. One can also find details of Besov space in [93, 25] and the references therein. In practice, one is generally interested in controlling a prescribed accuracy with a minimal number of parameters; we shall say that the adaptive algorithm is of optimal order $s > 0$ if whenever the solution is in B^s, then for all $\epsilon > 0$, there exists $j(\epsilon)$ such that

$$\|u - u_{\Lambda_j}\| \leq \epsilon, j \geq j(\epsilon) \tag{4.30}$$

and such that

$$\#(\Lambda_{j(\epsilon)}) \leq C\|u\|_{B^s}^{1/s} \epsilon^{-1/s}. \tag{4.31}$$

Such a property ensures an optimal memory size for the description of the approximate solution. An adaptive algorithm is called computationally optimal if, in addition to (4.30)-(4.31), the number of arithmetic operations required to derive u_{Λ_j} is proportional to $\#\Lambda_j$. Cohen, Dahmen, and DeVore [29] have developed and analyzed an adaptive scheme which for a wide class of operator equations, including those of negative order, is optimal with respect to best N-approximation and is also compuationally optimal in the above sense. A simplified version of this algorithm has been developed by Cohen and Masson (see reference above) and a more elaborate version of this has been studied and tested by Barinka et al. [7].

4.4 Evolution Equation

Let us consider the following *evolution partial differential equation*,

$$(\text{EPDE}) \begin{cases} \frac{\partial u}{\partial t} + Au = 0 \\ \text{with the periodic boundary conditions,} \\ u(x+1,t) = u(x,t) \\ \text{and initial condition} \\ u(x,0) = u_0(x), \end{cases}$$

where the operator A is linear or nonlinear of space variables.

Let V be a Hilbert space. Then u is a weak solution of (EPDE) in V of the following problem:

$$(\text{P}) \begin{cases} \text{Find } u \in V \text{ so that for all } v \in V \\ \langle u_t, v\rangle + \langle Au, v\rangle = 0 \\ \langle u(\cdot,0) - u_0, v\rangle = 0. \end{cases}$$

For the subspaces $\{V_j\}$ of the MRA, the problem (P) can be converted into the following approximate problem (P_J):

$$(\mathrm{P}_J) \begin{cases} \text{Find } u_J \in V_J \text{ so that for all } v_J \in V_J \\ \qquad\qquad \langle u_{Jt}, v\rangle + \langle Au_J, v\rangle = 0 \\ \qquad\qquad \langle u_J(\cdot, 0) - P_J u_0, v\rangle = 0 \end{cases}$$

for some positive integer J.

Now consider the Burger's equation with the boundary conditions studied in [78] and [79].

$$(\mathrm{BE}) \begin{cases} \frac{\partial u}{\partial t} + u\frac{\partial u}{\partial x} = \kappa \frac{\partial^2 u}{\partial x^2} \\ u(x+1, t) = u(x, t) \\ u(x, 0) = u_0(x), \end{cases}$$

where κ is a small positive real number. Let Δt be the time step and $u_n(x) = u(x, n\Delta t)$, where u is the solution of (BE).

The time discretization is given by

$$\frac{u_{n+1} - u_n}{\Delta t} + u_n \frac{\partial u_n}{\partial x} = \kappa \frac{\partial^2 u_{n+1}}{\partial x^2},$$

which can also be written as

$$\begin{cases} \left(\mathrm{I} - \kappa\Delta t \frac{\partial^2}{\partial x^2}\right) u_{n+1} = u_n - \Delta t u_n \frac{\partial u_n}{\partial x}, \\ u_{n+1}(0) = u_{n+1}(1) \end{cases}$$

where I stands for the identity operator. Liandrat, Perrier, and Tchamitchian [79] introduced a family of functions $\theta_{j,k},\ 0 \le j \le J-1,\ 0 \le k \le 2^j - 1$, as

$$\left(I - \kappa\Delta t \frac{\partial^2}{\partial x^2}\right) \theta_{j,k} = \psi_{j,k}$$

and computed the wavelet coefficients of u_{n+1} by

$$\langle u_{n+1}, \psi_{j,k}\rangle = \langle u_n, \theta_{j,k}\rangle + \frac{\Delta t}{2} \left\langle u_n^2, \frac{\partial \theta_{j,k}}{\partial x}\right\rangle.$$

The use of wavelets as basis functions for the discretization of PDEs has been of great success [12, 117, 125]. The properties of MRA seem to be a generalization of finite element methods with some characteristics of multigrid methods. It is the localizing ability of wavelet expansions that gives rise to sparse operators and good numerical stability to method.

For wavelets fundamental solutions to the heatlets, we refer to Shen and Strang [120].

5 Optimization Problems of Chemical Engineering

In chemical engineering, complex dynamic optimization problems formulated on moving horizons have to be solved on line. A multiscale approach based on wavelets has been presented in [13, 16], where a hierarchy of successively, adaptively refined problems are constructed. They are solved in the framework of nested iteration as long as the real-time restrictions are fulfilled. In [14], it is shown that by using properly scaled wavelets for the formulation of the problem, the condition number of the refinement level can be kept bounded independently. This scaling can be viewed as a scale dependent diagonal preconditioner.

We now illustrate the said optimization problem of chemical engineering and explain its discretization in wavelet domain. The dynamic behavior of the plant is often modeled by a system of differential-algebraic equations, which together with bounds on selected variables form the constraints of optimization. The *control functions* are denoted by $\mathbf{u}$, and the *parameters are denoted* by $\mathbf{p}$. For *state estimation*, both $\mathbf{u}$ and $\mathbf{p}$ are given. The goal is to estimate the output function $\mathbf{y}$, which corresponds to the signals, and the *state functions* $\mathbf{x}$ in the receding fixed time interval $[t_{k-m}, t_k]$. Here t_k denotes the current time. The underlying measurements are discrete and noisy, so they have to be transformed into denoised functions $\mathbf{z}$. Additive model correction terms are introduced into the model equation as functions $\mathbf{v}$ and $\mathbf{w}$, which have to be estimated as well. For given $\mathbf{z} \in (L_2)^{n_y}$, $\mathbf{p} \in R^{n_p}$ and $\mathbf{u} \in (L_2)^{n_u}$ and unknown functions

$$\mathbf{x} \in (H^1)^{n_x}, \ \mathbf{y} \in (L_2)^{n_y}, \ \mathbf{v} \in (L_2)^{n_v}, \text{ and } \mathbf{w} \in (L_2)^{n_w}$$

with $n_w \leq n_x$, $n_v \leq n_y$, and time-invariant indicator matrices $\mathbf{W}, \mathbf{V}$, the resulting optimization problem of size n_x has the form

$$\min_{\mathbf{x},\mathbf{y},\mathbf{v},\mathbf{w}} \int_{t_{k-m}}^{t_k} \left\{ (\mathbf{y} - \mathbf{z})^t \mathbf{Q} (\mathbf{y} - \mathbf{z}) \right\} d\tau, \tag{5.1}$$

subject to the constraints

$$\dot{\mathbf{x}} - \mathbf{A}x - \mathbf{W}\mathbf{w} = \mathbf{B}u \tag{5.2}$$

$$\mathbf{y} - \mathbf{C}x - \mathbf{V}\mathbf{v} = 0. \tag{5.3}$$

For the discretization of the optimization problem on a finite time horizon $[t_{k-m}, t_k]$, the horizon is scaled to $[0, 1]$, and the following equality is obtained:

$$\langle \dot{\mathbf{x}} - \mathbf{Ax} - \mathbf{Ww}, \boldsymbol{\nu}_1 \rangle = 0, \ \forall \boldsymbol{\nu}_1 \in (L_2)^{n_x} \tag{5.4}$$

$$\langle \mathbf{y} - \mathbf{Cx} - \mathbf{Vv}, \boldsymbol{\nu}_2 \rangle = 0, \ \forall \boldsymbol{\nu}_2 \in (L_2)^{n_y}. \tag{5.5}$$

As dicussed by Binder et al. [13], this problem may be ill-posed and to overcome this difficulty, that is, to guarantee continuity as well as uniqueness of solution, the Tikhonov regularization process is adopted. Thus, (5.1) takes the form

$$\min_{\mathbf{xy},\mathbf{vw}} \int_{t_k-m}^{t_k} \{(\mathbf{y}-\mathbf{z})^t \mathbf{Q}(\mathbf{y}-\mathbf{z}) + \mathbf{v}^t \mathbb{R}_v \mathbf{v} + \mathbf{w}^t \mathbb{R}_w\} d\tau. \tag{5.6}$$

(5.6) together with constraints (5.2) to (5.3).

In principle, the weights $Q, \mathbb{R}_v, \mathbb{R}_w$ could be the operator and may be chosen in such a way that the cost function is equivalent to the square of the Sobolev norm.

Let

$$\mathcal{Y} = n(H^1)^{nx} \times (L_2)^{n_y+n_v+n_w} \tag{5.7}$$

$$\mathcal{M} = (L_2)^{n_x+n_y}. \tag{5.8}$$

The weak formulation of the necessary conditions of the minimization problem (5.6), (5.2), (5.3) can be expressed as follows: find $v \in \mathcal{Y}$ and the Lagrange multiplier $\lambda = (\mu^t, v^t)^t \in \mathcal{M}$ such that

$$a(v, v') + b(v, v') = \langle 2\mathbf{Q}z, v' \rangle, \quad \text{for all } v' \in \mathcal{Y}, \tag{5.9}$$

$$b(\xi, v) = \langle \xi_1, \mathbf{B}u \rangle, \quad \text{for all } \xi \in \mathcal{M}, \tag{5.10}$$

where

$$a(v, v') = \langle y, \mathbf{Q}y' \rangle + \langle v, \mathbb{R}_v v' \rangle + \langle w, \mathbb{R}_w w' \rangle \tag{5.11}$$

$$b(\xi, v) = \langle \xi_1, \dot{x} - \mathbf{A}x - \mathbf{W}_w \rangle + \langle \xi_2, \mathbf{y} - c\mathbf{x} - V_v \rangle. \tag{5.12}$$

By the Riesz representation theorem, we seek the formulation problem (5.11)–(5.12) which is equivalent to the operator equation

$$\mathcal{L} \begin{bmatrix} v \\ \lambda \end{bmatrix} = \begin{bmatrix} h(z, u) \\ g(z, u) \end{bmatrix} \tag{5.13}$$

with unknown $v \in \mathcal{Y}$ and $\lambda \in \mathcal{M}$. Thus, in this setting, the solution λ of the dual optimization problem is automatically part of the solution. In view of the well posedness of the problem, the operator $\mathcal{L}$ is a topological isomorphism from $\mathcal{Y}\times\mathcal{M}$ to the dual $\mathcal{Y}^*\times\mathcal{M}^*$, that is, there exists positive constants $c, \widetilde{c}$ such that

$$c(\|v\|_{\mathcal{Y}}^2 + \|\lambda\|_{\mathcal{M}}^2) \leq \left\| \mathcal{L} \begin{bmatrix} v \\ \lambda \end{bmatrix} \right\|_{(H^1)^{n_x})^* \times (L_2^{n_x+2n_y+n_v+n_w})} \leq \widetilde{c}(\|v\|_{\mathcal{Y}}^2 + \|\lambda\|_{\mathcal{M}}^2). \tag{5.14}$$

Following the terminology of [14], let $\Psi = \{\psi_{j,k}(x)\}$ and $\widetilde{\Psi} = \{\widetilde{\psi}_{j,k}(x)\}$ satisfy $\langle \Psi, \widetilde{\Psi} \rangle = I$. The corresponding primal multiresolution is generated by the classical hat function

$$\varphi(x) = 1 - |x| \quad \text{for } x \in [-1, 1]$$
$$\varphi(x) = 0 \qquad \text{elsewhere.}$$

For the last component w of the space $\mathcal{Y}$, the Haar basis Ψ^H is chosen. The generating scaling function is $\varphi^H = \chi_{[0,1]}$, since the Haar basis is orthonormal $\ll \psi^H, \psi^H \gg = I$, where $\ll \cdot, \gg$ denotes the matrix given by

$$\left[\left\langle \psi_{j,k}(x), \widetilde{\psi}_{(j,k)}(x) \right\rangle \right]_{(j,k)\in\in\Lambda,(j',k')\in\Lambda'}.$$

Similarly, all components of the first group of Lagrange multipliers μ are discretized with the aid of the Haar basis Ψ^H, while the dual wavelets in $\widetilde{\Psi}$ are used for the second group of components of the Lagrange multipliers ν in $\mathcal{M}_\Lambda$. Thus, the state equations (5.2) and (5.3) are tested respectively by $\psi_{j,k}^H$ and $\widetilde{\psi}_{j,k}$ dual to the piece linear wavelets $\psi_{j,k}$ determined by a MRA order 2 adopted to $[0, 1]$. It may be noted that the chosen collection of wavelets forms bases for the relevant function spaces. Hence, we obtain an infinite-dimensional but discretized problem formulation which is still equivalent to (5.9) and (5.10). The restriction to a finite index set of employed basis functions

$$\Lambda \subseteq \{(j, k)/j \geq j_0,\ k \in \Lambda_j\}$$

leads to the finite-dimensional problem essential for the numerical treatment. By Binder et al. ([14], P.507–509), the restriction of (5.9) and (5.10) equivalently (5.13) to $(\mathcal{Y}_\Lambda, M_\Lambda)$ yields the following Karush-Kuhn-Tucker

(KKT) system

$$\begin{bmatrix} & & & & X_\Lambda^T & -C_\Lambda^T \\ & 2Q_\Lambda & & & & I_\Lambda \\ & & 2R_{r_\Lambda} & & & -V_\Lambda^T \\ & & & 2R_{W_\Lambda} & -W_\Lambda^T & \\ X_\Lambda & & & -W_\Lambda & & \\ -C_A & I_\Lambda & -V_\Lambda & & & \end{bmatrix} \begin{bmatrix} d_x \\ d_y \\ d_v \\ d_w \\ d_\mu \\ d_\nu \end{bmatrix} = \begin{bmatrix} 0 \\ 2Q_\Lambda d_z \\ 0 \\ 0 \\ B_\Lambda d_u \\ 0 \end{bmatrix}, \quad (5.15)$$

where the Kronecker product of a matrix $A \in R^{n\times m}$ with a matrix B is defined by

$$A \otimes B := (a_{i,j} B)_{\substack{i=1,\ldots,n \\ j=1,\ldots,m}}$$

and

$$Q_\Lambda = Q\otimes \ll \Psi_{\Lambda_y}, \Psi_{\Lambda_y} \gg, \quad R_{v_\Lambda} = R_v \otimes \ll \Psi_{\Lambda_v}, \Psi_{\Lambda_v} \gg, \quad (5.16)$$

$$R_{w_\Lambda} = R_w \otimes I_{\Lambda_w},$$

$$X_\Lambda = I_{n_x} \otimes \ll \Psi^H_{\Lambda'_x}, \Psi_{\Lambda_x} \gg - A\otimes \ll \Psi_{\Lambda'_x}, \Psi_{\Lambda_x} \gg, \quad (5.17)$$

$$W_\lambda = W\otimes \ll \Psi^H_{\Lambda'_x}, \Psi^H_{\Lambda_x} \gg,$$

$$I_\Lambda = I_{n_y} \otimes I_{\Lambda_y}, C_\Lambda = C\otimes \ll \widetilde{\Psi}_{\Lambda_y}, \Psi_{\Lambda_x} \gg, \quad (5.18)$$

$$V_\Lambda = V\otimes \ll \widetilde{\Psi}_{\Lambda_y}, \Psi_{\Lambda_y} \gg,$$

$$B_\Lambda = B\otimes \ll \Psi^H_{\Lambda'_x}, \Psi^H_{\Lambda_x} \gg . \quad (5.19)$$

The linear system (5.15) is nothing but the Galerkin discretization of (5.13) with respect to the finite-dimensional spaces $\mathcal{Y}_\Lambda, \mathcal{M}_\Lambda$. The system in (5.15) has the following block structure:

$$\mathcal{L}_\Lambda \begin{bmatrix} v_\Lambda \\ \lambda_\Lambda \end{bmatrix} = \begin{bmatrix} \mathcal{A}_\Lambda & \mathcal{B}^t_\Lambda \\ \mathcal{B}_\Lambda & 0 \end{bmatrix} \begin{bmatrix} v_\Lambda \\ \lambda_\Lambda \end{bmatrix} = \begin{bmatrix} h_\Lambda \\ g_h \end{bmatrix}. \quad (5.20)$$

Numerical algorithms and wavelet-based preconditioning of (5.20) are discussed in ([14], pp.510–524).

It has been shown ([14], Theorem 5.1) that if $\mathcal{L}$ is well-posed with respect to $\mathcal{Y}\times\mathcal{M}$ and the used Galerkin scheme is stable, then for diagonal matrix T with diagonal entries $2^j, (j,k) \in \Lambda$, $T_\Lambda^{-1}\mathcal{L}_\Lambda T_\Lambda^{-1}$ is uniformly bounded.

6 Black-Scholes Model of Option Pricing

In 1973, Black and Scholes [15] and Merton [100] published their papers on the theory of option pricing. Since then the growth of the field of derivative securities has been phenomenal. The option price theory, widely known as the Black-Scholes model, is acclaimed to be the most successful theory in finance and economics. In recognition of their fundamental contributions to the pricing theory of derivatives, Scholes and Merton received the 1997 Nobel Prize in economics. Unfortunately, Black was not alive to receive the award.

Willmott, Dewynne, and Howison [141, 142] have popularized option pricing, especially among mathematicians, and elaborately presented its numerical methods. Since numerical methods are of vital importance for finding and visualizing solutions of European and American options, the Newton Institute of Mathematical Sciences [118], Cambridge University, took the initiative to organize a conference on this theme. The papers of Barles [6], Broadie, and Detemple ([118], pp. 43–66) survey the numerical methods for European options, while the papers of Aithlia and Carr ([118], pp. 67–68) and Zhong [118], pp. 93–114 present numerical methods for American options. Siddiqi, Manchanda, and Kocvara [124] have studied the application of an iterative two-step algorithm developed by Kocvara and Zowe [85] to American option pricing. They have demonstrated that the new algorithm SSORP-PCG outperforms the SORP method, even though the problem data favor the latter. The concept of wavelets has been used in finance and economics by Gencay and Seluk [69]. Time domain decomposition for European options has been carried out by Crann et al. [35]. A finite element approach to the pricing of discrete lookbacks with stochastic volatility has been discussed by Forsyth, Vetzal, and Zvan [63]. Barucci, Polidoro, and Vespri [8] have analyzed partial differential equations arising in the evaluation of Asian options which are strongly degenerate. They have shown that the solution of the no-arbitrage partial differential equation is sufficiently regular and so classical finite difference methods had been applied to approximate the solution.

Now we examine the wavelet-based methods for European and American options based on [125].

6.1 Wavelet-Based Method for European Option

It is well known [141, 142] that a European call option $C(S,t)$ is a solution of the following boundary value problem:

$$
\text{(EC)}\begin{cases} \frac{\partial C}{\partial t}+\frac{1}{2}\sigma^2 S^2\frac{\partial^2 C}{\partial S^2}+rS\frac{\partial C}{\partial S}-rC=0 \\ C(S,T)=\max(S-E,0) \\ C(0,t)=0 \\ C(S,t)\to S \text{ as } S\to\infty, \end{cases}
$$

where S, σ, r, E, and T are underlying asset, volatility, interest rate, exercise price, and expiry time, respectively. On the other hand, a European put option $P(S,t)$ is a solution of the following boundary value problem:

$$
\text{(EP)}\begin{cases} \frac{\partial P}{\partial t}+\frac{1}{2}\sigma^2 S^2\frac{\partial^2 P}{\partial S^2}+rS\frac{\partial P}{\partial S}-rP=0 \\ P(S,T)=\max(E-S,0) \\ P(0,t)=Ee^{-r(T-t)} \text{ if } r \text{ is independent of time} \\ P(0,t)=Ee^{-\int_t^T r(\tau)d\tau} \text{ if } r \text{ is dependent of time.} \end{cases}
$$

As $S\to\infty$, the option is unlikely to be exercised, and so, $P(S,t)\to 0$ as $S\to\infty$.

Equations (EC) and (EP) are known as the Black-Scholes models [15] for call and put options, respectively. The Black-Scholes call option model can be transformed into the linear diffusion equation

$$
\text{(DE)} \qquad \frac{\partial u}{\partial \tau}=\frac{\partial^2 u}{\partial x^2} \quad \text{for } -\infty<x<\infty, \tau>0
$$

with

$$
u_0(x)=u(x,0)=\max(e^{\frac{1}{2}(k+1)x}-e^{\frac{1}{2}(k-1)x},0)
$$

$$
S=Ee^x,\ \ t=T-\tau/\frac{1}{2}\sigma^2
$$

and

$$
C(S,t)=Ee^{-\frac{1}{2}(k-1)x-\frac{1}{4}(k+1)^2\tau}u(x,\tau)
$$

where $k=\frac{r}{1/2\sigma^2}$.

Similarly, the Black-Scholes put option model can be written in the form of a linear diffusion equation with approximate boundary conditions.

It has been shown that [56, 71] a class of wavelet-based algorithms for linear evolution equations are faster than classical methods such as finite differences. The solution of (DE) is given by

$$
P(x,\tau)=\frac{1}{\sqrt{4\pi\tau}}\int_{\mathbb{R}} e^{-|x-y|^2/4t}u_0(y)dy. \tag{6.1}
$$

Greengard and Strain [73, 74] have developed an algorithm to evaluate the discretized version of (6.1), which is several thousand times faster than direct methods. An improvement of this algorithm based on wavelet transformation has been carried out.

The variational formulation of (DE) is given as

$$\langle \frac{\partial u}{\partial t}, v\rangle + \int_0^1 \frac{\partial u}{\partial x}\frac{\partial v}{\partial x}dx = 0, \forall v \in H^1(\mathbb{R}). \tag{6.2}$$

The approximate problem of (DE) which is equivalent to (6.2) can be formulated as follows.

Find a function $u_n(t)$, for $t > 0$, such that

$$\begin{aligned} \int_0^1 \frac{\partial u_n(t)}{\partial t} v dx + \int_0^1 \frac{\partial u_n(t)}{\partial x}\frac{\partial v}{\partial x}dx &= 0 \\ u_n(0) &= u_{0n}, \end{aligned}$$

where u_{0n} is the L_2 projection $P_n(u_0)$ of the initial data u_0 on V_n. This is equivalent to a system of first order ordinary differential equations and can be handled by the well-known Lax-Wendroff scheme (see, for example, Quateroni and Valli [115], pp. 469–472).

7 Wavelets for Financial Time-Series Analysis

Capobianco [23] presented the application of wavelets to nonstationary time series with an aim of detecting the dependence structure which is typically found to characterize dependence intraday stock index financial returns. With very high-frequency data one must consider these features in order to build statistical models which are able to achieve reliable inference results.

The wavelets basically adopt a flexible degree of smoothing; by increasing the resolution level j we decrease smoothing and vice versa. We have the choice of working with so-called decimated wavelets instead of stationary, i.e., undecimated, ones. When we try to link wavelet and financial time-series models, the principle of *parsimony* in model parameters may be matched to that of sparsity of signal representation, i.e., the ability to approximate a function by using relatively few coefficients. Given that a better reconstruction might be crucial for financial time series in order to capture the underlying volatility structure, data denoising can be useful. A widely employed threshold for denoising the financial time series which adapt to each resolution is given in [24]. It is known as the Stein unbiased risk estimator (SURE) and is defined as

$$\lambda_j = \arg\min_{t \geq 0} SURE(d_j, t)$$

$$SURE(d_j, t) = K - 2\sum_{k=1}^{K} I_{[|d_{j,k}| \leq t\sigma_j]} + \sum_{k=1}^{K} \min\left[\left(\frac{d_{j,k}}{\sigma_j}\right)^2, t^2\right]$$

$$\widetilde{f}(x) = \sum_k \widetilde{c}_{j',k}\phi_{j',k} + \sum_{j>j'}\sum_k \operatorname{sgn}\left(\widetilde{d}_{j,k}\right)\left(|\widetilde{d}_{j,k}| - \lambda_j\right) + \psi_{j,k}(x).$$

where

$$\widetilde{d}_{j,k} = (1/n)\sum_{i=1}^{n} \psi_{j,k}(x_i), \quad \widetilde{c}_{j',k} = (1/n)\sum_{i=1}^{n} \phi_{j',k}(x_i).$$

For literature on denoising we refer to [33, 51, 54, 122].

8 Maxwell's Equations

The concept of wavelets has been introduced in the applied mathematics literature as a new mathematical object for performing localized *time-frequency* characterization. In recent years a growing attention has been paid to the development of schemes using wavelets for solving Maxwell's equations [41, 68, 67, 75, 80, 111, 119, 131, 132]. The utilization of a wavelet-type basis has the advantage that the condition number of the system matrix does not increase rapidly with an increase in the number of unknowns, unlike the original version of finite element methods. Very sparse coefficient matrices have been obtained due to the vanishing moments, localization, and MRA of the wavelets.

Fujii and Hoefer [68] described the application of wavelets through the *Wavelet-Galerkin scheme* of the time dependent Maxwell's equations. They have used the Deslauriers-Dubuc interpolating function as the scaling function, and the wavelet is the shifted and contracted version of the scaling function. These functions constitute non-L_2 biorthogonal bases that are smooth, symmetric, compactly supported, and exactly interpolating. Unlike the Daubechies orthogonal wavelets [45] of which the interpolation property is limited to the bases of low regularity, the proposed basis set yields a scheme of an arbitrary order of regularity as well as saves the computation cost. For discretization of Maxwell's equations in the wavelet domain, we refer to Fujii and Hoefer [68, 67], Grivet-Talocia [75], and references therein.

We present here briefly the wavelet-based numerical simulation of Maxwell equations [132, 131].

For a given function ρ (the charge density) and σ (electric conductivity of the medium in some domain $\Omega \subset R^3$), as well as the given vector field J_s (the suppressed current density), one is interested in the electric field E, the electric displacement $\mathbf{D}$, the magnetic field $\mathbf{H}$, and the magnetic flux $\mathbf{B}$, which are related by the following set of equations (called Maxwell equations):

$$\text{div } \mathbf{D} = \rho \tag{8.1}$$

$$\text{div } \mathbf{B} = 0 \tag{8.2}$$

$$\frac{\partial \mathbf{B}}{\partial t} + \text{curl } E = 0 \tag{8.3}$$

$$\frac{\partial}{\partial t} - \mathbf{D} \text{ curl } H + \sigma E = -J_s \tag{8.4}$$

If dielectric materials or insulation are linear and isotropic then $\mathbf{B} = \mu\mathbf{H}$, $\mathbf{D} = \epsilon\mathbf{E}$, where μ is the magnetic permeability and ϵ is the electric permeability. Then (8.1)-(8.4) can be written as

$$\frac{\partial}{\partial t}\begin{pmatrix}\mathbf{B}\\ \mathbf{C}\end{pmatrix} + \mathcal{A}\begin{pmatrix}\mathbf{B}\\ \mathbf{D}\end{pmatrix} = \begin{pmatrix}0\\ -J_s\end{pmatrix}, \tag{8.5}$$

where $\mathcal{A} = \begin{pmatrix} 0 & \text{curl } \dfrac{1}{\epsilon(\cdot)} \\ -\text{curl } \dfrac{1}{\mu(\cdot)} & \dfrac{\sigma}{\epsilon} \end{pmatrix}$.

It can be seen that (8.5) is equivalent to

$$\mathbf{u} + \nu \text{ curl } wru = f. \tag{8.6}$$

For details, see Urban [132].

A variational formulation of (8.6) is as follows:

$$a(u, v) = \langle f, v \rangle, \tag{8.7}$$

where $a(u, v)$ is given by

$$a(u, v) = \langle u, v \rangle_{L_2} + \nu \langle \text{ curl } u, \text{curl } v \rangle_{L_2}.$$

The operator $\mathcal{L}$ induced by $a(\cdot, \cdot)$ is a mapping from $\mathbf{H}(\text{curl}; \Omega)$ to the dual space of $\mathbf{H}(\text{curl } ; \Omega)$. It may be recalled that $\mathbf{H}(\text{curl}; \Omega) = \{\xi \in L_2(\Omega)/\text{curl } \xi \in L_2(\Omega)\}$. It is a Hilbert space with respect to the norm

$$\|\xi\|^2_{H(\text{curl};\Omega)} = \|\xi\|^2_{L_2(\Omega)} + \|\text{curl}\xi\|^2_{L_2}$$

$$H_0(\text{curl}; \Omega) = \text{clos}_{H(\text{curl};\Omega)} C_0^\infty(\Omega).$$

Urban [131] has studied wavelet bases in $H(\text{curl};\Omega)$ denoted Ψ^{curl} and applied those concepts to prove the following theorem.

Theorem 8.1 ([132] Theorem 15, p. 117). *The operator $D_{curl} = I + \nu\, curl^* curl)^{1/2}$ and wavelet basis Ψ^{curl} fulfill the assumptions of Theorem 4.3, that is, for $A = a(\Psi^{curl}, \Psi^{curl})$, then the matrix $\mathbf{D}_{curl}^{-t}\mathbf{A}_\Lambda\mathbf{D}_{curl}^{-1}$ is uniformly bounded as $\#\Lambda \to \infty$, independent of ν.*

9 Turbulence Analysis

The disordered fluid flow or chaotic dynamics of motion are usually called *turbulence.* The dynamics of turbulent flows depend not only on different length scale but also on different positions and directions. Consequently, the physical quantities such as energy, velocity, pressure, etc. become highly intermittent. The Fourier transform cannot give the local description of turbulent flows, but the wavelet transform analysis has the ability to provide a wide variety of local information of the physical quantities associated with turbulence.

The classical theory of turbulence was developed in Fourier transform space by introducing the *Fourier energy spectrum* $\widetilde{E}(\xi)$ of a function $f(x)$ in the form

$$\widetilde{E}(\xi) = |\widetilde{f}(\xi)|^2.$$

However, $\widetilde{E}(\xi)$ does not give any local information on turbulence. Since $\widetilde{f}(\xi)$ is a complex function of a real variable ξ, it can be expressed in the form

$$\widetilde{f}(\xi) = |\widetilde{f}(\xi)| \exp\{i\widetilde{\theta}(\xi)\}.$$

The phase spectrum $\widetilde{\theta}(\xi)$ is totally lost in the Fourier transform analysis of turbulent flows, and only the modulus of $\widetilde{f}(\xi)$ is utilized. This is possibly another major weakness of the Fourier energy spectrum analysis of turbulence since it cannot take into consideration any organization of the turbulent field. Therefore, the wavelet transform is adopted to define the space-scale energy density by

$$\widetilde{E}(l,x) = \frac{1}{l}|\widetilde{f}(l,x)|^2,$$

where $\widetilde{f}(l,x)$ is the wavelet transform of a given function $f(x)$. The local energy spectrum $\widetilde{E}(l,x_0)$ in the neighborhood of x_0 is given by

$$\widetilde{E}(l,x_0) = \frac{1}{l}\int_{-\infty}^{\infty} \widetilde{E}(l,x)\chi\left(\frac{x-x_0}{l}\right) dx, \tag{9.1}$$

where the function χ is considered as a filter around x_0. In particular, if χ is a *Dirac delta function*, then the local wavelet energy spectrum becomes

$$\widetilde{E}(l, x_0) = \frac{1}{l}|\widetilde{f}(l, x_0)|^2.$$

The local energy density can be defined by

$$\widetilde{E}(x) = C_\psi^{-1} \int_0^\infty \widetilde{E}(l, x)\frac{dl}{l}.$$

On the other hand, the global wavelet spectrum is given by

$$\widetilde{E}(l) = \int_{-\infty}^{\infty} \widetilde{E}(l, x)dx.$$

This can be expressed in terms of the Fourier energy spectrum $\widetilde{E}(\xi) = |\widetilde{f}(\xi)|^2$, so that

$$\widetilde{E}(l) = \int_{-\infty}^{\infty} \widetilde{E}(\xi)|\widetilde{\psi}(l\xi)|^2 d\xi, \tag{9.2}$$

where $\widetilde{\psi}(l\xi)$ is the Fourier transform of the analyzing wavelet ψ. Thus, the global wavelet energy spectrum corresponds to the Fourier energy spectrum smoothed by the wavelet spectrum at each scale.

Another significant feature of turbulence is the so-called intermittency phenomenon. Farge et al. [58] used the wavelet transform to define the local intermittency as the ratio of the local energy density and the space averaged energy density in the form

$$I(l, x_0) = \frac{|\widetilde{f}(l, x_0)|^2}{\int_{-\infty}^{\infty} |\widetilde{f}(l, x)|^2 dx}. \tag{9.3}$$

If $I(l, x_0) = 1$ for all l and x_0, then there is no intermittency, that is, the flow has the same energy spectrum everywhere, which then corresponds to the Fourier energy spectrum. If $I(l, x_0) = 10$, the point at x_0 constitutes ten times more than average to the Fourier energy spectrum at scale l [58]. This shows a striking contrast with the Fourier transform analysis, which can describe a signal in terms of wave numbers only but cannot give any local information. For more details we refer to [57, 59, 60, 61, 62, 94, 95].

10 Limitations, Ridgelets, and Curvelets

In many important imaging applications, images exhibit edge discontinuities across curves. As, for example, in biological imagery this occurs whenever two different organs or tissue structures meet. [21] showed in his thesis that ridgelets and curvelets are appropriate tools to study these problems. We present these concepts briefly and refer to [22, 50, 52, 53, 127] for updated details.

10.1 Ridgelets

A *ridgelet* is a function of the form

$$\psi_{a,b,u}(x) = \frac{1}{\sqrt{a}}\psi\left(\frac{u \cdot x - b}{a}\right), \tag{10.1}$$

where a and b are scalar and u is a vector of unit length in R^2 (more generally, one can consider ridgelets in R^n). In the sequel, we assume that ψ satisfies $\int_R \frac{|\psi(\zeta)|^2}{|\zeta|^2} d\zeta = 1$. Of course, a ridgelet is a ridge function whose profile displays an oscillatory behavior (like a wavelet).

A ridgelet has a scale a, an orientation u, and a location parameter b. Ridgelets are concentrated around *hyperplanes*; roughly speaking, the ridgelet (10.1) is supported near the strip $\{x \ : \ |u \cdot x - b| \leq a\}$. Like wavelets, one can represent any function as a superposition of these ridgelets. The ridgelet coefficients are defined as

$$\mathbb{R}_f(a, u, b) = \int_R f(x) a^{-1/2} \psi\left(\frac{u \cdot x - b}{a}\right) dx,$$

and then, for any $f \in L_1 \cap L_2(R^d)$, we have

$$f(x) = \left(\frac{1}{2\pi}\right)^{-1} \int \mathbb{R}_f(a, u, b) a^{-1/2} \psi\left(\frac{u \cdot x - b}{a}\right) d\mu(a, u, b),$$

where $d\mu(a, u, b) = da/a^{d+1} du db$.

10.2 Curvelets

As remarked by Candés and Donoho [22], the visual appearance of curvelets does not match the name given to it. The curvelets waveforms look like brushtrokes; brushlets would have been an appropriate name, but it was given by Y. Meyer and R. Coifman in a different context of Gabor analysis. Curvelets exemplify a certain curve scaling low-width = length^2 - which is naturally associated with curves. One can think of a curve in the plane as distribution supported on the curve in the same way that a point in the plane can be thought of as a Dirac distribution supported at that point. The curvelets scheme can be used to represent that distribution as a superposition of functions of various lengths and widths obeying the scaling law-width $\cong$ length^2.

We now briefly describe the curvelet construction which is based on the following ideas:

(a) ridgelets in two dimension,

(b) multiscale ridgelets, and

(c) bandpass filtering or subband filtering.

In two dimensions a system of analysis could be developed based on ridge functions

$$\psi_{a,b,\theta}(x) = a^{-1/2}\psi((x_1 \cos\theta + x_2 \sin\theta - b)/a), \qquad (10.2)$$

where $x = (x_1, x_2) \in R^2$.

For details, see Candés and Donoho [53, 22]. For definitions of ortho ridgeltes, and multiscale ridgelets, monoscale ridgelets, and we refer to Donoho [53].

10.3 Subband Filtering

To remedy the "energy blowup", one decomposes f into subbands using standard filterbank ideas. Then one assigns one specific monoscale dictionary $\mathcal{M}_s$ to analyze one specific (and specially chosen) subband.

We define coronae of frequencies $|\xi| \in [2^{2s}, 2^{2s+2}]$ and subband filters D_s extracting components of f in the indicated subbands; a filter P_0 deals with frequencies $|\xi| \leq 1$. The filters decompose the energy exactly into subbands:

$$\|f\|_2^2 = \|P_0 f\|_2^2 + \sum_s \|D_s f\|_2^2.$$

The construction of such operators is standard; the coronization oriented around powers 2^{2i} is nonstandard – and essential for us. Explicitly, we build a sequence of filters Φ_0 and $\Psi_{2s} = 2^{4s}\Psi(2^{2s}\cdot)$, $s = 0, 1, 2, \ldots$, with the following properties: Φ_0 is a low-pass filter concentrated near frequencies $|\xi| \leq 1$; Ψ_{2s} is bandpass, concentrated near $|\xi| \in [2^{2s}, s^{2s+2}]$; and we have

$$|\widehat{\Phi}_0(\xi)|^2 + \sum_{s\geq 0} |\widehat{\Psi}(2^{-2s}\xi)|^2 = 1, \qquad \forall\, \xi.$$

Hence, D_s is simply the convolution operator $D_s f = \Psi_{2s} * f$.

10.4 Definition of Curvelet Transform

Assembling the above ingredients, we are able to sketch the definition of the curvelet transform. We let M' consist of M merged with the collection of integral triples (s, k_1, k_2, ϵ), where $s \leq 0, (s, k_1, k_2)$ indexes coarse scale dyadic squares in the plane of side $2^{-s} \geq 1$, and $\epsilon \in \{01, 10, 11\}^2$ is a gender indicator.

The curvelet transform is a map $L^2(\mathbf{R}^2) \mapsto \ell^2(\mathbf{M}')$, yielding curvelet coefficients $(\alpha_\mu : \mu \in M')$. These come in two types.

At *coarse scales* we have wavelet coefficients,

$$\alpha_\mu = \langle W_{s,k_1,k_2,\dots,\epsilon}, P_0 f \rangle, \quad \mu = (s, k_1, k_2) \in M' \setminus M,$$

where each $W_{s,k_1,k_2,\epsilon}$ is a Meyer wavelet. At *fine scales* we have multiscale ridgelet coefficients of the bandpass filtered object:

$$\alpha_\mu = \langle D_s f, \psi_\mu \rangle, \quad \mu \in M_s, s = 1, 2, \dots .$$

Note well that for $s > 0$, each coefficient associated to scale 2^{-s} derives from the subband filtered version of $f - D_s f-$ and not from f.

Several properties are immediate.

- Tight frame:

$$\|f\|_2^2 = \sum_{\mu \in M'} |\alpha_\mu|^2.$$

- Existence of coefficient represents (frame elements): There are $\gamma_\mu \in L^2(\mathbf{R}^2)$ so that

$$\alpha_\mu \equiv \langle f, \gamma_\mu \rangle.$$

- L^2 reconstruction formula:

$$f = \sum_{\mu \in M'} \langle f, \gamma_\mu \rangle \gamma_\mu.$$

- Formula for frame elements: For $s \leq 0, \gamma_\mu = P_0 W_{s,k_1,k_2,\epsilon}$, while for $s > 0$,

$$\gamma_\mu = D_s \psi_\mu, \quad \mu \in M_s.$$

  In short, fine-scale curvelets are obtained by bandpass filtering of multiscale ridgelets coefficients where the passband is rigidly linked to the scale of spatial localization.

- Anistrotopy scaling law: By linking the filter passband $|\xi| \approx 2^{2s}$ to the scale of spatial localization, 2^{-s} imposes that (1) most curvelets are negligible in norm (most multiscale ridgelets do not survive the bandpass filtering D_s); (2) the non-negligible curvelets obey *length* $\approx 2^{-s}$ while *width* $\approx 2^{-2s}$. In short, the system obeys approximately the scaling relationship

$$\text{width} \;\approx\; \text{length}^2.$$

  Note: It is at this last step that the 2^{2s} coronization scheme comes fully into play.

- Oscillatory nature: Both for $s > 0$ and $s \leq 0$, each frame element has a Fourier transform supported in an annulus away from 0.

11 Concluding Remarks and Open Problems

In the previous sections we have presented a review of current researches, specially related to applications of wavelet methods to different areas of science and technology. We mention here some of the problems which may be studied and resolved in the near future. In [93] several open problems related to wavelet packets are discussed. In a plenary lecture [46], Professor Ingrid Daubechies emphasized the role of wavelets in a nonlinear approximation framework with special reference to compression and/or sparse representation of images and large data sets. She talked about limitations of wavelet theory and about the need for further research of wavelet packets. Several theoretical and applied aspects could be investigated. For example, we may study the pointwise convergence of wavelet packet expansion on the lines of Manchanda et al. [92, 93] and Kelly, Kon, and Raphael [81]. Investigation of the relationship between the size of wavelet packet coefficients and function spaces like Lip_α, Zygmund class of functions, Sobolev, and Besov spaces could be an interesting theme.

Application of the wavelet method to image processing, particularly compression and denoising, has been discussed by DeVore, Jawerth, and Popov [49], DeVore, Jawerth, and Lucier [48], and Chambolle et al. [25]. One may examine the results of these papers for wavelet packets, initially for Walsh-type wavelet packets.

Capobianco [24] argues that with wavelet packets, because of the presence of an oscillation index b related to a periodic behavior in the series and because of a richer combination of wavelet functions, we obtain a better domain of functions from which to select a basis to represent a signal. We can still select an orthogonal transform from the so-called wavelet packet transform (WPT) perfectly equivalent to the DWT employed before, but we can do more: from the wavelet packets we can try to choose combinations of them according to procedures which adapt the transform, as in the case of best basis algorithm, to match the characteristics of the signal by minimizing an additive cost function computed over the wavelet coefficient and equivalent to a minimum entropy transform. Nielson [105] has studied a class of highly nonstationary wavelet packets. We may study the problems replacing wavelet packets by highly nonstationary wavelet packets.

Applications of wavelets and wavelet packets for solutions of partial differential equations provide a very promising direction for future research, see [146, 41] and many other papers available on the website of Dahmen. Furati and Siddiqi are investigating the error estimation of Maxwell's equations by the Galerkin method where the wavelet/wavelet packet is taken as the basis function. The work is on the lines of Monk [98, 99]. It is expected that the error estimation in this case would be sharper than the classical methods.

In the recent past, the Hilbert transform has been used for signal analysis of data from nonstationary nonlinear processes. The Hilbert transform $(H\psi)_{j,k}(x)$ is a wavelet if $\psi_{j,k}(x)$ is a wavelet [137]. One may examine the effectiveness of wavelet basis by the Hilbert transform.

For more applications of wavelets we refer to the very recent papers of Kim, Yang, and Kwon [82], Kumar and Shen [83], and Palfner, Mali, and Miller [110] published in the proceedings of "The 6th World Multiconference on Systemics, Cybernetics and Informatics", Orlando, FL, July 14–18, 2002.

References [2, 9, 17, 18, 20, 37, 47, 64, 65, 66, 72, 77, 86, 87, 88, 91, 96, 102, 112, 116, 130, 134, 136, 143, 144] contain either significant applications or lucid presentation essential for proper understanding of wavelet methods, particularly we refer to [17, 77, 96, 136] for a simple introduction. Reference [126] deals with a generalized Black-Scholes model of option pricing which may be studied by wavelet techniques. References [2, 97, 112] provide a basic concept of image processing and are useful references for those who are interested to pursue applications of wavelets in medical sciences. In the recent past, the study of applications of wavelets to nonlinear variational problems has been initiated in research papers [30, 34]. In our opinion, there is a wide scope for further research in view of huge existing literature on variational inequalities. Different aspects of applications of wavelet to physics can be found in Van den Berg [10]. Besides applications of wavelets to turbulence mentioned earlier, some other applications of current interest are also discussed in different chapters of Debnath [87]. It is clear that due to rapid progress in the field, it is practically impossible to keep track of every single application of wavelet, but we are quite hopeful that the citation of most of the published work may be found in one of the references of this chapter.

For updating information about applications of wavelets, ridgelets, and curvelets, researchers are advised to visit the following websites:

Cohen: http://www.ann.jussieu.fr/~cohen/
Dahmen: http://www.igpm.rwth-aachen.de/~dahmen/
DeVore: http://www.math.sc.edu.~devore/
Donoho: http://www-stat.standford.edu/~donho
Siddiqi: http://www.kfupm.edu.sa/cs/

Acknowledgment. K.M. Furati and A.H. Siddiqi would like to express their gratitude to the King Fahd University of Petroleum & Minerals for financial support via Project No. # MS/Safing Sensor/234.

References

[1] M.K. Ahmad and A.H. Siddiqi, Image classification and comparative study of compression techniques, *Journal of Sampling Theory in Signal and Image Processing*, **1**(2) (2002), 155–180.

[2] A. Aldroubi, The wavelet transform: A surcing guide, in *Wavelets in Medicine and Biology*, Edited by A. Aldroubi and M. Unser, CRC Press, Boca Raton, FL, pp. 3–36, 1991.

[3] A. Aldroubi and M. Unser, *Wavelets in Medicine and Biology*, CRC Press, Boca Raton, FL, 1996.

[4] M. Antonini, M. Barlaud, P. Mathieu, and I. Daubechies, Image coding using wavelet transform, *IEEE Trans. on Image Processing*, **1**(2) (1992), 205–220.

[5] E. Bacry, S. Mallat, and G. Papanicolaou, Wavelet based space-time adaptive numerical method for partial differential equations, *Math. Anal.*, **26**(7) (1992), 793–834.

[6] G. Barles, Convergence of numerical schemes for degenerate parabolic equations arising in finance theory, in *Numerical Methods in Finance*, Edited by C. Rogers and D. Talay, Publications of the Newton Institute, Cambridge University Press, London; New York, pp. 1–21, 1997.

[7] A. Barinka et al. Adaptive wavelet schemes for elliptic problems - Implication and numerical experiments, *SIAM J. Sci. Comput.*, **30** (2001), 910–939.

[8] E. Barucci, S. Polidoro, and V. Vespri, Some results on partial differential equations and asian options: Math models and methods, *Appl. Sci.*, **11** (2001), 475–497.

[9] J.J. Benedetto, *Harmonic Analysis and Applications*, CRC Press, Boca Raton, FL, 1996.

[10] J.C. Van den Berg (Editor), *Wavelets in Physics*, Cambridge University Press, Cambridge, 1999.

[11] G. Beylkin, R. Coifman, and V. Rokhlin, Fast wavelet transforms and numerical algorithms, *Comm. Pure Appl. Maths.*, **44** (1993), 141–183.

[12] G. Beylkin and J.M. Keiser, On the adaptive numerical solution of nonlinear partial differential equations in wavelet bases, *J. Comput. Physics*, **132** (1997), 233–259

[13] T. Binder, L. Blank, W. Dahmen, and W. Marquardt, Multiscale concepts for moving horizon optimization, in *Online Optimization of Large Scale Systems*, Edited by M. Grötschel, S.O. Krumke, and J. Rambau, Springer-Verlag, Berlin, pp. 341–361, 2001.

[14] T. Binder, L. Blank, W. Dahmen, and W. Marquardt, Iterative algorithm for multiscale state estimation, Part I: Concepts, *J. Optim. Theory Appl.*, **111** (2001), 501–528, Part II: pp. 529–551.

[15] F. Black and M. Scholes, The pricing of options and corporate liabilities, *J. Pol. Econ.*, **81** (1973), 637–659.

[16] L. Blank, Preconditioning via a Schur complement method on application in state estimation, *Institut für Geometrie und Praktische Mathematik*, RWTH Aachen, Germany, Bericht Nr. 209, Nov. 2001.

[17] C. Blatter, *Wavelets A Primer*, AK Peters, Massachussetts, 1998.

[18] O. Bratteli and P. Jorgenson, *Wavelets Through a Looking Glass*, Birkhäuser, Basel, 2002.

[19] C.M. Brislawn, Finger prints go digital, *Notices Amer. Math. Soc.*, **42**(11) (1995), 1278–1283. Website: http://www.c3.lanl.gov/brislawn.

[20] C. Burrus, R. Gopinathan and H. Guo, *Wavelets and Wavelet Transforms: A Primer*, Prentice Hall, New York, 1998.

[21] E.J. Candés, Ph.D. Thesis, Stanford University, 1998.

[22] E.J. Candés and D.L. Donoho, Curvelets and curvilinear integrals, *J. Approx. Theory*, **113** (2001), 59–90.

[23] E. Capobianco, Wavelet transforms for the statistical analysis of returns generating stochastic processes, *Int. J. Theor. Appl. Finance*, **4** (2001), 511–534.

[24] E. Capobianco, Multiresolution approximation for volatility processes, *Quantitative Finance*, **2** (2002), 91–110.

[25] A. Chambolle, R.A. DeVore, N.Y. Lee, and B. Lucier, Nonlinear wavelet image processing: Variational problems, compression and noise removal through wavelet shrinkage, *IEEE Transactions on Image Processing*, **7**(3) (1998), 319–335.

[26] C.K. Chui, *An Introduction to Wavelets*, Academic Press, Inc., Boston, 1992.

[27] A. Cohen, Wavelets in numerical analysis, in *The Handbook of Numerical Analysis*, Edited by P.G. Ciarlet and J.L. Lions, Elsevier, Amsterdam, Vol. VII, pp. 417–711, 1999.

[28] A. Cohen et al., *Amer. J. Math.*, **121** (1999), 587–628.

[29] A. Cohen, W. Dahmen, and R. DeVore, Adaptive wavelet methods for elliptic operator equations, convergence rates, *Math. Comp.*, **70** (2001), 27–75.

[30] A. Cohen, W. Dahmen, and R. DeVore, Adaptive wavelet methods – Beyond the elliptic operator case, *Found. Comput. Math.*, **2** (2002), 203–245.

[31] A. Cohen and J. Froment, Image compression and multiscale approximation, in *Wavelets and Applications*, Proceedings of the Int. Conf., Marseilles, France, May 1989, Edited by Y. Meyer, Springer-Verlag, Berlin, New York, 1992.

[32] A. Cohen and R. Masson, Wavelet adaptive methods for second order elliptic problems, boundary conditions and domain decomposition, *Numer. Math.*, **86** (2000), 193–238.

[33] R.R. Coifman and D.L. Donoho, Translation invariant denoising, in *Wavelets and Statistics*, Edited by A. Antoniadis, Springer-Verlag, Heidelberg, 1995.

[34] V. Commincoli, T. Scapolla, G. Naldi, and P. Venini, A wavelet-like Galerkin method for numerical solution of variational inequalities arising in elastoplasticity, *Comm. Numer. Math. Engg.*, **16** (2000), 133–144.

[35] D. Crann, A. Davies, C.H. Lai, and S.H. Leong, Time domain decomposition for european options in financial modeling, *Contemporary Mathematics*, **218** (1998), 486–491.

[36] S. Dahlke, *Wavelets: Construction Principles and Application to the Numerical Treatment of Operator Equations*, Aachen, Shaker Verlag, 1997.

[37] S. Dahlke, W. Dahmen, and I. Weinreich, Multiresolution analysis and wavelets on S^2 and S^3, *Numer. Funct. Anal. Optim.*, **16**(1&2) (1995), 19–41.

[38] S. Dahlke, P. Maass, and G. Teschke, Reconstruction of wideband reflectivity densities by wavelet transforms, *Adv. Comput. Math*, **18** (2003), 189-209..

[39] W. Dahmen, Wavelet and multiscale methods for operator equations, *Acta Numer.*, **6** (1997), 55–228.

[40] W. Dahmen, Wavelet methods for PDEs - Some recent developments, *J. Comput. Appl. Maths.*, **128** (2001), 133–185.

[41] W. Dahmen, T. Klint, and K. Urban, On fictitious formulation for Maxwell's equations, Preprint, Aachen, IGPM; Germany (website of Professor Dahmen), 2003.

[42] W. Dahmen and A. Kunoth, Multilevel preconditioning, *Numer. Math.*, **63** (1992), 315–344.

[43] W. Dahmen, S.Prössdorf, and S. Schneider, Wavelet approximation methods for pseudo-differential equations II: Matrix compression and fast solution, *Adv. Comput. Math.*, **1** (1993), 259–335.

[44] I. Daubechies, Orthonormal bases of compactly supported wavelets, *Comm. Pure Appl. Maths.*, **41** (1988), 909–996.

[45] I. Daubechies, *Ten Lectures on Wavelets*, SIAM, Philadelphia, 1992.

[46] I. Daubechies, Wavelets, nonlinear approximation and compression, Invited Plenary Lecture, *SIAM 50th Anniversary & 2002 Annual Meeting*, July 8-12, 2002, Philadelphia Mariot Hotel.

[47] R.A. DeVore, Nonlinear approximation, *Acta Numer.*, **7** (1998), 51–150.

[48] R.A. DeVore, B. Jawerth, and B.J. Lucier, Image compression through wavelet transform coding, *IEEE Trans. Inform. Theory*, **38**(2) (1992).

[49] R.A. DeVore, B. Jawerth, and V. Popov, Compression of wavelet decompositions, *Amer. J. Math.*, **114** (1992), 737–785.

[50] M.N. Do, Contourlets, in *Beyond Wavelets*, Edited by J. Stoekler and G.V.Welland Academic Press, New York, pp. 1–27, 2001.

[51] D.L. Donoho, Denoising by soft-thresholding, *IEEE Trans. Inform. Theory*, **IT-41** (1995), 613–627.

[52] D.L. Donoho, Orthogonal ridgelets and linear singularities, *SIAM J. Math. Anal.*, **31** (2000), 1062–1099.

[53] D.L. Donoho, Ridge functions and orthonormal ridgelets, *J. Approx. Theory*, **111** (2001), 143–179.

[54] D.L. Donoho and I.M Johnstone, Ideal spatial adaptation via wavelet shrinkage, *Biometrika*, **81** (1994), 425–455.

[55] G. Efi-Foufoula and P. Kumar (Editors), *Wavelets in Geophysics*, Academic Press, New York, 1994.

[56] B. Enguist, S. Osher, and S. Zhong, Fast wavelet-based algorithms for linear evolution equations, *SIAM J. Sci. Comput.*, **15** (1994), 755–775.

[57] M. Farge, Wavelet transforms and their applications to turbulence, *Ann. Rev. Fluid Mech.*, **24** (1992), 395–457.

[58] M. Farge, E. Goirand, Y. Meyer, F. Pascal, and M.V. Wickerhauser, Improved predictability of two dimensional turbulent flows using wavelet packet compression, *Fluid Dyn. Res.*, **10** (1992), 229–242.

[59] M. Farge, N. Kevlahan, V. Perrier, and E. Goirand, Wavelets and turbulence, *Proc. IEEE*, **84**(4) (1996), 639–669.

[60] M. Farge, N. Kevlahan, V. Perrier, and K. Schneider, Turbulence analysis, modeling and computing using wavelets in physics, in *Wavelets in Physics*, Edited by J.C. van den Berg, Cambridge University Press, Cambridge, pp. 117–200, 1999.

[61] M. Farge and G. Rabreau, Wavelet transform to analyze coherent structures in two-dimensional turbulent flows, *Proceedings on Scaling, Fractals and Nonlinear Variability in Geophysics*, Paris, 1989.

[62] M. Farge, K. Schneider, and N. Kevlahan, Non-Gaussianity and coherent vortex simulation for two-dimensional turbulence using an adaptive orthonormal wavelet basis, *Phys. Fluids*, **11**(8) (1999), 2187–2201.

[63] P.A. Forsyth, K.R. Vetzal, and R. Zvan, A finite element approach to the pricing of discrete lookbacks with stochastic volatility, *Appl. Math. Finance*, **6** (1999), 87–106.

[64] W. Freeden, D. Glockner, and R. Litzenberger, A general Hilbert space approach to wavelets and its applications in geopotential determination, *Numer. Funct. Anal. Optim.*, **20**(9&8) (1999), 853–879.

[65] W. Freeden and M. Schreiner, Orthogonal and nonorthogonal multiresolution analysis, scale discrete and exact fully discrete wavelet transform on the sphere, *Const. Approx.*, **14** (1998), 493–515.

[66] J. Fröhlich and K. Schneider, An adaptive wavelet-vaguelette algorithm for the solution of PDEs, *J. Comput. Physics*, **130** (1997), 174–190.

[67] M. Fujii and W.J.R. Hoefer, A three-dimensional Haar-wavelet based multiresolution analysis smilar to the FDTD method – Derivation and application, *IEEE Trans. on Microwave Theory and Techniques*, **46**(12) (1998), 2463–2475.

[68] M. Fujii and W.J.R. Hoefer, Application of biorthogonal interpolating wavelets to the Galerkin schemes of time-dependent Maxwell's wquations, *IEEE Microwave and Wireless Components Lett.*, **11**(1) (2001), 22–24.

[69] R. Gencay and F. Seluk, *An Introduction to Wavelets and Other Filtering Methods in Finance and Economics*, Academic Press, New York, 2001.

[70] R. Glowinski, W. Lawton, M. Ravachol, and E. Tenenbaum, Wavelets solution of linear and nonlinear elliptic, parabolic and hyperbolic problems in one space dimension, in *Computing Methods in Applied Sciences and Engineering*, Edited by R. Glowinski and A. Lichnewsky, SIAM, Philadelphia, pp. 55–120, 1990.

[71] S. Goedecker, Wavelets and their application for the solution of partial differential equations in physics, Lecture Notes, M.D. Reymond University, Lausane, 1998.

[72] R.C. Gonzales and R.E. Wood, *Digital Image Processing*, Addison-Wesley Publishing Company, Reading, MA, 1993.

[73] L. Greengard and J. Strain, A fast algorithm for the evolution of heat potentials, *Comm. Pure Appl. Math.*, **43** (1990), 949–964.

[74] L. Greengard and J. Strain, The fast Gauss transform, *SIAM J. Sci. Statis. Comput.*, **12** (1991), 79–94.

[75] S. Grivet-Talocia, On the accuracy of Haar-based multiresolution time-domain schemes, *IEEE Microwave and Guided Wave Lett.*, **10**(10) (2000), 397–399.

[76] A. Grossman and J. Morlet, Decomposition of Hardy functions into square integrable wavelets of constant shape, *SIAM J. Math. Anal.*, **15**(1984), 723–736.

[77] B.B. Hubbard, *The World According to Wavelets*, A.K. Peters, Wellesley, 1996.

[78] S. Jaffard and Ph. Laurencot, Orthonormal wavelets, analysis of operators and application to numerical analysis, in *Wavelets - A Tutorial in Theory and Applications*, Edited by C.K. Chui, pp. 543–601, 1992.

[79] J. Liandrat, V. Perrier, and Ph. Tchamitchian, Numerical resolution of the regularized Burgers equation using the wavelet transform, in *Wavelets and Applications*, Edited by Y. Meyer, LMA/CNRS, Masson, pp. 420–433, 1992.

[80] G. Kaiser, *A Friendly Guide to Wavelets*, Berkhäuser, Boston-Basel-Berlin, 1994.

[81] S.E. Kelly, M.A. Kon, and L. Raphael, Pointwise convergence of wavelet expansions, *J. Funct. Anal.*, **126** (1994), 102–138.

[82] E-jae Kim, Sung-il Yang, and Y. Kwon, Adaptive wavelet denoising for a car interior noise, in the *Proceedings of The 6th World Multiconference on Systemics, Cybernetics and Informatics*, Orlando, Florida, July, pp. 14–18, 2002.

[83] A. Kumar and H.C. Shen, Recognition of palmprints using wavelet based features, in the *Proceedings of The 6th World Multiconference on Systemics, Cybernetics and Informatics*, Orlando, Florida, July, pp. 14–18, 2002.

[84] M. Kobyashi (Editor), *Wavelets and Their Applications, Case Studies*, SIAM, Philadelphia, 1998.

[85] M. Kocvara and J. Zowe, An iterative two-step algorithm for linear complementarity problems, *Numer. Math.*, **68** (1994), 95–106.

[86] J. Kovacevic and I. Daubechies (Editors), Special issue on Wavelets, Proc. IEEE, **84** (1996), 507–614.

[87] L. Debnath (Editor), *Wavelet Transforms and Their Applications*, Birkhäuser, Boston, 2002.

[88] A.K. Louis, P. Maass, and A. Reider, *Wavelet Theory and Applications*, John Wiley & Sons, Chichester, New York, 1997.

[89] S. Mallat, A theory of multiresolution signal decomposition, the wavelet representation, *IEEE Trans. Patt. Analy. and Mach. Intell.*, **11**(7) (1989), 674–693.

[90] S. Mallat, *A Wavelet Tour of Signal Processing*, Academic Press, New York, 1998.

[91] Y. Mallat, D. Coomans, J. Kantsky, and O. DeVel, Classification using adaptive wavelets for feature extraction, *IEEE Trans. Patt. Anal. and Mach. Intell.*, **19**(10) (1997).

[92] P. Manchanda, A.S. Mukheimer, and A.H. Siddiqi, Pointwise convergence of wavelet expansions associated with Dilation matrix, *Appl. Anal.*, **76**(3–4) (2000), 301–308.

[93] P. Manchanda and A.H. Siddiqi, Role of functional analysis methods in imaging science during the 21st century, in *Current Trends in Industrial and Applied Mathematics*, Edited by P. Manchanda, K. Ahmad, and A.H. Siddiqi, Anamaya Publishers, New Delhi, pp. 1–28, 2002.

[94] C. Meneveau, Analysis of turbulence in the orthonormal wavelet representation, *J. Fluid Mech.*, **232** (1991), 469–520.

[95] C. Meneveau, Wavelet analysis of turbulence: The mixed energy cascade, in *Wavelets, Fractals and Fourier Transforms*, Edited by M. Farge, J.C.R. Hunt, and J.C. Vassilicos, Oxford University Press, Oxford, pp. 251–264, 1993.

[96] Y. Meyer, *Wavelets, Algorithms and Applications*, SIAM, Philadelphia, 1993.

[97] Y. Meyer, Lecture Notes, *School on Mathematical Problems in Image Processing*, 4–22 September 2000, *Abdus Salam International Centre for Theoretical Physics*, SMR 1237/3, 4, Trieste, Italy, 2000.

[98] P.B. Monk, A mixed method for approximating Maxwell's equation, *SIAM J. Numer. Anal.*, **28**(6) (1991), 1610–1634.

[99] P.B. Monk, A finite element method for approximating the time-harmonic Maxwell's equation, *Numer. Math.*, **63** (1992), 243–321.

[100] R.C. Merton, Theory of rational option pricing, *Bell J. Econom. Management Sci.*, **4** (1973), 141–183.

[101] J. Morlet, G. Arens, E. Fourgeau, and D. Giard, Wavelet propagation and sampling theory, Part I: Complex signal and scattering in multilayer media, *J. Geophys.*, **47** (1982), 203–221.

[102] J. Morlet, G. Arens, E. Fourgeau, and D. Giard, Wavelet propagation and sampling theory, Part II: Sampling theory and complex waves, *J. Geophys.*, **47** (1982), 222–236.

[103] M. Nielson, *Size Properties of Wavelet Packets*, Ph.D. Thesis, Washington University, St. Louis, 1999.

[104] M. Nielson, Walsh-type wavelet packet expansions, *Appl. Comput. Harmonic Anal.*, **9** (2000), 265–285.

[105] M. Nielson, Highly non-stationary wavelet packets, *Appl. Comput. Harmonic Anal.*, **12** (2002), 209–229.

[106] M. Nielson, On convergence of wavelet packet expansions, *Approx. Theory Appl.*, **18** (2002), 34–50.

[107] M. Nielson, Size properties of wavelet packets generated using finite filters, *Rev. Mat. Iberoamericuna*, **18** (2002), 249–265.

[108] M. Nielsen and D.-X. Zhou, Mean size of wavelet packets, *Appl. Comput. Harmon. Anal.*, **13** (2002), 22–34.

[109] H. Neunzert and A.H. Siddiqi, *Industrial Mathematics - Case Studies and Mathematical Methods*, Kluwer Academic Publishers, Dordrecht-Boston-London, 2000.

[110] T. Palfner, A. Mali, and E. Müller, Optimized progressive coding of stereo images using discrete wavelet transform, in the *Proceedings of The 6th World Multiconference on Systemics, Cybernetics and Informatics*, Orlando, Florida, July, pp.14–18, 2002.

[111] G. Pan, Orthogonal wavelets with applications in electromagnetics, *IEEE Trans. on Magnetics*, **32**(3) (1996), 975–983.

[112] D.B. Percival and A.T. Walden, *Wavelet Methods for Time-Series Analysis*, Cambridge University Press, London, New York, 2000.

[113] T. von Petersdorff and C. Schwab, Wavelet approximations for first kind boundary integral equations on polygons, *Numer. Math.*, **74** (1996), 479–519.

[114] S. Qian and J. Weiss, Wavelets and numerical solution of partial differential equations, *J. Comp. Physics*, **106** (1993), 155–175.

[115] A. Quarteroni and A. Valli, *Numerical Approximation of Partial Differential Equations*, Springer-Verlag, New York, 1994.

[116] F. Rashid-Farrokhi, R. Liu, C. Bertein, and D. Walnut, Wavelet based multiresolution local tomography, *IEEE Trans. on Image Processing*, **6** (1997), 1412–1430.

[117] H.L. Resnikoff and R.O. Wells Jr., *Wavelet Analysis*, Springer-Verlag, New York, 1998.

[118] C. Rogers and D. Talay (Editors), *Numerical Methods in Finance*, Publications of the Newton Institute, Cambridge University Press, London, New York, 1997.

[119] G. Rubinacci, A. Tamburrino, S. Ventre, and F. Villone, Interpolating wavelets for the solution of Maxwell equations in the time domain, *IEEE Trans. on Magnetics*, **34**(5) (1998), 2275–2278.

[120] J. Shen amd G. Strang, On wavelet fundamental solutions to the heat equation-heatlets, *J. Diff. Eqn.*, **16** (2000), 403–421.

[121] R.H. Shumway and D.S. Stoffer, *Time Series Analysis and Its Applications*, Springer-Verlag, New York, 2000.

[122] A.H. Siddiqi, *Applied Functional Analysis*, Marcel-Dekker, New York, 2003.

[123] A.H. Siddiqi, M.K. Ahmad, and A. Mukheimer, Current developments in fractal image compression, in *Functional Analysis with Current Applications in Science, Technology and Industry*, Edited by M. Brokate and A.H. Siddiqi, Pitman Res. Notes Math. Ser. 377, Addison Wesley Longman Limited, England, pp. 304–331, 1998.

[124] A.H. Siddiqi, P. Manchanda, and M. Kocvara, An iterative two-step algorithm for american option pricing, *IMA J. Business and Industry*, **11** (2000), 71–84.

[125] A.H. Siddiqi, P. Manchanda, and M. Kocvara, Fast wavelet based algorithm for option pricing, in *The 6th World Multiconference on Systemics, Cybernetics and Informatics*, Edited by N. Callaos, G. Whymark, and W. Lesso, Orlando, FL, Vol. XIII, 2002, pp. 141–146.

[126] K.R. Sircar and G. Papanicolaou, General black-scholes models accounting for increased market volatility from hedging strategies, *Appl. Math. Finance*, **5** (1998), 45–82.

[127] J.L. Starke, E.J. Candes, and D.L. Donoho, The curvelet transform for image denboising, *IEEE Trans. on Image Processing*, **11** (2002), 670–684.

[128] G. Strang and T. Nguyen, *Wavelets and Filter Banks*, Cambridge University Press, Wellesley, 1996.

[129] A. Teolis, *Computational Signal Processing with Wavelets*, Birkhäuser, Boston, Dordrecht, 1998.

[130] P. Topiwala (Editor), *Wavelet Image and Video Compression*, Kluwer Academic Publishers, Dordrecht, 1998.

[131] K. Urban, Wavelet bases in $H(\mathrm{div})$ and $H(\mathrm{curl})$, *Math. Comp.*, **70** (2000), 739–760.

[132] K. Urban, *Wavelets in Numerical Simulation*, Springer-Verlag, New York, 2002.

[133] B. Torresani, Time-frequency representations: Wavelet packets and optimal decomposition, *Ann. Inst. Henri Poincare, Phy. Theory*, **56**(2) (1992) 215–234.

[134] O.V. Vasilyev, S. Paolucci, and M. Sen, A multilevel wavelet collocation method for solving partial differential equations in a finite domain, *J. Comput. Phys.*, **120** (1995), 33–47.

[135] M. Vetterli and J. Kovačevic, *Wavelets and Subband Coding*, Prentice-Hall, Englewood Cliffs, NJ, 1995.

[136] J. Walker, *A Primer of Wavelets and Their Scientific Applications*, CRC Press, Boza Raton, FL, 1999.

[137] D.F. Walnut, *An Introduction to Wavelet Analysis*, Birkhäuser, Boston, 2002.

[138] J. Weiss, Application of compactly supported wavelets to the numerical solution of parabolic differential equations, in *Proceedings of Summer Study Program in Geophysical Fluid Dynamics*, 1990.

[139] M.V. Wickerhauser, *Applied Wavelet Analysis from Theory to Software*, A.K. Peters, Wellesley, 1994.

[140] M.V. Wickerhauser, Some problems related to wavelet packet bases and convergence, (Special issue on Wavelet and Fractal Methods in Science and Engineering, Lead Editor, A.H. Siddiqi), *Arabian J. Sci. Engg.*, **28** (2003), 45–59.

[141] P. Willmott, J.N. Dewynne, and S.D. Howison, *Option Pricing: Mathematical Models and Computation*, Oxford Financial Press, 1993.

[142] P. Willmott, S.D. Howison, and J.N. Dewynne, *The Mathematics of Financial Derivatives*, Cambridge University Press, London; New York, 1995.

[143] P. Wunsch and A.F. Laine, Wavelet descriptors for multiresolution recognition of handprint characters, *Pattern Recognition*, **28**(8) (1995), 1237–1249.

[144] S. Zafar, Multiscale video representation using multiresolution motion compensation and wavelet decomposition, *Argonne Wavelets Workshop*, 1994.

[145] C.J. Zarowski, An approach to initializing the wavelet packet transform, *IEEE Signal Process. Lett.*, **4**(5) (1997), 132–134.

[146] C.J. Zarowski, Wavelet packet transform initializing for piecewise polynomial and bandlimited inputs, *IEEE Trans. Signal Process*, **47** (1999), 224–226..

Chapter 7

WAVELET METHODS FOR INDIAN RAINFALL DATA

J. Kumar and P. Manchanda

Gurunanak Dev University

N. A. Sontakke

Indian Institute of Tropical Meteorology

Abstract

Wavelet and wavelet-based multifractal methods have been applied to understand real-world problems represented by time series [1–5, 7, 9–11]. India is one of the largest countries whose economy and political system hinge on agricultural production. The agricultural industry and hydroelectric power industry in India are mainly dependent on rainfall, particularly during the rainy season from June to September (June, July, August, and September – JJAS) and post monsoon season from October to December (October, November and December – OND). The study of behavior of rainfall in India has been a topic of intense research for more than 100 years, applying statistical and Fourier analytic methods, see for example [13–16]. In the present study, wavelet methods, especially the MATLAB Wavelet Toolbox, are used to examine the Indian rainfall series from 1813 to 1995.

1 Introduction

The time series analysis approach for analyzing and understanding real-world problems such as climatic and financial data is quite popular in the scientific world. Till very recently, statistical and Fourier analytic methods were applied, but today wavelet and wavelet-based multifractal methods are becoming more popular and reliable. In Section 2 we present the main ingredients of wavelet and multifractal formalism which are quite helpful in revealing the basic structure of the rainfall time series of Indian rainfall between 1813 and 1995. In Section 3 we present MATLAB wavelet tool box analysis of the Indian rainfall data between 1813 and 1995, applying Haar, Daubechies, Coifman, Mexican hat, Meyer, Marlet, biorthogonal, and symmetric wavelets at different levels. In Section 4 we discuss the main features of the results obtained in Section 3 and indicate the possibility of more realistic study by applying the concept of wavelet transform modulus maximum [5] developed by Arneodo et al.

2 Wavelet and Wavelet-Based Multifractal Formalism

The concept of wavelet methods and their applications to various fields has been developed by Coifman, Daubechies, and Meyer, Mallat et al.; updated references can be found in Mallat [5], Siddiqi [8], Addison [1], Arneodo et al. [2], Kumar and Foufoula-Georgiou [4], and Percival and Walden [7]. We present here basic ingredients of wavelet and wavelet-based multifractal formalism.

2.1 Wavelet Methods

The name wavelet means small waves (the sinusoids used in Fourier analysis are big waves), and in brief, a wavelet is an oscillation that decays quickly. Equivalent mathematical conditions are

$$\int_{-\infty}^{\infty} | \psi(t) |^2 \; dt < \infty, \tag{2.1}$$

$$\int_{-\infty}^{\infty} \psi(t)\, dt = 0. \tag{2.2}$$

The admissibility condition is

$$\int_{-\infty}^{\infty} \frac{|\hat{\psi}(\xi)|^2}{|\xi|} d\xi < \infty, \tag{2.3}$$

where $\hat{\psi}(\xi)$ denotes the Fourier transform of $\psi(t)$. In wavelet theory a function is represented by infinite series expansion in terms of dilated and translated versions of the basis function $\psi(t)$, called mother wavelet, satisfying the above conditions

$$\psi_{a,b}(t) = a^{-\frac{1}{2}} \psi \left(\frac{t-b}{a} \right) \quad \text{where} \quad a > 0 \tag{2.4}$$

$$\begin{aligned} T_\psi f(a,b) &= a^{-\frac{1}{2}} \int_{-\infty}^{\infty} f(t) \psi \left(\frac{t-b}{a} \right) dt = \langle f, \psi_{a,b} \rangle \quad (2.5) \\ &= \text{inner product of } f \text{ and } \psi_{a,b}. \end{aligned}$$

$T_\psi f(a,b)$ is called the wavelet transform of $f(t)$.

A wavelet transform decomposes a signal into several groups (vectors) of coefficients. Different coefficient vectors contain information about characteristics of the sequence at different scales.

It may be observed that the wavelet transform is a prism which exhibits properties of signal such as points of abrupt changes, seasonality, or periodicity. The wavelet transform is a function of a and b, where

a = the scale or frequency and

b = spatial position or time.

The plane defined by the variables (a, b) is called the scale-space or time-frequency plane. The wavelet transform $T_\psi f(a,b)$ measures the variation of f in a neighbourhood of b.

For a compactly supported wavelet (wavelet vanishing outside a closed and bounded interval), the value of $T_\psi f(a,b)$ depends upon the value of f in a neighbourhood of b of size proportional to the scale a. At small scales, $T_\psi f(a,b)$ provides localized information such as localized regularity (smoothness) of f.

The local regularity of a function (or signal) is often measured with a Lipschitz exponent (Hurst parameter, also fractal dimension). The global and local Lipschitz regularity can be characterized by the asymptotic decay of wavelet transformation at small scales. For example, if f is differentiable at $b, T_\psi f(a,b)$ has the order $a^{3/2}$ as $a \to 0$.

$\psi_{j,k}(t) = 2^{\frac{j}{2}}\psi(2^j t - k)$ is a discretized version of $\psi_{a,b}(t)$, where j, k are integers.

$\psi(t)$ is called an orthonormal wavelet if

$$\langle \psi_{j,k}(t), \psi_{m,n}(t) \rangle = \int_{-\infty}^{\infty} \psi_{j,k}(t)\psi_{m,n}(t)\, dt = \begin{cases} 1, & j = m,\ k = n \\ 0 & \text{otherwise.} \end{cases}$$

$d_{j,k} = \int_{-\infty}^{\infty} f(t)\psi_{j,k}(t)\, dt$ are called wavelet coefficients of $f(t)$, and $\sum_{j,k} d_{j,k}\psi_{j,k}(t)$ is called the wavelet series of f. With every orthonormal wavelet there is an associated function $\phi(t)$ called the scaling function such that

$$\phi_{j,k}(t) = 2^{\frac{j}{2}}\phi(2^j t - k).$$

$c_{j,k} = \int_{-\infty}^{\infty} f(t)\phi_{j,k}(t)$ are called scaling coefficients, and $\sum_{j,k} c_{j,k}\phi_{j,k}(t)$ is called the scaling series.

Mallat Decomposition Algorithm

$$c_{j,k} = \sum_{l \in Z} h_{l-2k} c_{j+1}, l \tag{2.6}$$

$$d_{j,k} = \sum_{l \in Z} (-1)^l h_{-l+2k+1} c_{j+1,l}, \tag{2.7}$$

where $\{h_j\}$ is a sequence characteristic of the associated wavelet. For example, for Haar wavelet

$$h_j = \begin{cases} \dfrac{1}{\sqrt{2}}, & j = 0, 1 \\ 0, & \text{otherwise.} \end{cases}$$

It is important to note that given scaling coefficients at any level j, all lower level scaling function coefficients $i < j$ can be computed recursively using (2.6) and all lower wavelet coefficients $i < j$ can be computed from the scaling function coefficients applying (2.7).

Reconstruction Algorithm

$$c_{j,k} = \sum_{l \in Z} h_{k-2l} c_{j-1,l} + (-1)^k h_{2l-k+1} d_{j-1,l}. \tag{2.8}$$

The scaling function coefficients at any level can be computed from only one set of lower level scaling function coefficients and all the intermediate. For wavelets with compact support, the sequences $\{d_{j,k}\}$ and $\{c_{j,k}\}$ will contain only finitely many non-zero elements. For wavelets with support on all of R, each sequence element h_k is, in general, non-zero, but elements decay exponentially.

Discrete wavelet transform (DWT) is commonly introduced using a matrix or a computation form. In matrix form we can represent the DWT through an orthogonal matrix

$$W = [W_1^t, W_2^t, \cdots, W_j^t, V_j]^t, \tag{2.9}$$

where V_j is a scaling function,

J = the largest level of the transform

t = indicates transpose

A DWT is applied to a vector X of observation as $d = WX$ and decomposes the data into sets of wavelets coefficients

$$d = [d_1^t, d_2^t, \cdots, d_j^t, c_j^t]^t \tag{2.10}$$

with

$$d_j = W_j X \quad \text{and} \quad c_j = V_j X. \tag{2.11}$$

Multiresolution Analysis

A wavelet transform leads to an additive decomposition of a signal into a series of different components describing smooth and rough features of the signal. In fact, we have

$$X = W^t d = \sum_{j=1}^{J} W_j^t d_j + V_j^t C_j = \sum_{j=1}^{J} D_j + C_j, \tag{2.12}$$

with D_j as the detail of the signal describing changes at τ_j and C_j as the smooth component associated with variations τ_{j+1} and higher. Mallat algorithms are used for computation of the DWT. It is faster than the Fourier transform (FT), being of complexities order $0(n)$ and $0(n \log_2 n)$, respectively.

- Sharp signal transitions create large amplitude wavelet coefficients.
- Singularities are detected by following across scales the local maxima of the wavelet transform.

Time Series

Wavelet analysis of a time series is the study of the above-mentioned properties by the breaking up of the signal (time series) into shifted and scaled versions of the original (mother) wavelet.

2.2 Fractals and Multifractals

Fractals describe objects that are too irregular to fit into traditional geometric settings. Various phenomena display fractal features when plotted as functions of time. Examples include atmospheric pressure, level of reservoir, and prices of the stock market, usually when recorded over fairly long time spans. The zooming capability of the wavelet transform not only locates isolated singular events, but can also characterize more complex multifractal signals having non-isolated singularities. Multifractals are fractal objects which cannot be completely described using a single fractal dimension (monofractals). They have, in fact, an infinite number of dimension measures associated with them.

The wavelet transform takes advantage of multifractal self-singularities in order to compute the distribution of the singularities. This singularity spectrum is used to analyze multifractal properties. Signals that are singular at almost every point are multifractals, and they appear in the maintenance of economic records, physiological data including heart records, electromagnetic fluctuations in galactic radiation noise, textures in images of natural terrain, variation of traffic flows etc.

Singularity Spectrum

Definition 2.1 (Self-Similar Set) *A set $S \subset R^n$ is said to be self-similar if it is the union of disjoint subsets $S_1, S_2, \cdots, S_R$ that can be obtained from S with a scaling, translation, and rotation. It may be observed that the self-similarity often implies an infinite multiplication of details, which create irregular structures.*

The triadic Cantor set and the Von Koch curve are well-known examples.

Definition 2.2 (Spectrum of singularity) *Let S_α be the set of all points $t \in R$, where the pointwise Lipschitz regularity of f is equal to α. The spectrum of singularity $D(\alpha)$ of f is the fractal dimension of S_α. The support of $D(\alpha)$ is the set of α such that S_α is not empty.*

The singularity spectrum gives the proportion of Lipchitz singularities that appear at any scale a. Fractal dimension is a first order parameter of complexity which can degenerate [6].

Partition Function. One cannot compute the Lipschitz regularity of a multifractal because its singularities are not isolated, and the finite numerical resolution is not sufficient to discriminate them. It is, however, possible to measure the singularity spectrum of multifractals from the wavelet transform modulus maximum, using a global partition function introduced by Arneodo, Bacry and Muzy, see, for example, [5].

Let ψ be a wavelet with n vanishing moments. Mallat ([5] Theorem 6.5) states that if f has pointwise Lipschitz regularity $\alpha_0 \leq n$ at v, then the wavelet transform $T_\psi f(a,b)$ has a sequence of modulus maxima that converges towards v at fine scales. The set of maxima at the scale 'a' can thus be interpreted as a cover of the support of f with wavelets at scale a. At these maxima locations

$$| T\psi f(a,b) | \approx a^{\alpha_0+\frac{1}{2}}.$$

Let $\{u_p(a)\}_{p\in Z}$ be the position of all local maxima of $| T\psi f(a,b) |$ at a fixed scale a. The partition function Z measures the sum at a power q of all these wavelet modulus maxima:

$$Z(q,a) = \sum_p | T\psi f(a,u_p) |^q .$$

For each $q \in R$, the scaling $\tau(q)$ measures the asymptotic decay of $Z(q,a)$ at fine scales:

$$\tau(q) = \lim_{a\to 0} \inf \frac{\log Z(q,a)}{\log a}.$$

This typically means that

$$Z(q,a) \approx a^{\tau(q)}.$$

The following theorem relates $\tau(q)$ to the Legendre transform of $D(\alpha)$ for self-similar signals.

Theorem 2.3 *(Arneodo, Bacry, Jaffard, Muzy, see [5]) Let $\wedge = [\alpha_{\min}, \alpha_{\max}]$ be the support of $D(\alpha)$. Let ψ be a wavelet with $n > \alpha_{\max}$ vanishing moments. If f is a self-similar signal, then*

$$\tau(q) = \min_{\alpha\in\wedge}(q(\alpha + \frac{1}{2}) - D(\alpha)). \tag{2.13}$$

Theorem 2.4 *[5] The scaling exponent $\tau(q)$ is a convex and increasing function of q.*

The Legendre transform (2.13) is invertible if and only if $D(\alpha)$ is convex, in which case

$$D(\alpha) = \min_{q \in R}(q(\alpha + \frac{1}{2}) - \tau(q)). \tag{2.14}$$

The spectrum $D(\alpha)$ of self-similar signals is convex.

We first calculate $Z(q,a) = \sum_q \mid T_\psi f(a,u_p) \mid^q$, then derive the decay scaling exponent $\tau(q)$, and finally compute $D(\alpha)$ with Legendre transform. If $q < 0$, then the value of $Z(q,a)$ depends mostly on the small amplitude $\max \mid T_\psi f(a,u_p) \mid$.

Procedure for Numerical Calculation

- Maxima: Compute $T_\psi f(a,u_p)$ and the modulus maxima at each scale a. Chain the wavelet maxima across scales.
- Partition function: Compute
$$Z(q,a) = \sum_p \mid T_\psi f(a,u_p) \mid^q .$$
- Scaling: Compute $\tau(q)$ with a linear regression of $\log_2 Z(a,q)$ as a function of $\log_2 a$:
$$\log_2 Z(q,a) \approx \tau(q) \log_2 a + C(q).$$
- Spectrum: Compute
$$D(\alpha) = \min_{q \in R}(q(\alpha + \frac{1}{2}) - \tau(q)).$$

Smooth Perturbation. Let f be a multifractal whose spectrum of singularity $D(\alpha)$ is calculated from $\tau(q)$. If a regular signal g is added to f, then the singularities are not modified and the singularity spectrum of $\tilde{f} = f + g$ remains unchanged. We study the effect of this smooth perturbation on the spectrum calculation. The wavelet transform of $\tilde{f}$ is

$$T\tilde{f}(a,u) = Tf(a,u) + Tg(a,u).$$

Let $\tau(q)$ and $\tilde{\tau}(q)$ be the scaling exponent of the partition functions $Z(q,a)$ and $\tilde{Z}(q,a)$ calculated from the modulus maxima, respectively.

Theorem 2.5 *[Arneodo, Bacry, Muzy] Let ψ be a wavelet with exactly n vanishing moments. Suppose that f is, self-similar function.*

- *If g is a polynomial of degree $p < n$, then*

$$\tau(q) = \tilde{\tau}(q) \quad \text{for all} \quad q \in R.$$

- *If $g^{(n)}$ is almost everywhere non-zero, then*

$$\tilde{\tau}(q) = \begin{cases} \tau(q) & q > q_c \\ (n + \frac{1}{2})q & q \leq q_c, \end{cases}$$

where q_c is defined by $\tau(q_c) = (n + \frac{1}{2})q_c$.

3 Wavelet Based Time Series Analysis of Indian Rainfall Data

Indian rainfall is governed by the summer monsoon ubiquitous over the country spanning from June through September due to its major contribution (70 to 90% of the annual rainfall). It is also a major component and driver of the global circulation system. As such, it has significant social and economic impacts. Agriculture, industry, hydroelectric power, irrigation, reservoirs drinking water, and many such areas over the country are directly linked with the summer monsoon performance. However, this seasonal rainfall exhibits large temporal and spatial variability.

Rainfall activity during the rest of the seasons though has limited spatial extent, is less intense compared to the summer monsoon rainfall, and is regionally important and noteworthy. The northeast winter monsoon during October to December is mainly active over southeast India and is the major rainfall season over the area. It is also spatially variable with its gradient from the southeast to northwest direction. During March to May, thunderstrom activity accounts for rainfall over the Southwest peninsula and northeast India. In northern latitudes western disturbances pass throughout the year, giving, good amount of rainfall.

Owing to large variability, understanding and prediction of Indian rainfall has been a practical necessity and therefore an important area of research. Long period instrumental series are vital in studies on climate variability and its modelling, monitoring, and prediction. Longest seasonal and annual Indian rainfall series have been reconstructed from the past records over different spatial rainfall zones and for the country as a whole. For constructional details we refer to Singh and Sontakke [12] and Sontakke and Singh [17].

In the present study, wavelet methods, especially the MATLAB Wavelet Toolbox, are used to examine the Indian rainfall series from 1813 to 1995.

Singh and Sontakke [13] have studied large-scale variations in the post monsoon rainfall over India to determine their association with the large-scale variations of the preceeding summer monsoon season. By applying the MATLAB Wavelet Toolbox, we look into the same problem for the same data. We also study the long-term behavior of Indian rainfall data at different levels by applying well-known wavelets such as Haar, Daubechies, Coiflet, Morlet, and Mexican hat. The wavelet transform of Indian rainfall data reveals that this data has multifractal structure due to the presence of self-similarity. Therefore, the study of spectrum singularity by applying the method of modulus maximum of wavelet transform developed by Arneodo is better suited for this study. This will be studied separately.

Experiment Results

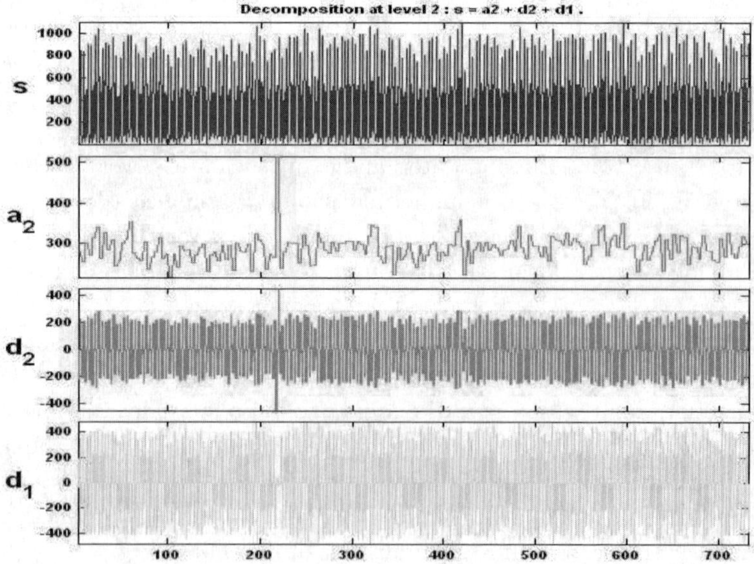

Figure 1: Analysis of Indian rainfall data for 1813–1995 (annual) by Haar wavelet 1-D, level 2.

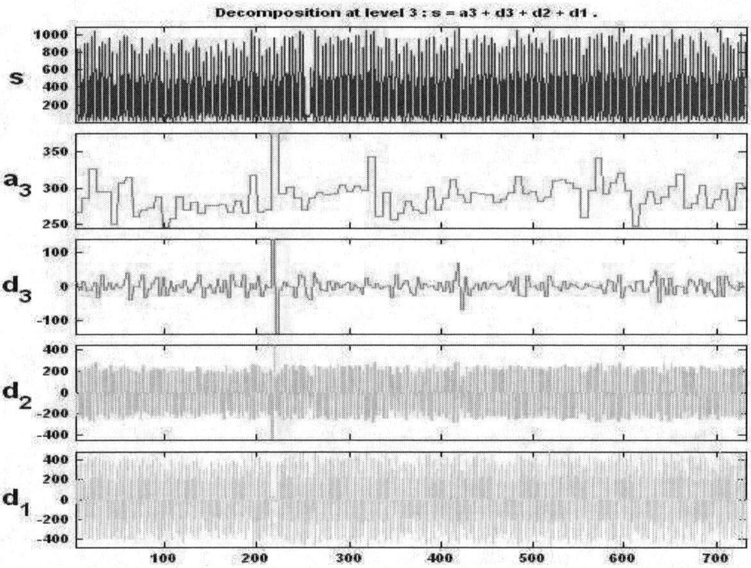

Figure 2: Analysis of Indian rainfall data for 1813–1995 (annual) by Haar wavelet 1-D, level 3.

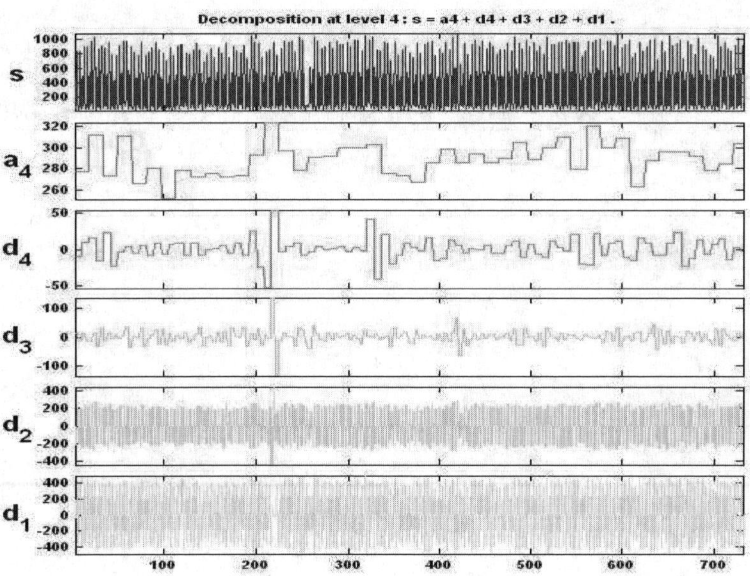

Figure 3: Analysis of Indian (annual average) rainfall data for 1813–1995 by Haar wavelet 1-D, Level 4

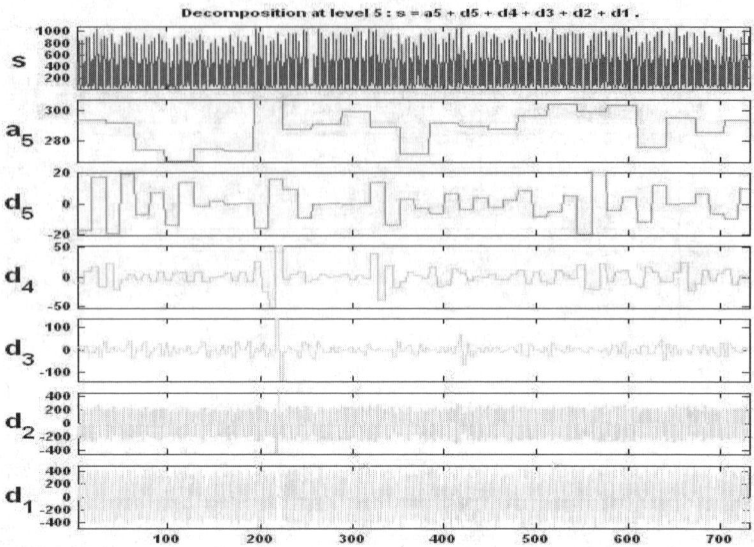

Figure 4: MATLAB analysis of annual Indian rainfall data by Haar wavelet 1-D, Level 5. We note that the result starts deteriorating after level 5. The best possible result is available at level 5.

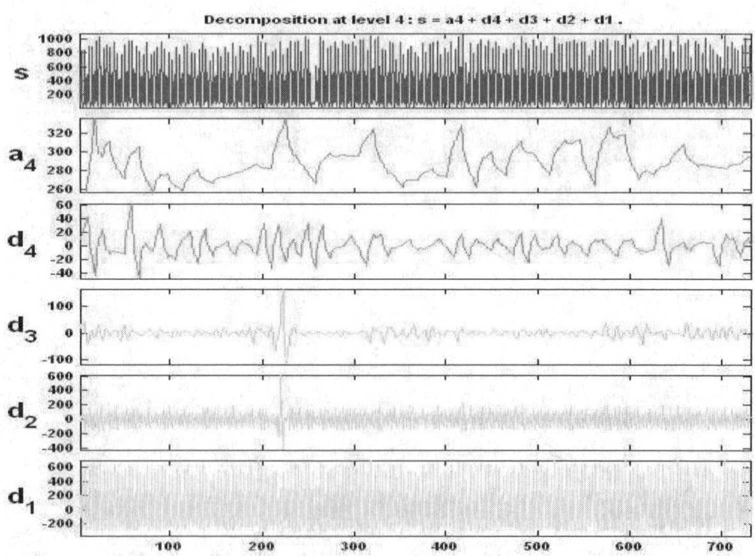

Figure 5: MATLAB analysis of annual Indian rainfall data for the year 1813–1995 by Daubechies 2 wavelet 1-D, level 4.

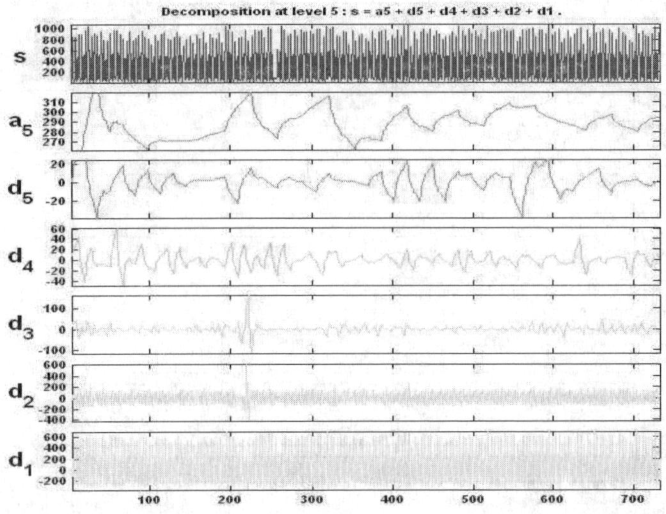

Figure 6: MATLAB analysis of annual Indian rainfall data for the year 1813–1995 by Daubechies 2 wavelet 1-D, level 5. We remark that it starts deteriorating after level 5.

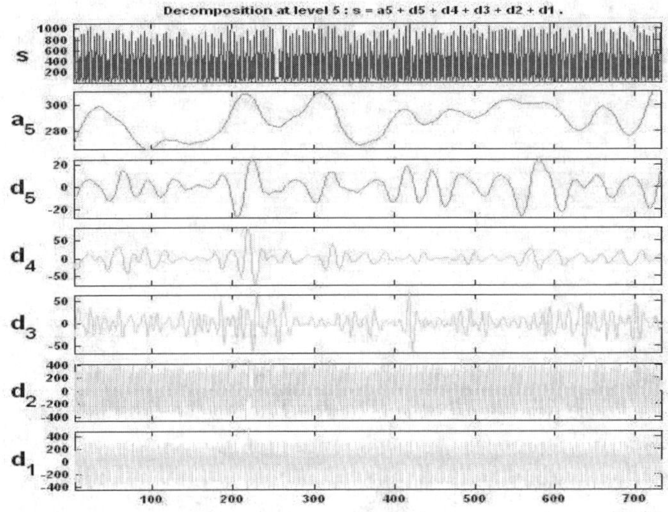

Figure 7: MATLAB analysis of annual Indian rainfall data for the year 1813–1995 by Daubechies 5 wavelet 1-D, level 5. Deterioration is observed after level 5.

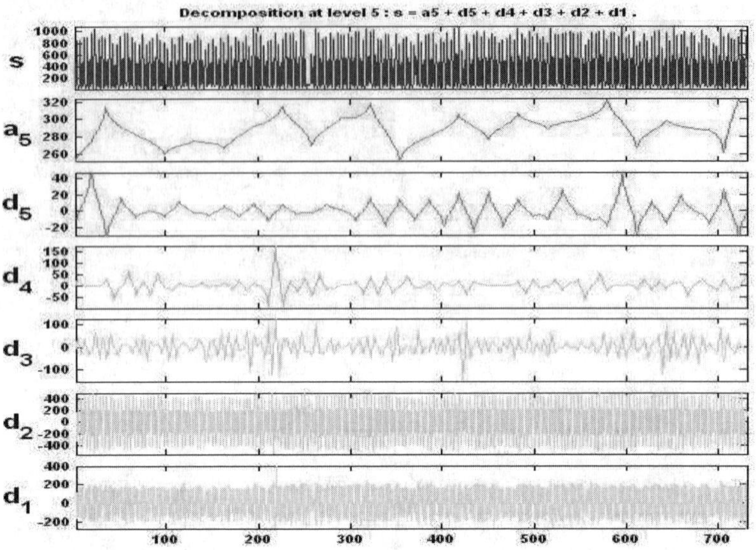

Figure 8: MATLAB analysis of Indian rainfall data for the period 1813–1995 by Coifman 1 wavelet 1-D, level 5

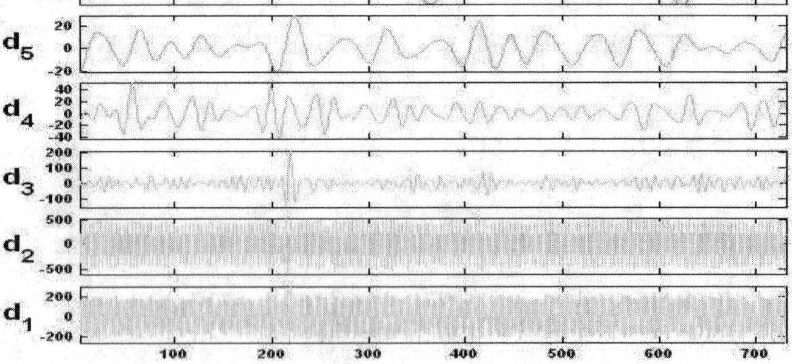

Figure 9: MATLAB analysis of Indian rainfall data for the period 1813–1995 by Coifman 4 wavelet 1-D, level 6. Clear distortion is observed.

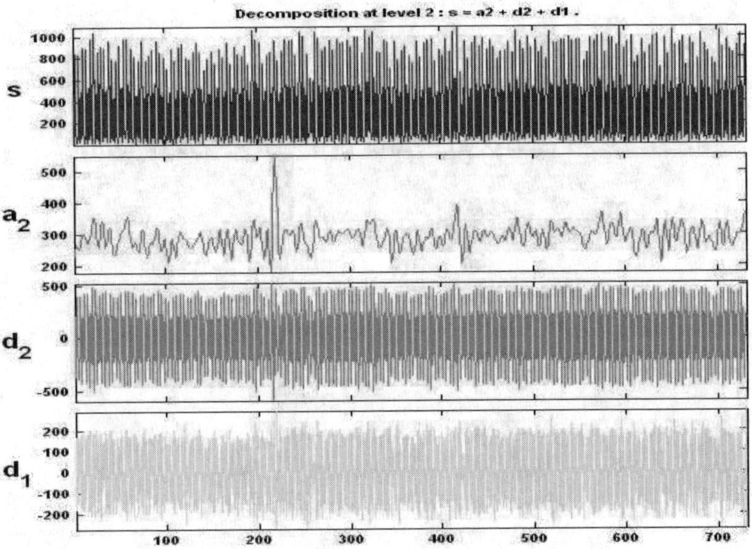

Figure 10: MATLAB analysis of Indian rainfall data for the period 1813–1995 by Coifman 5 wavelet 1-D, level 2

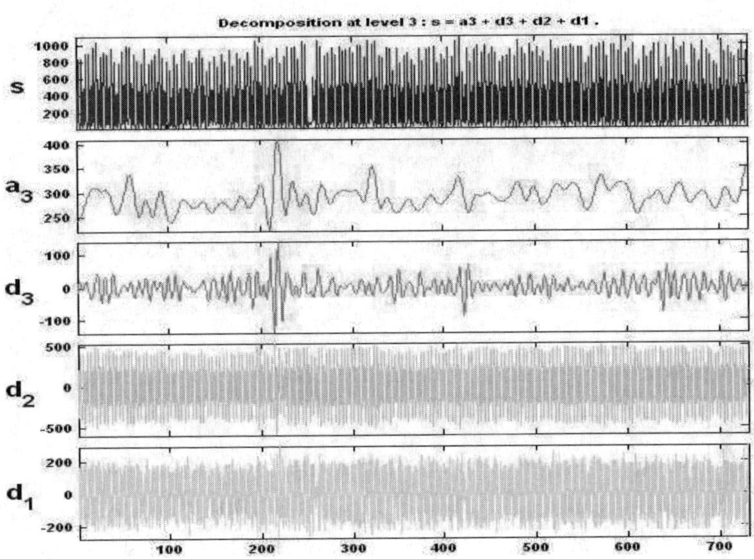

Figure 11: MATLAB analysis of Indian rainfall data for the period 1813–1995 by Coifman 5 wavelet 1-D, level 3

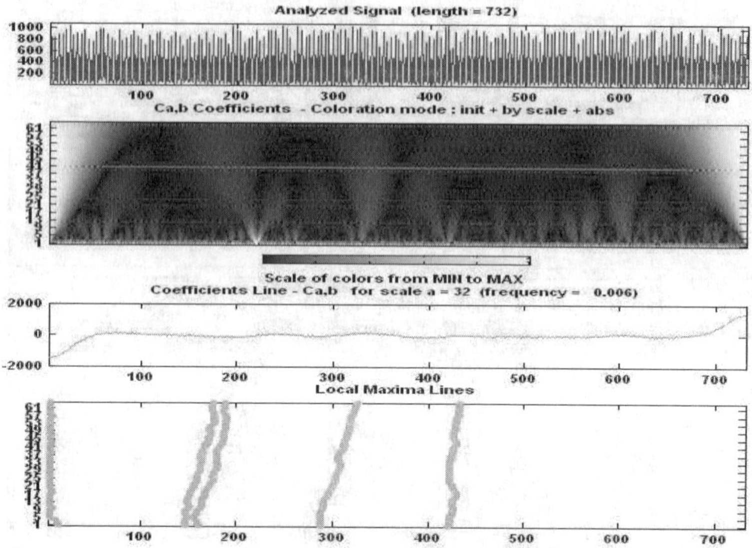

Figure 12: MATLAB analysis of Indian rainfall data for the period 1813–1995 by continuous Gauss wavelet 1-D, sampling period 1.

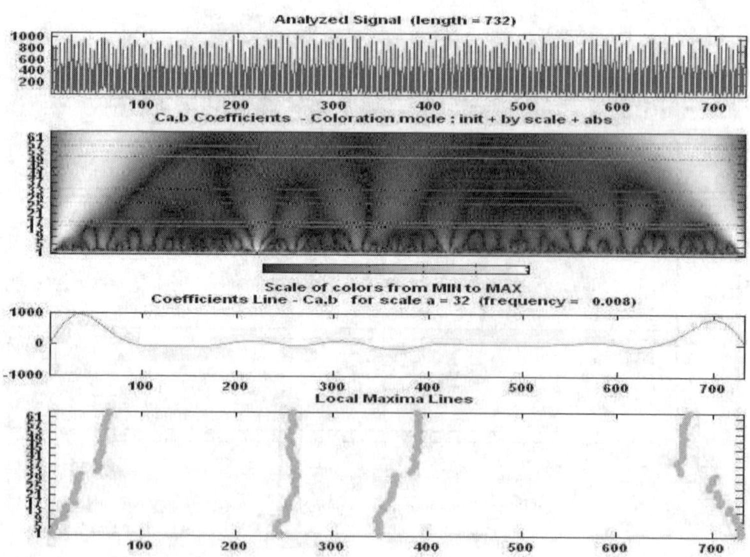

Figure 13: MATLAB analysis of Indian rainfall data for the period 1813–1995 by continuous Mexican hat wavelet 1-D, sampling period 1.

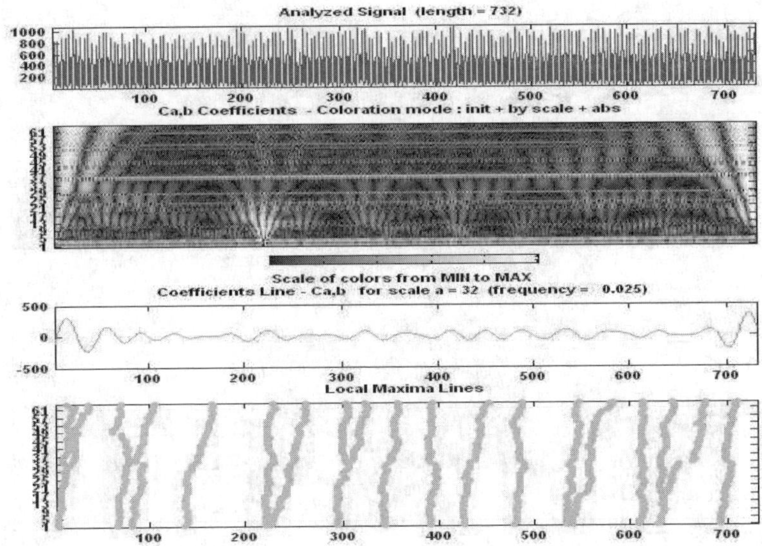

Figure 14: MATLAB analysis of Indian rainfall data for the period 1813–1995 by continuous Morlet wavelet 1-D, sampling period 1.

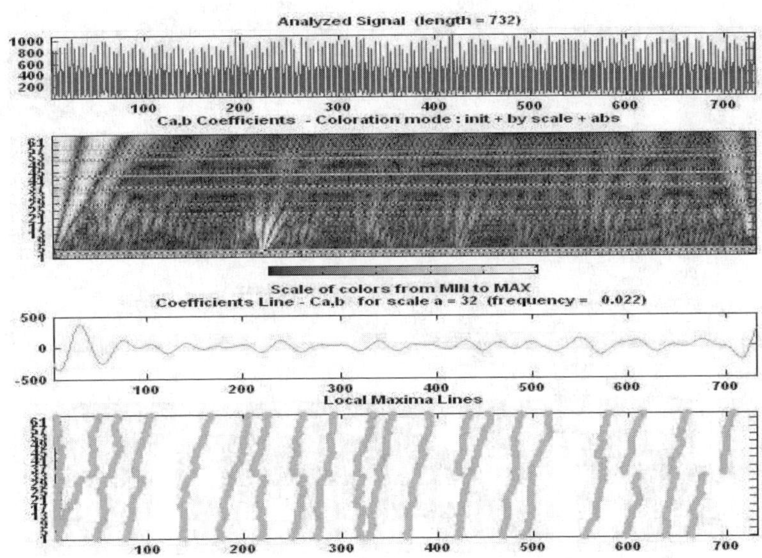

Figure 15: MATLAB analysis of Indian rainfall data for the period 1813–1995 by continuous Meyer wavelet 1-D, sampling period 1.

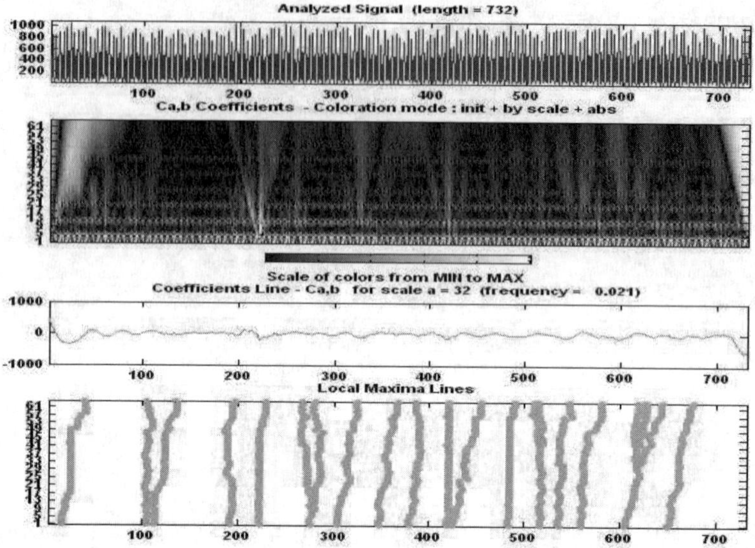

Figure 16: MATLAB analysis of Indian rainfall data for the period 1813–1995 by continuous symmetric 2 wavelet 1-D, sampling period 1.

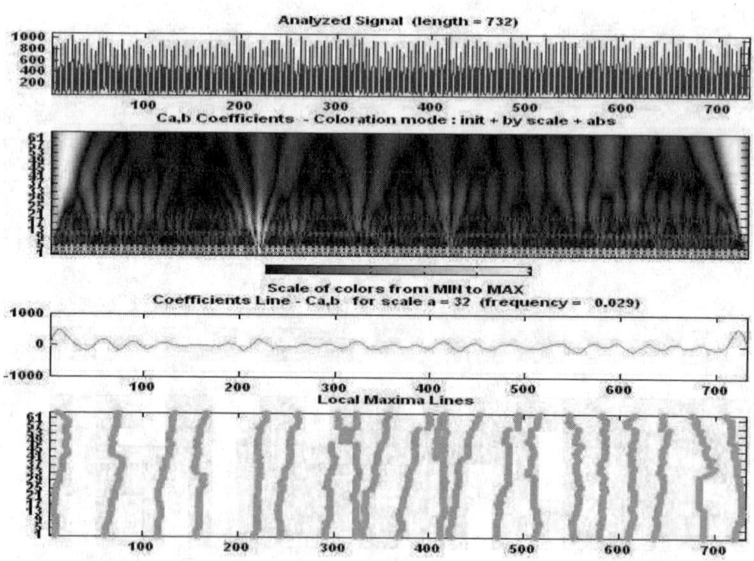

Figure 17: MATLAB analysis of Indian rainfall data for the period 1813–1995 by biorthogonal (continuous) 2.6 wavelet 1-D, sampling period 1.

4 Main Features

We have observed the following facts during our experiment with the MATLAB Wavelet Toolbox on Indian rainfall data of the period 1813–1995.

(i) It has been noted by several scholars that the Haar wavelet is more suited for detection of singularities in different types of signals. We have studied whether this fact holds true by the MATLAB Toolbox analysis of the Indian rainfall data from 1813–1995. The Haar wavelet analysis exhibits discontinuities clearly.

(ii) The main purpose of this paper is to look into the over all trend of the rainfall signal of the yearly and seasonal rainfall during 1813–1995 at different levels using different wavelets available in the MATLAB Wavelet Toolbox.

The rainfall signal of the above-mentioned years is a hazy picture obscured by noise. There is so much noise in the original signal that the overall shape is not apparent on visual inspection. In the experiment with different wavelets available in the MATLAB Wavelet Tool-box, we observe that the trends become more and more clearer with each approximation A_1 to A_4 or A_5. After A_5 the trend begins to distort. Coiflet 5 (see levels 3–5 in Figure 18–20, 23–28, and 30–35) provides the most appropriate trend and resembles with the earlier studies using different methodologies. It is just like finding an appropriate image through the zooming of a camera, Xerox machine, or projector.

It may be noted that the trend is the slowest part of the signal.

(iii) As the scale increases the resolution decreases, producing a better estimate of the unknown trend. Another way to think of this is in terms of frequency. Successive approximations possess progressively less high frequency information with higher frequencies removed, what is left is the overall trend of the signal.

(iv) The MATLAB Wavelet Tool-box is useful in revealing the pattern of rainfall in India annually and during the two most relevant seasons, an object that is complementary to the one of revealing a signal hidden in noise.

(v) If the signal itself inludes sharp changes, then the successive approximations look less and less similar to the original signal. Our experiment with the MATLAB Wavelet Toolbox indicates that there are only a few sharp changes.

(vi) The MATLAB Wavelet Toolbox can detect a self-similar or fractal nature of a signal representing a real-world problem. Our analysis of the Indian rainfall date with a continuous wavelet one-dimensional graphical tool reveals self-similarity and consequently the fractal nature of the Indian rainfall signal. Using the methodology discussed in Section 2, one can examine whether the behavior of the signal is persistent in a certain interval of time. We will pursue this theme separately along with spectrum singularity of annual and seasonal Indian rainfall date for the period 1813–1995.

(vii) It has been observed [Website (ii)] that $H = 2 - D$, where H is the Hurst exponent and D is a fractal dimension.

A Hurst exponent value H, $0.5 < H < 1$ indicates a persistant behavior, that is, positive auto-correlation; while a Hurst exponent value H, $0 < H < 0.5$ indicates a time series with an anti persistant behavior, that is, negative auto-correlation. A value of Hurst exponent $H = 0.5$ indicates a true random walk (a Brownian time series).

We are persuing these properties of time series of seasonal and regional Indian rainfall data, and they will be reported separately.

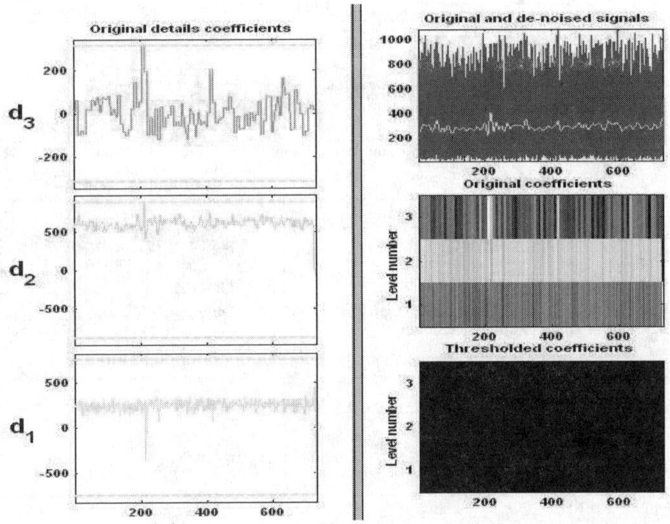

Figure 18: Indian rainfall data for the period 1813–1995 is analyzed by Coif 5 wavelet 1-D, level 3 of the MATLAB Wavelet Toolbox. One part indicates the noise data, while the other part (figure) provides pure signal.

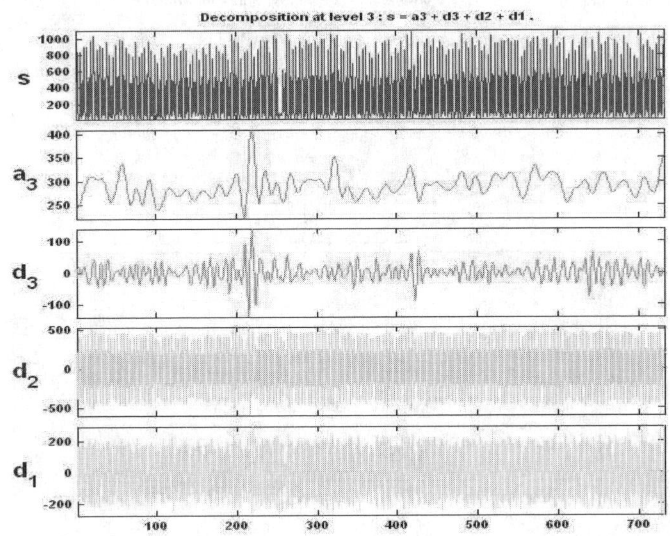

Figure 19: Denoising by Coif wavelet 1-D, level 3 of annual Indian rainfall data of the period 1813–1995.

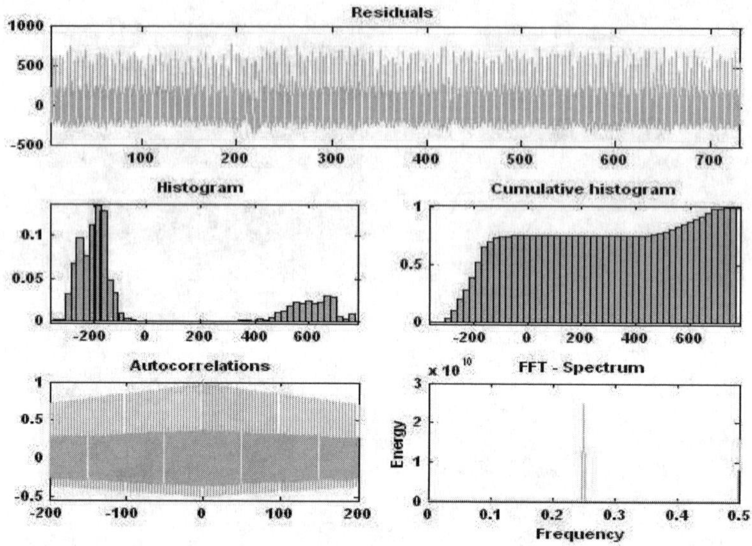

Figure 20: statistical properties of annual Indian rainfall data for the period 1813–1995. Coif wavelet 1-D, level 3 has been used.

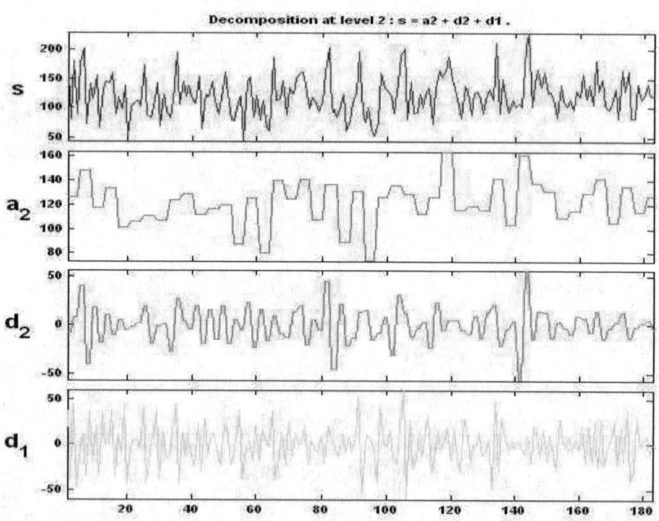

Figure 21: MATLAB analysis of Indian rainfall data for the month of October, November, and December for the period 1813–1995 by the Haar wavelet 1-D, level 2.

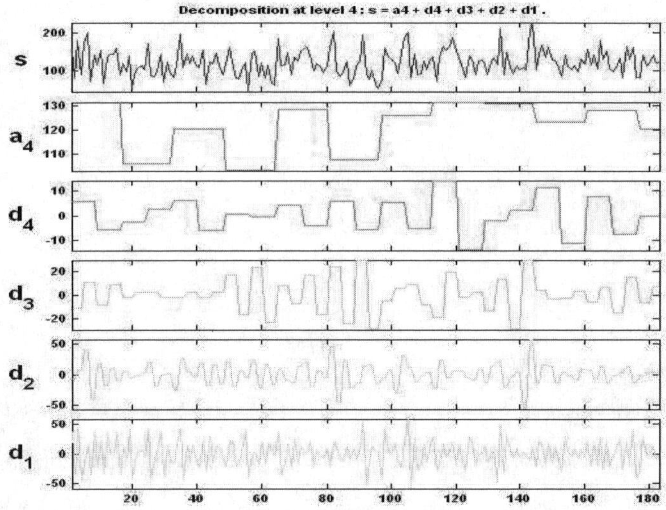

Figure 22: MATLAB analysis of Indian rainfall data for the months of October, November and December for the period 1813–1995 by the Haar Wavelet 1-D, Level 4

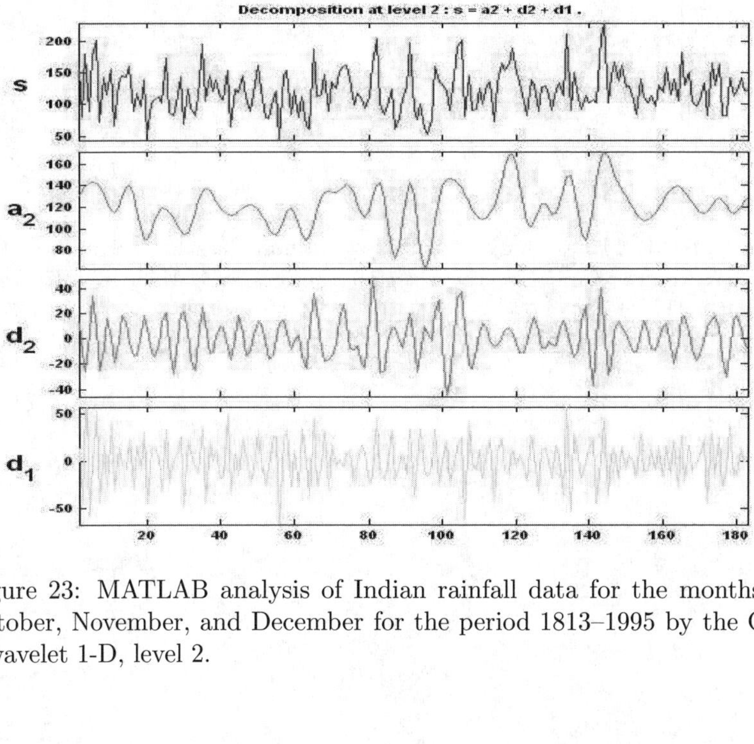

Figure 23: MATLAB analysis of Indian rainfall data for the months of October, November, and December for the period 1813–1995 by the Coif 5 wavelet 1-D, level 2.

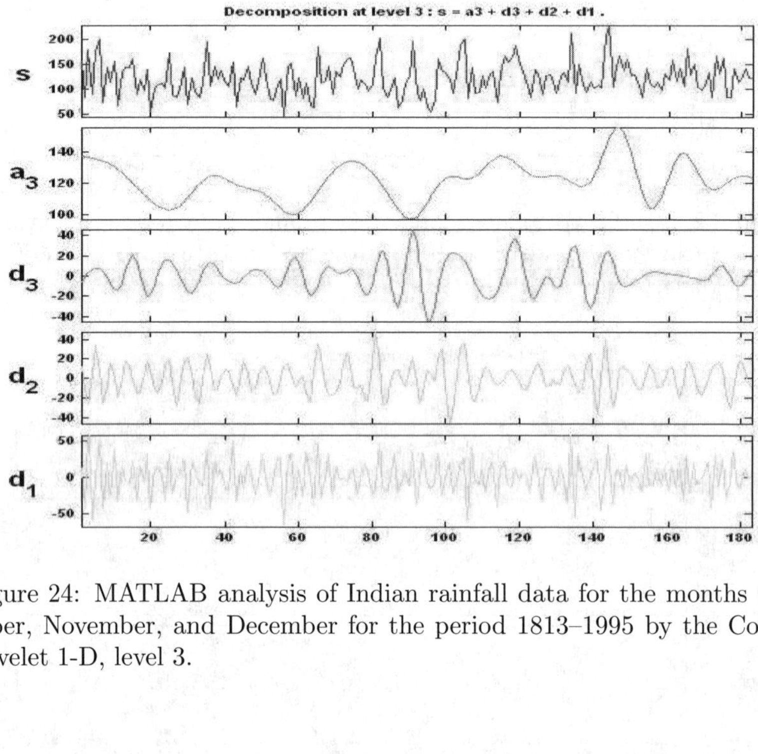

Figure 24: MATLAB analysis of Indian rainfall data for the months October, November, and December for the period 1813–1995 by the Coif 5 wavelet 1-D, level 3.

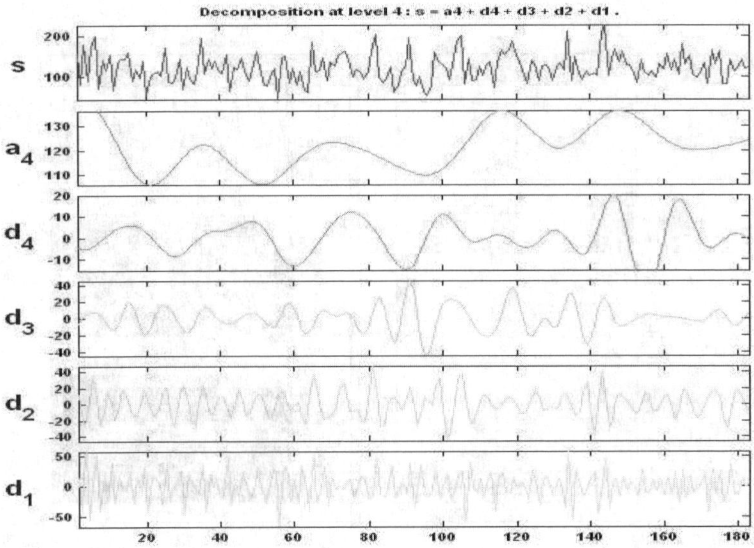

Figure 25: MATLAB analysis of Indian rainfall data for months October, November, and December for the period 1813–1995 by Coif 5 wavelet 1-D, level 4.

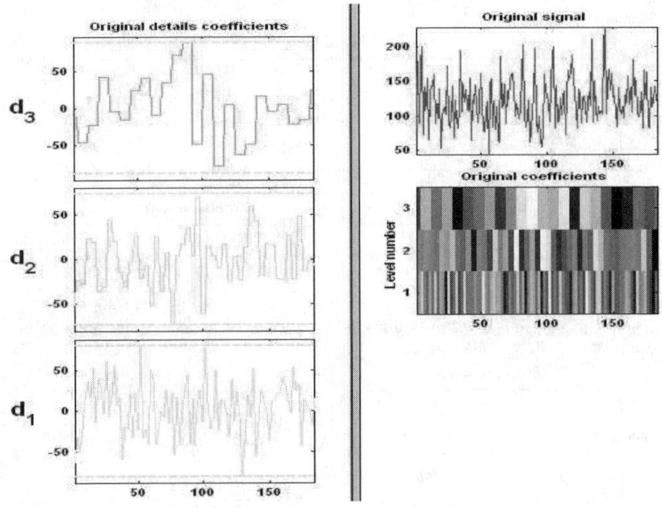

Figure 26: MATLAB analysis of Indian rainfall data for months October, November, and December for the period 1813–1995 by Coif 5 wavelet 1-D, level 3.

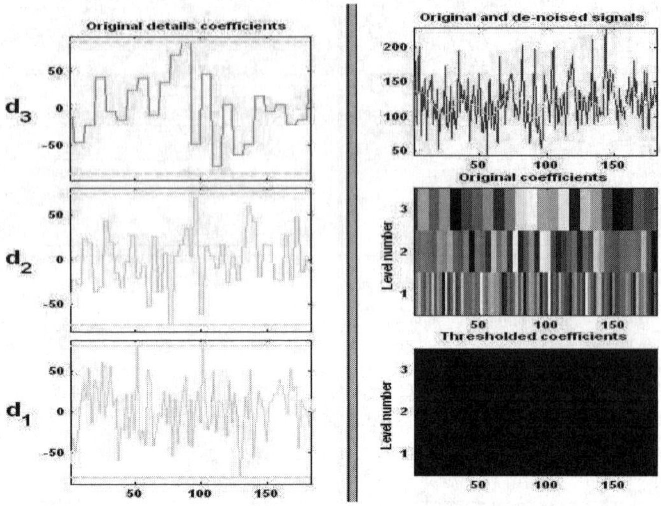

Figure 27: MATLAB wavelet toolbox (Coif 5, level 3) analysis of seasonal rainfall data for the Month of October, November, and December 1813–1995 is presented in this figure.

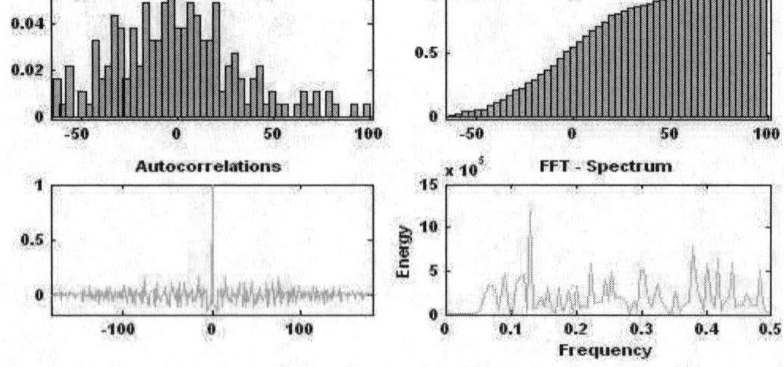

Figure 28: Wavelet 1-D: Coif 5, level 3.

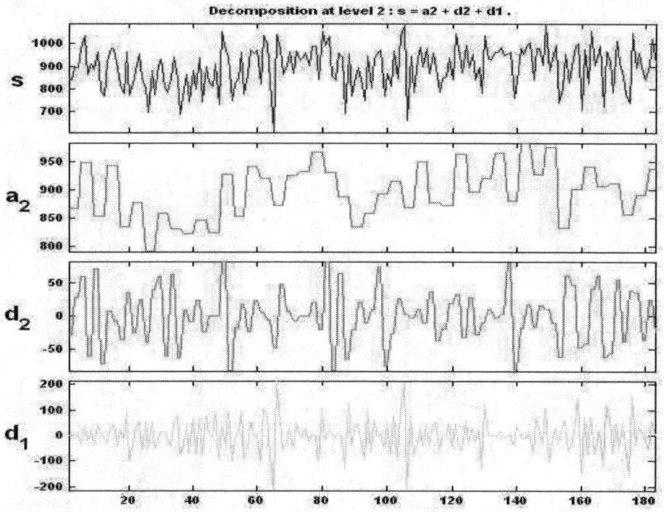

Figure 29: MATLAB analysis of the Indian rainfall data for the months of June, July, August, and September for the period 1813–1995 by the Haar wavelet 1-D, level 2.

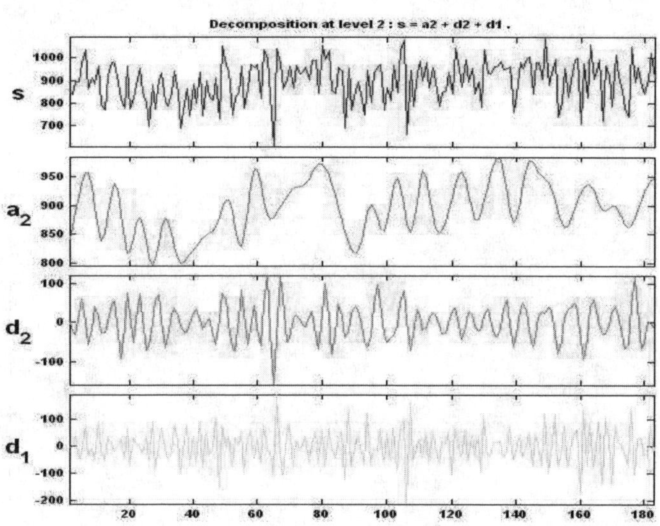

Figure 30: MATLAB analysis of the Indian rainfall data for the months of June, July, August, and September for the period 1813–1995 by the Coif 5 wavelet 1-D, level 2.

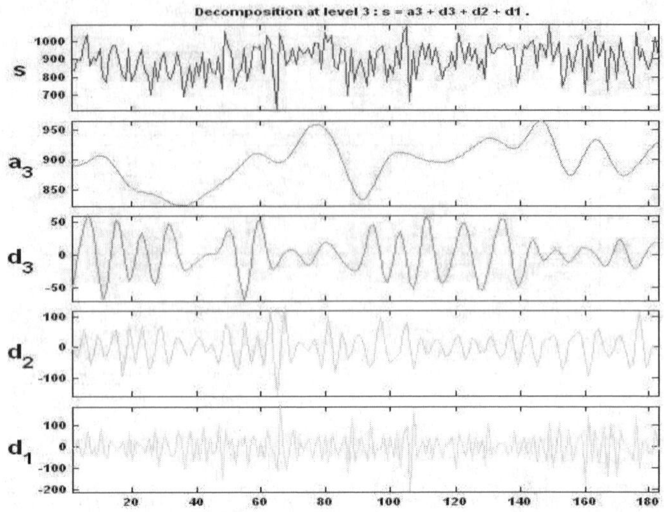

Figure 31: MATLAB analysis of the Indian rainfall data for the month of June, July, August, and September for the period 1813–1995 by the Coif 5 wavelet 1-D, level 3.

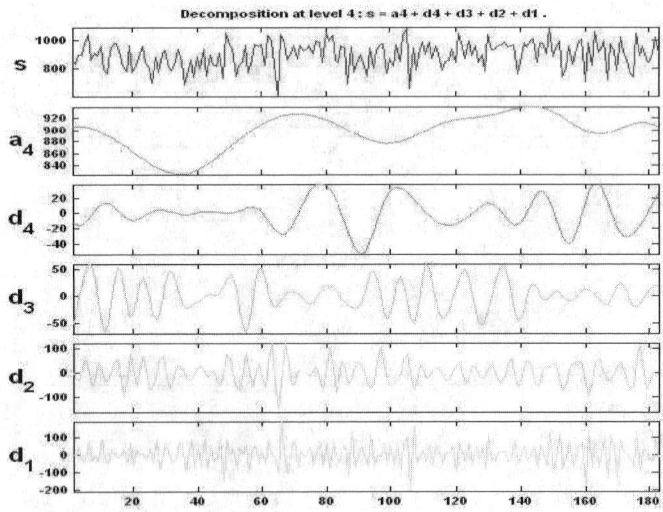

Figure 32: MATLAB analysis of the Indian rainfall data for the month of June, July, August and September for the period 1813-1995 by Coif 5 Wavelet 1-D, Level 4. Deterioration is apparent at level 4.

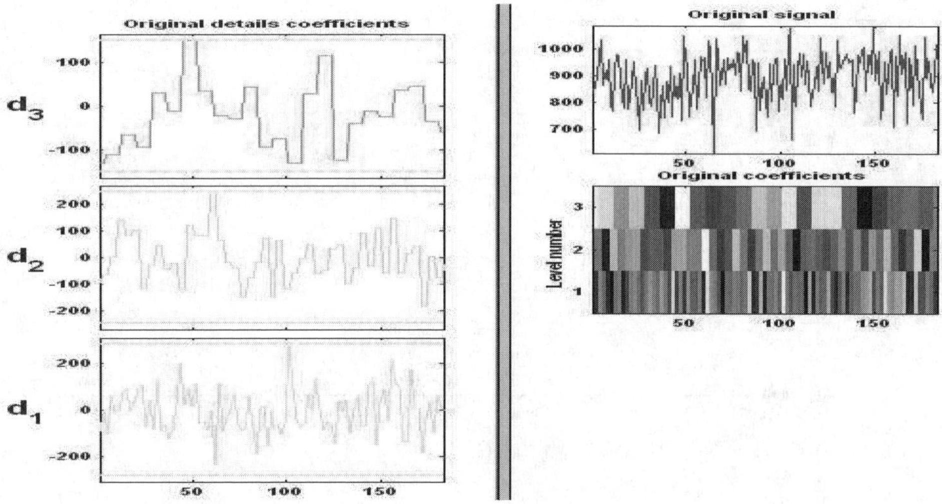

Figure 33: MATLAB analysis of the Indian rainfall data for the months of June, July, August, and September for the period 1813–1995 by the Coif 5 wavelet 1-D, level 3.

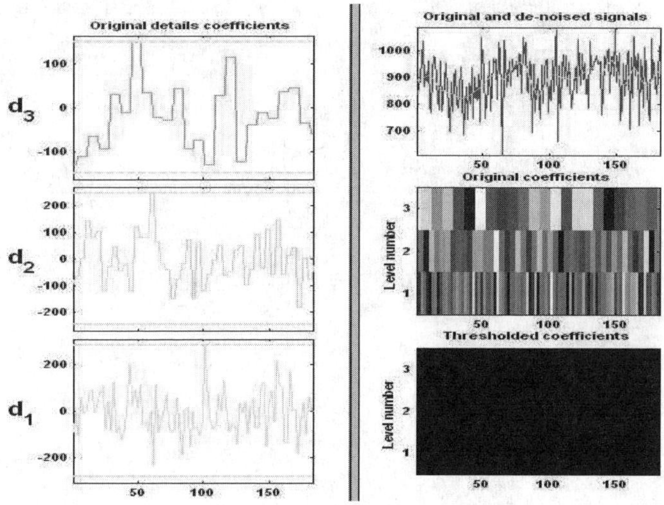

Figure 34: The red part indicates the original signal of the rainfall data of India for the months of June, July, August, and September for the period 1813–1995. The yellow graph shows pure signal (denoised signal) Coif 5 wavelet 1-D, level 3.

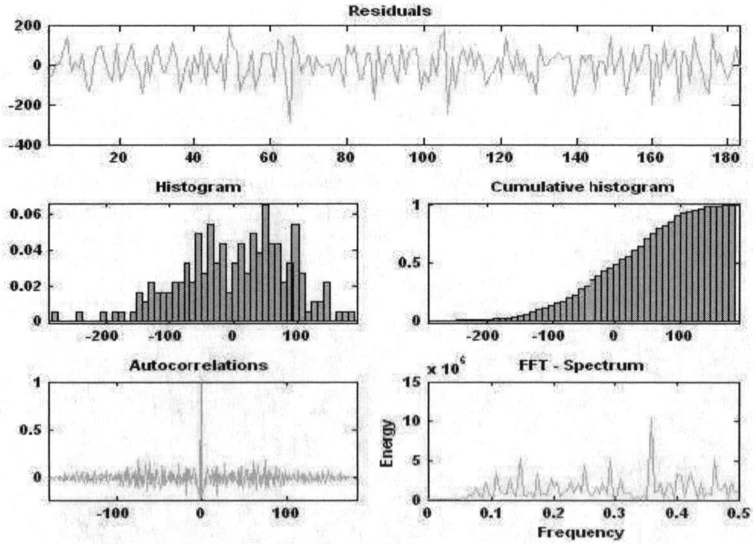

Figure 35: Coif 5 wavelet 1-D, level 3.

Acknowledgment. The authors would like to thank Prof. A. H. Siddiqi, King Fahd University of Petroleum & Minerals for his valuable suggestions and guidance during the preparation of this chapter.

References

[1] P. S. Addison, *The Illustrated Wavelet Transform Handbook: Introductory Theory and Applications in Science and Engineering, Medicine and Finance*, Institute of Physics Publishing, Bristol and Philadelphia, 2002.

[2] A. Arneodo, B. Audit, N. Decuster, P. Kestener, and S. C. Roux, A wavlet based method for multifractal image analysis: From traditional concepts to experimental applications, *Adv. in Imaging and Electronic Physics*, **126** (2003), 1–92.

[3] Z.Z. Hu and T. Nitta, Wavelet analysis of summer rainfall over North China and India, *J. Meteorological Society of Japan*, **6** (1996), 833–844.

[4] J. Kumar and F. Foufaula-Georgiou, Wavelet analysis for geophysical applications, *Rev. Geo-Physics*, **35** (1997), 385–412.

[5] S. Mallat, *A Wavelet Tour of Signal Processing*, Academic Press, New York, 1999.

[6] F. Nekka and J. Li, A continuous translation based method to reveal the fine structure of fractal sets, *Arabian J. Sci. Engg.*, **28** (2003), 169–188 (theme issue, Wavelet and Fractal Methods in Science and Engineering Part I).

[7] D. B. Percival and A. T. Walden, *Wavelet Methods for Time Series Analysis*, Cambridge University Press, London; New York, 2000.

[8] A. H. Siddiqi, *Applied Functional Analysis*, Marcel Dekker, New York, 2004.

[9] A. H. Siddiqi, *Lecture Notes Short Course on Wavelets and Applications*, Sept. 2003 to Dec. 2003, Department of Mathematical Sciences, King Fahd University of Petroleum and Minerals, Dhahran, Saudia Arabia; Website: http:www.edu.sa/Math/special events.

[10] A. H. Siddiqi, S. Khan, and S. Rehman, Wavelet based computer simulation for wind speed on Saudia Arabia, *Proc. First Saudia Conference*, KFUPM, Dhahran, Saudia Arabia, pp. 313–326, April 2001.

[11] A. H. Siddiqi, Z. Aslan, and A. Tokagozlu, Wavelet computer simulation of some meteorological parameters, in *Trends in Industrial and Applied Mathematics*, Edited by A.H. Siddiqi and M. Kocvara, Kluwer Academic Publishers, Boston/Dordrecht/London, pp. 95–115, 2002.

[12] N. Singh and N. A. Sontakke, The instrumental peiod rainfall series of the Indian region: A documentation, *Reserach Report No. RR-067*, Indian Institute of Tropical Meteorology, Pune, p. 79, 1996.

[13] N. Singh and N. A. Sontakke, On the variability and prediction of rainfall in the post monsoon seasons over India, *International J. Climatology*, **19** (1999), 309–339.

[14] N. Singh and N. A. Sontakke, On climatic fluctuations and environmental changes of the Indo-gangetic plains, India Climate Change, **52** (2002), 287–313.

[15] N. Singh, N. A. Sontakke, and S. K. Patwardhan, Rainfall prediction for rice growing areas: Potential and prospects, in *RICE-in a Variable Climate*, Edited by Y.P. Abrol and S. Gadgil, APC Publications Pvt. Ltd., New Delhi, pp. 63–105, 1999.

[16] N. Singh and S. K. Patwardhan, Prediction of terrestial and extraterrestrial parameter by modelling and extrapolating their natural regularities, *MAUSAM*, **52**(1) (2001), 117–132.

[17] N. A. Sontakke and N. Singh, Longest instrumental regional and all-India summer monsoon rainfall series using optimum observations: Reconstruction and update, *The Holocene*, **6** (1996), 315–331.

Websites

(i) Matlab Wavelet Toolbox: http://www.math.works.com

(ii) http://www.bearcav.com/misl/misl_tech/wavelets/hurst

Chapter 8

WAVELET ANALYSIS OF TROPOSPHERIC AND LOWER STRATOSPHERIC GRAVITY WAVES

O. Oğuz
Istanbul Commerce University

Z. Can
Yildiz Technical University

Z. Aslan
Beykent University

A. H. Siddiqi
King Fahd University of Petroleum & Minerals

Abstract

To define the wave packets, wavelet analysis has been applied on horizontal wind speed data. This is the more objective methodology to detect wave packets. Horizontal wind speed, meridional and zonal components based on radiosonde launches at Göztepe (Istanbul) between 1993 and 1997 are analyzed. The vertical profiles of meridional and zonal wind components at 850, 700, 500, 300 and 100 hPa pressure heights are transformed by using the Daubechies wavelet and the Morlet wavelet. High wind speed variance observations are identified as gravity wave packets. A strong seasonal cycle and maximum wave variance have been observed in winter.

1 Introduction

Gravity waves play a crucial role in determining the large-scale structure of the middle atmosphere. There are several techniques to detect gravity waves. Recently, wavelet analysis of vertical profiles of meridional and zonal components of horizontal wind has been used to obtain some parameters of gravity waves [8]. From another point of view, one can conclude that it is essential to consider the effects of gravity waves in climate change simulations. A relationship between gravity waves and meteorological parameters like pressure, temperature, and wind speed has been investigated [3].

McLandress [5] discusses the problem of parametrising unresolved gravity waves in general circulation models (GCMs) of the middle atmosphere. The results of this paper show the northward wind component at local solar noon for March and April as a function of latitude and height.

On the basis of the data series of the length of day (LOD), the atmospheric angular momentum (AAM), and the Southern Oscillation Index (SOI) between January 1970 and June 1999, the relationships among inter-annual LOD, AAM, and the El Nino/Southern Oscillation (ENSO) are analyzed by the wavelet transform method [7]. The results suggest that they have similar time-variation of spectral structures. The signals of 1997–98 El Nino and 1998–99 La Nina events can be detected from the analysis of LOR or AAM.

A calendar of the negative and positive phases of the North Sea-Caspian Pattern (NCP) for the period 1958–1998 was used to analyze the implication of the NCP upper level teleconnections on the regional climate of the eastern Mediterranean basin [4].

This chapter presents the basic dynamics of linear internal gravity waves. Wavelet analysis is used to transform vertical profiles of horizontal wind speeds based on radiosonde data observed at Göztepe (Istanbul) between 1993 and 1996.

2 Material and Method

2.1 Material

The data about horizontal wind speed analyzed in this chapter was recorded by the radiosonde system at Göztep (Istanbul) between 1993 and 1997. In particular, the meridional and zonal components of the horizontal winds at

850, 700, 500, 300 and 100 hPa pressure heights are analyzed to determine the vertical profiles of gravity waves.

2.2 Method

Recently, several works have been devoted to methods to extract gravity wave parameters from vertical profiles of the horizontal wind speed. In this chapter a wavelet-based method is used to decompose vertical profiles of horizontal wind into the gravity wave packets. It is assumed that only a single gravity wave is observed at a given layer. However, there are some other studies on the existence of the multiple gravity waves [1, 2, 6].

3 Analysis and Results

3.1 Time Series Analysis of Horizontal Wind Speed Components

Monthly variation of zonal and meridional wind speed standard deviations at the 100 hPa pressure level between 1996 and 1997 shows higher standard deviation values in spring that can be associated to the gravity waves activities and lower deviation values in summer and winter for the zonal wind speed analysis at the 100 hPa pressure level. On the other hand, the analysis of meridional wind speed at the 100 hPa pressure level indicates higher standard deviation values in winter and lower standard deviation in summer. When we repeat similar analysis techniques on zonal wind speed at 500 and 850 hPa pressure levels, we obtain the following results. It can be seen at the 500 hPa pressure level that lower deviation values are in summer, and higher deviation values are observed in winter. At the 850 hPa pressure level lower and higher deviations are observed in summer and spring, respectively. In general, processed data with moving averages shows similar trends at 850 and 500 hPa pressure levels. The extreme standard deviation is observed at the 300 hPa level. Standard deviations at pressure levels in 1993 and 1997 are higher than the values observed in other periods. The standard deviations of the zonal and meridional wind speeds at the pressure levels 100, 300, 500, 700 and 850 hPa for years 1993, 1994, 1995, 1996, and 1997 are presented in the Tables I–V.

The amount of wave energy propagating from 100 hPa to 850 hPa is maximum in winter. These results are similar to the analysis of gravity wave packets over Macquarie Island [8].

Table 1a: Monthly standard deviations of zonal wind speed (100 hPa)

Month	1993	1994	1995	1996	1997
1	7.27	7.52	8.13	6.98	7.83
2	12.34	4.25	7.42	6.41	6.95
3	9.30	9.42	6.37	6.30	7.35
4	8.15	5.91		6.18	8.14
5	5.52	6.01	4.46	6.40	4.24
6	5.69	7.75		11.01	6.58
7	5.64	5.23	4.78	7.24	10.53
8	4.62	5.28	5.99	4.53	5.43
9	9.15	5.68	4.97	7.20	6.62
10	5.50	4.94	4.25	5.45	8.92
11	10.03		5.9	6.35	8.97
12	9.26		5.57	5.42	11.15

Table 1b: Monthly standard deviations of meridional wind speed (100 hPa)

Month	1993	1994	1995	1996	1997
1	5.90	5.27	8.31	7.36	6,77
2	12.77	5.55	4.63	7.17	5.13
3	7.65	5.45	6.94	4.79	6.92
4	7.36	5.63		5.48	6.69
5	2.67	7.82	4.52	3.86	5.97
6	7.17	6.95		6.17	4.20
7	6.7	7.17	4.46	7.18	7.57
8	5.9	5.54	5.48	5.07	5.72
9	7.81	6.64	6.59	6.20	4.73
10	4.94	6.08	4.70	5.07	4.29
11	9.72		7.60	5.99	8.93
12	5.76		6.73	6.95	8.50

Table 2a: Monthly standard deviations of zonal wind speed (300 hPa)

Month	**1993**	**1994**	**1995**	**1996**	**1997**
1	14.91	14.19	10.62	11.01	14.86
2	17.41	9.36	11.21	11.17	12.52
3	14.43	13.42	14.91	11.05	14.06
4	12.36	10.87		10.89	11.74
5	9.15	9.18	5.94	10.99	8.92
6	9.33	13.21		11.07	10.18
7	9.18	8.48	7.44	10.05	7.66
8	9.05	8.18	8.76	6.53	10.60
9	12.55	6.69	9.69	10.14	7.84
10	7.14	8.70	11.52	7.40	13.25
11	13.22		8.29	11.10	9.78
12	9.96		9.78	9.68	13.55

Table 2b: Monthly standard deviations of meridional wind speed (300 hPa)

Month	**1993**	**1994**	**1995**	**1996**	**1997**
1	12.97	11.50	17.19	12.35	12.13
2	16.6	11.59	13.67	11.10	14.11
3	13.90	13.28	14.39	10.86	15.00
4	13.29	13.52		12.93	15.18
5	10.26	10.72	12.00	8.07	11.36
6	10.45	13.34		11.38	7.15
7	12.97	12.63	5.83	12.33	12.12
8	8.77	7.56	8.85	7.66	10.08
9	11.67	10.27	11.18	9.68	9.43
10	6.84	9.70	15.14	10.96	12.96
11	14.64		13.79	10.52	13.10
12	14.68		11.49	13.67	15.32

Table 3a: Monthly standard deviations of zonal wind speed (500 hPa)

Month	1993	1994	1995	1996	1997
1	10,32	8.82	10.04	7.09	10.41
2	13.87	6.00	7.483	7.03	9.22
3	9.34	9.40	7.52	6.89	9.47
4	11.56	6.73		8.15	7.74
5	5.77	6.54	6.35	7.11	8.53
6	7.71	10.64		6.75	7.34
7	7.40	3.57	5.11	4.624	4.69
8	7.23	5.73	5.17	4.38	4.99
9	7.61	3.76	5.60	4.78	6.17
10	5.847	5.42	7.00	5.26	8.74
11	8.22		5.78	7.47	6.90
12	7.99		7.84	8.71	12.44

Table 3b: Monthly standard deviations of meridional wind speed (500 hPa)

Month	1993	1994	1995	1996	1997
1	10.85	7.23	11.33	7.13	9.83
2	15.90	7.04	10.08	9.53	11.45
3	10.66	10.02	10.37	8.976	11.41
4	8.79	9.20		9.67	11.00
5	5.12	7.42	8.43	6.86	7.60
6	6.79	9.74		7.85	5.22
7	8.19	6.24	4.16	6.91	7.13
8	4.77	5.43	5.87	4.79	6.21
9	7.21	7.34	9.03	6.30	5.69
10	5.36	6.21	10.17	7.54	6.73
11	9.59		10.78	9.35	8.33
12	8.55		9.08	8.68	9.31

Table 4a: Monthly standard deviations of zonal wind speed (700 hPa)

Month	1993	1994	1995	1996	1997
1	9.92	6.93	8.27	4.63	7.69
2	10,32	5.09	7.14	7.01	6.80
3	6.73	5.93	6.79	5.87	6.30
4	9.10	5.60		5.32	5.76
5	4.65	3.83	4.98	5.26	6.81
6	5.67	6.66		4.08	5.31
7	4.83	3.22	5.05	3.54	4.69
8	4.79	5.16	3.52	3.98	4.27
9	5.32	3.90	5.02	4.40	6.33
10	3.89	4.03	4.82	3.42	6.18
11	6.46		5.00	5.28	6.64
12	5.60		6.81	7.15	7.12

Table 4b: Monthly standard deviations of meridional wind speed (700 hPa)

Month	1993	1994	1995	1996	1997
1	7.01	5.29	7.69	4.72	7.15
2	12.59	4.96	7.93	9.92	8.91
3	7.34	5.92	8.35	6.13	7.77
4	6.46	6.71		5.95	8.06
5	4.35	5.92	6.90	5.47	5.47
6	5.00	5.72		4.97	4.80
7	5.46	4.34	3.78	4.40	6.15
8	3.71	4.28	4.06	3.88	4.59
9	4.68	5.61	6.40	5.33	3.88
10	4.11	6.22	5.45	4.77	7.42
11	6.83		7.25	7.46	6.14
12	6.54		9.86	7.74	7.01

Table 5a: Monthly standard deviations of zonal wind speed (850 hPa)

Month	1993	1994	1995	1996	1997
1	8.33	5.14	6.46	3.35	5.83
2	8,17	4.69	5.60	6.60	4.83
3	5.94	4.92	5.24	4.69	4.76
4	7.21	4.74		3.53	4.69
5	4.04	2.97	4.57	4.61	5.53
6	4.07	4.93		3.89	3.93
7	4.00	3.41	2.71	3.32	3.98
8	4.20	4.38	3.52	3.71	5.71
9	4.10	3.77	4.01	3.55	3.95
10	4.62	3.46	4.02	2,97	6.53
11	6.32		4.05	4.62	6.56
12	5.80		5.05	5.03	7.74

Table 5b: Monthly standard deviations of meridional wind speed (850 hPa)

Month	1993	1994	1995	1996	1997
1	6.59	4.86	6.78	4.28	5.50
2	8.97	3.97	6.18	6.52	7.18
3	5.72	4.78	6.66	5.54	5.65
4	5.56	5.21		3.92	6.26
5	3.18	4.10	4.96	4.25	4.75
6	3.95	4.14		3.61	3.22
7	3.34	2.57	1.85	2.65	3.85
8	2.54	2.73	3.21	2.661	3.89
9	3.58	3.40	3.90	4.37	3.23
10	4.19	4.15	3.49	3.88	6.74
11	7.34		6.46	5.80	7.66
12	6.88		7.24	6.85	8.46

3.2 Wavelet Analysis of Horizontal Wind Speed Components Using Daubechies Wavelet

Wavelet analysis of zonal wind speeds are performed for the year 1997 by using the Daubechies wavelet. The large-scale effect at the 500 hPa pressure level is observed in April, May, June, and July is shown in Figure 1. Other periods have been affected by small-scale fluctuations. During the summer period, easterly winds are dominant. At the 850 hPa pressure level large-scale effects play a crucial role on zonal wind speed variations in general (Figure 2). Between October and December large- and small-scale fluctuations have more deviations on zonal wind speed components.

In a similar way wavelet analysis of meridional wind speed for the year 1997 is analyzed. From January to April combined effects of large- and small-scale fluctuations are observed at the 500 hPa pressure level as shown in Figure 3. Next section is under the effect of small-scale events in late spring, and autumn northerly flows are dominant. In the late spring, and early summer terms small-scale fluctuations have been observed at the 850 hPa pressure level in Figure 4. Larger deviations were recorded at the end of the year.

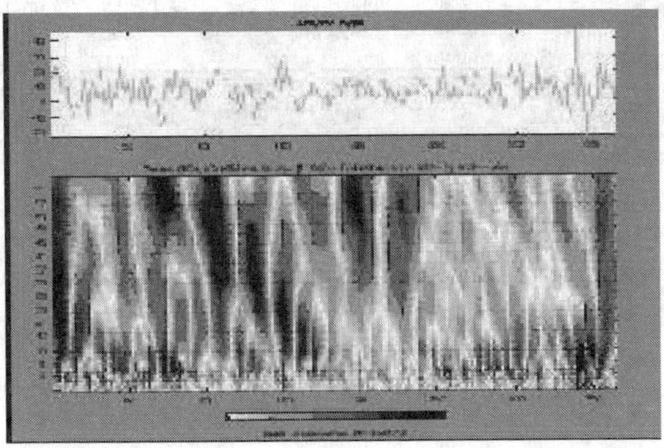

Figure 1: Continuous wavelet 1-D of zonal wind speed 1997 (500 hPa).

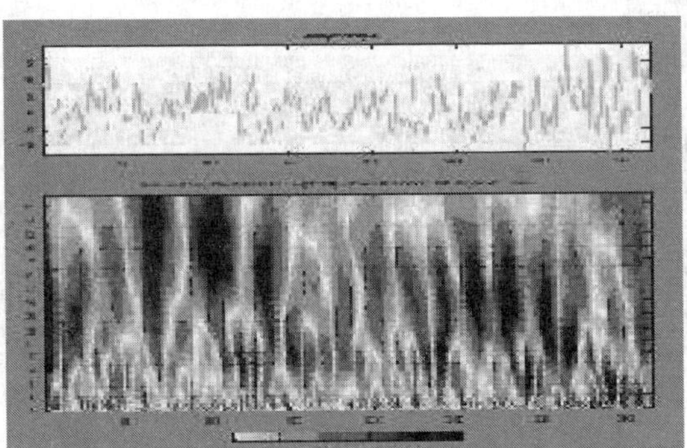

Figure 2: Continuous wavelet 1-D of zonal wind speed 1997 (850 hPa).

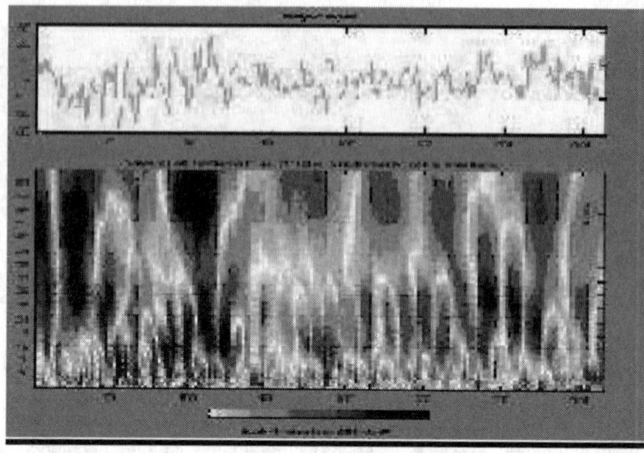

Figure 3: Continuous wavelet 1-D of meridional wind speed 1997 (500 hPa).

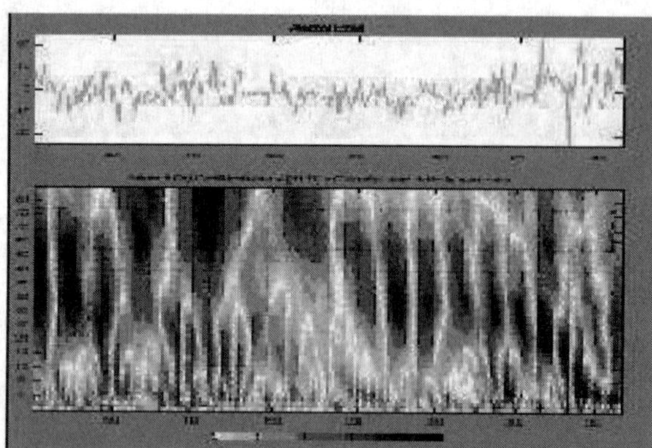

Figure 4: Continuous wavelet 1-D of meridional wind speed 1997 (850 hPa).

3.3 Wavelet Analysis of Zonal Wind Speeds Performed for the Year 1997 by Using Morlet Wavelet

In this work the Morlet wavelet is also used to analyze zonal wind speeds for the year 1997. The results are compatible with the results derived by using the Daubechies wavelet. In Figure 5 small-scale fluctuations are observed in July and August at the 500 hPa pressure level. This analysis is repeated for the 850 hPa pressure level (Figure 6), and small-scale fluctuations are observed in February and March and large-scale fluctuations in October and November. There are some similarities between the two analyses for the meridional wind speeds at 500 and 850 hPa pressure levels (see Figure 7 and Figure 8).

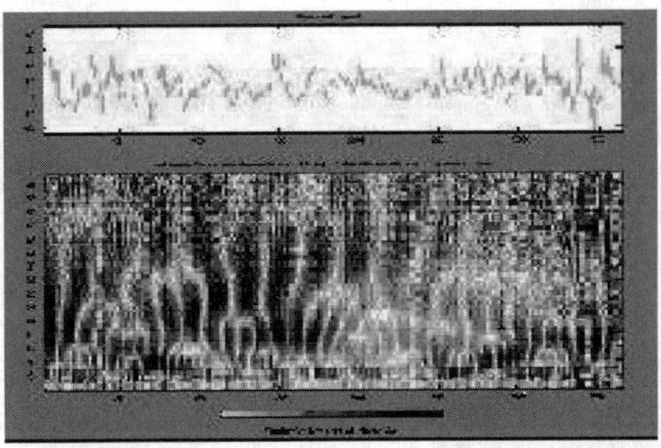

Figure 5: Continuous wavelet 1-D of zonal wind speed 1997 (500 hPa).

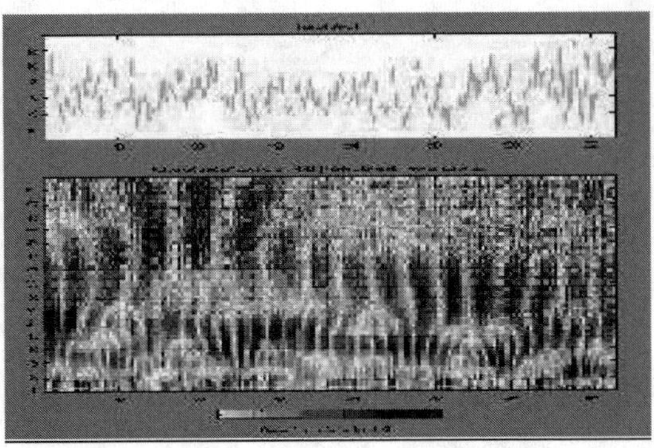

Figure 6: Continuous wavelet 1-D of zonal wind speed 1997 (850 hPa).

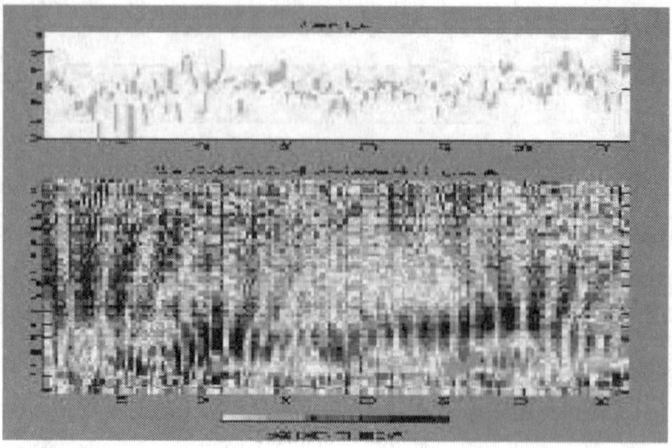

Figure 7: Continuous wavelet 1-D of meridional wind speed 1997 (500 hPa).

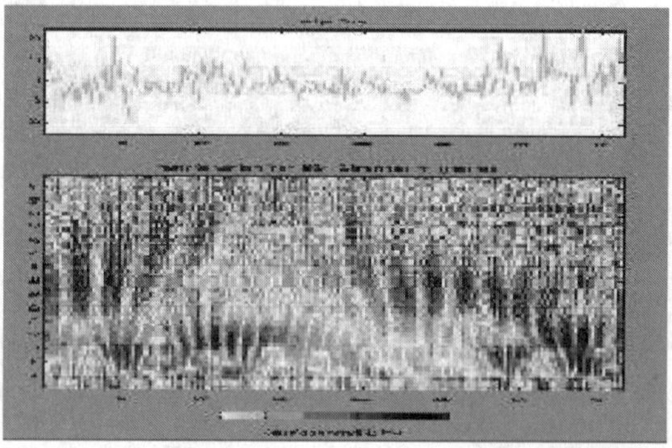

Figure 8: Continuous wavelet 1-D of meridional wind speed 1997 (850 hPa).

4 Conclusions

Time series analysis of raw and smoothed data (wind speed components) based on moving averages (1 = 5) shows similar results at 850 and 500 hPa.

In order to define large- and small-scale effects on wind speed component deviations, continuous 1D-Daubechies wavelet analysis of zonal and meridional wind speed deviations were analyzed.

Large-scale effects on zonal wind speed deviations have been observed in winter, spring, and autumn at all pressure levels. Generally, small-scale fluctuations are observed during four seasons at 700 and 850 hPa. At the 500 hPa pressure level small-scale events do not play an important role in the gravity wave variations in spring.

Large-scale variations play an important role in the meridional wind speed deviations in autumn, winter, and spring. These results have a similar structure with the variation of zonal wind speed components for all pressure levels in the observation period.

It is observed that small-scale events are not effective in the meridional wind speed deviations in summer and winter.

Continuous wavelet 1-D analysis of the Morlet wavelet were also applied at the same data. Both results are similar. Application of the method to once daily radiosonde soundings of lower atmosphere over Istanbul (Göztepe) explains the higher wave variance with a maximum in winter.

Acknowledgment. The authors acknowledge support from Beykent University, Istanbul Commerce University, and Yildiz Technical University.

The study has been supported by the research foundation of Yildiz Technical University with project No. 21-01-01-02, 2001-2004.

References

[1] S. J. Allen and R. A. Vincent, Gravity wave activity in the lower atmosphere: Seasonal and latitudinal variations, *J. Geophysical Res.*, **100** Issue D1 (1995), 1327–1350.

[2] Z. Aslan and O. Oguz, Time variation of gravity waves in lower atmosphere, IC/IR/98/22 Internal report/Trieste/Italy, 1998.

[3] Z. Can, Z. Aslan, and O. Oguz, The variations of temperature, pressure and wind speed values: Effects on gravity waves, *Il Nuovo Cim.*, **25** (2002), 137.

[4] H. P. Kutiel, M. Maheras, M. Turkes, and S. Paz, North Sea-Caspian Pattern (NCP)-a upper level atmospheric teleconnection affecting the eastern Mediterranean-implications on the regional climate, *Theor. Appl. Climatology*, **72** (2002), 173–192.

[5] C. McLandress, On the importance of gravity waves in the middle atmosphere and their parameterization in general circulation models, *J. Atmospheric and Solar-Terrestrial Phys.*, **60** (1998), 1357–1383.

[6] S.-Y. Ogino, M. D. Yamanaka, and S. Fukao, Meridional and temporal variations of lower-stratospheric gravity wave activity over Japan, West Pasific ocean and East Indian ocean, *Advances in Space Res.*, **27** (8) (2001), 1475–1478.

[7] Y. H. Zhou, D. W. Zheng, and X. H. Liao, Wavelet analysis of interannual LOD, AAM, and ENSO: 1997-98 El Nino and 1998–99 La Nina signals, *J. Geodesy*, **75** (2001), 164–168.

[8] F. Zink and R. A. Vincent, Wavelet analysis of stratospheric gravity wave packets over Macquarie Island, *J. Geophysical Res.*, **106** (2001), 10273.

Chapter 9

ADVANCED DATA PROCESSES OF SOME METEOROLOGICAL PARAMETERS

A. Tokgozlu
Süleyman Demirel University

Z. Aslan

İstanbul Commerce University

Abstract

This chapter provides a case study on the application of curve fitting and harmonic analysis for some meteorological parameters in Isparta. Future projections of air temperature, pressure, rainfall, rate and wind speed data based on polynomial, final intervals and trigonometric functions have been evaluated for the study area. To define the model parameters, data observed in Isparta between 1929–2001 have been considered. Results of the harmonic analysis (Fourier series) and analysis of amplitude and phase values explain the small-, meso-, and large-scale effects on the parameters.

The results of this chapter are evaluated and compared with wavelet analysis for this pilot project. This study has a key role to define the climate changing effects on meteorological parameters.

1 Introduction

In recent years, harmonic analysis has emerged as a useful tool in studying annual patterns of some meteorological parameters [2, 10]. The spatial distributions of temperature, precipitation, and pressure have been examined by Kirkyla and Hameed [5], Currie and Hameed [4], Oku [9], and Aslan et al. [1]. The results of harmonic analysis enable us to distinguish different precipitation regimes and transition regions [6–11].

2 Material and Method

2.1 Study Area

The study area (Isparta; latitude : $37^\circ\ 46^{'}$ N, longitude: $30^\circ\ 33^{'}$ E, height, 997 m. above mean sea level) is in southwestern Anatolia. This area is called the Area of Lakes. This region is under the effect of central and southwestern Anatolian climatological conditions. It is under the combined effects of Mediterranean and terrestrial climate conditions with hot and dry summers and cold and wet winters. The annual rainfall rate is 568 mm in Isparta. Complex topography causes orographic and convective rain formation in the study area.

2.2 Data

In this study, wind speed and air pressure, air temperature, and rainfall rate observations in Isparta between 1929 and 2001 have been analyzed.

2.3 Methodology

In the first part of this chapter, statistical data analysis techniques (statistical descriptive and time variation of parameters) have been analyzed by using EXCEL and SPSS packet programs.

Statistical characteristics such as mean, standard deviation, variance, extremes, and skewness of meteorological parameters (air temperature (T), air pressure (P), horizontal wind speed (V) and precipitation (R)) have been studied in this chapter.

In the second part of this chapter FORTRAN-77 , DBOS program were used to calculate first and higher order harmonics of parameters.

Harmonic analysis is based on the series of trigonometric functions given in the following equation [5]:

$$X = X_0 + \sum_{i=1}^{N/2} A_i \cos(360it/p + \phi_i) \tag{2.1}$$

where

X: value at time t,
X_0 : arithmetic mean,
A_i: amplitude of the harmonics,
Φ_i: phase angles,
N: number of observations,
p: period of observations,
t : time.

In this study, p is 12 months. A large first harmonic indicates strong annual variation, and a comparatively large second harmonic amplitude points to a strong semi-annual variation. The phase angle determinates the year when the maximum or minimum of a given harmonic occurs. Short-period harmonics indicate influences of local phenomena. Amplitudes and phase determine various boundaries and areas of transition.

The ratio of the amplitudes of the first and second harmonics determines the relative importance of the two harmonic components.

The relative contribution of the first three harmonics to the seasonal variance is given by the ratio of the sum of squares of their amplitudes to the sum of squares of all six amplitudes. Hence, a ratio value close to unity suggests that the first three harmonics account for most or all of the seasonal variation in the curve; on the other hand, a smaller fraction implies a more complex annual curve with a greater amount of variability contained by the high-frequency harmonics [5].

3 Analysis

3.1 Statistical Analysis of Meteorological Parameters

Statistical characteristics (monthly mean, standard deviation, variance, extremes, and skewness) of meteorological parameters (air temperature (T), air pressure (P), horizontal wind speed (V), and precipitation (R)) have been presented in Tables 3.1.a–3.1.d.

Table 3.1.a. Statistical Characteristics of Air Temperature (T, ^{O}C) In Isparta Between 1929 and 2001.

	January	February	March	April	May	June	July	August	September	October	November	December
Mean	1,7	2,7	5,7	10,6	15,4	19,6	23,2	22,9	18,5	13,0	7,6	3,6
SD.	2,121	2,110	1,925	1,548	1,402	1,433	1,142	1,222	1,280	1,545	1,702	1,962
Max.	5,2	6,8	11,0	14,7	18,7	22,2	26,1	25,7	21,7	17,6	11,7	6,7
Min.	-3,2	-2,8	1,2	6,4	13,0	12,9	20,2	19,6	15,4	9,8	3,6	-1,8
Var.	4,497	4,451	3,705	2,397	1,965	2,054	1,303	1,492	1,639	2,386	2,898	3,851
Skew.	-0,445	-0,526	-0,067	-0,139	0,446	-1,810	-0,078	-0,282	0,264	0,209	-0,062	-0,678

Table 3.1.b. Statistical Characteristics of Air Pressure (P, hPa) In Isparta Between 1939 and 1999.

	January	February	March	April	May	June	July	August	September	October	November	December
Mean	899,18	898,20	897,80	897,72	898,62	898,31	896,87	897,53	900,13	901,91	901,87	900,56
SD.	4,36	3,76	3,61	2,48	2,55	2,36	2,38	2,46	2,51	2,57	2,79	3,65
Max.	907,7	906,1	907,9	902,2	903,1	902,0	901,1	901,4	904,0	906,5	907,3	909,0
Min.	890,7	890,8	888,2	892,1	892,8	891,9	892,2	891,8	893,9	895,8	895,6	893,8
Var.	19,00	14,16	13,02	6,14	6,50	5,59	5,65	6,05	6,28	6,61	7,77	13,36
Skew.	-0,04	-0,10	0,18	-0,13	-0,13	-0,16	0,10	-0,15	-0,15	-0,16	-0,18	-0,09

Table 3.1.c. Statistical Characteristics of Horizontally Wind (V, m/s) In Isparta Between 1939 and 2001.

	January	February	March	April	May	June	July	August	September	October	November	December
Mean	1,9	2,2	2,5	2,4	1,9	1,7	1,8	1,6	1,5	1,5	1,6	1,9
SD.	0,690	0,697	0,839	0,693	0,599	0,542	0,517	0,492	0,495	0,575	0,692	0,758
Max.	3,4	4,0	4,7	4,3	3,4	3,2	3,3	2,9	2,8	3,1	3,2	4,0
Min.	0,5	0,6	0,7	0,9	0,7	0,4	0,7	0,7	0,4	0,4	0,5	0,3
Var.	0,4760	0,4857	0,7038	0,4797	0,3591	0,2938	0,2673	0,2425	0,2449	0,3302	0,4795	0,5744
Skew.	0,0838	0,2365	0,5923	0,2846	0,4171	0,3573	0,3324	0,5467	0,4011	0,5995	0,2156	0,4885

Table 3.1.d. Statistical Characteristics of Precipitation (R, mm) In Isparta Between 1929 and 2001.

	January	February	March	April	May	June	July	August	September	October	November	December
Mean	79,4	70,7	60,9	51,9	55,1	32,7	13,3	11,4	16,3	36,8	45,8	92,6
SD.	56,858	45,781	39,321	31,765	35,367	26,008	17,905	15,337	18,430	29,086	33,525	66,538
Max.	308,6	222,0	183,4	188,8	197,4	112,3	87,4	58,6	112,1	124,2	202,5	360,1
Min.	1,7	7,9	9,7	3,4	0,8	0,0	0,0	0,0	0,0	0,0	3,2	0,0
Var.	3232,78	2095,94	1546,11	1009,02	1250,86	676,425	320,573	235,209	339,671	846,001	1123,92	4427,24
Skew.	1,550	1,050	1,146	1,607	1,035	0,871	2,127	1,630	2,667	0,963	2,059	1,521

Table 3.2 Contribution of first three harmonics and, relative importance (A_2/A_1) of first two harmonics

		T(°C)		P(hPa)		V(m/s)		R(mm)	
		*	A2/A1	*	A2/A1	*	A2/A1	*	A2/A1
1929-1930	Ave.	0,999932	0,010223					0,679456	0,549554
	Max.	0,999995	0,010396					0,928644	0,782913
	Min.	0,99987	0,010051					0,430267	0,316195
1931-1960	Ave.	0,999641	0,011874	0,462479	0,451651	0,831157	0,339426	0,706155	0,480369
	Max.	0,999996	0,041931	0,686065	0,977737	0,991751	1,652174	0,998445	1,982621
	Min.	0,99581	4,82E-05	0,100617	0,08254	0,09133	0,021661	0,167627	0,007642
1961-1990	Ave.	0,999833	0,01199	0,512889	0,405643	0,845524	0,407111	0,834454	0,80734
	Max.	0,999999	0,056063	0,859469	1,180317	0,99965	3,275	0,999453	8,209815
	Min.	0,999198	0,001695	0,163875	0,080043	0	0	0,178267	0,000813
1990-1999	Ave.	0,999291	0,023679	0,640403	0,348273	0,908084	0,41714	0,446442	3,097814
	Max.	0,999991	0,089931	0,727093	0,973227	0,984361	2,169231	0,857854	24,51041
	Min.	0,994688	0,003057	0,476107	0,129217	0,76801	0,035088	0,042501	0,065252
1929-1999	Ave.	0,999686	0,013373	0,513522	0,413772	0,726692	0,952263	0,849572	0,38418
	Max.	0,999999	0,089931	0,859469	1,180317	0,999453	24,51041	0,99965	3,275
	Min.	0,994688	4,82E-05	0,100617	0,080043	0,042501	0,000813	0	0

*......... $(A_1^2+A_2^2+A_3^2)/(A_1^2+A_2^2+A_3^2+A_4^2+A_5^2+A_6^2)$

Maximum (25.7°C) and minimum (–3.2°C) air temperature values have been observed in August and January, respectively in the study area. Maximum standard deviation values have been observed in February and January (Table 3.1.a).

Maximum (909 hPa) and minimum (888,2 hPa) air pressure values have been observed in December, and March, respectively in study area. Maximum standard deviation values have been observed in January (Table 3.1.b).

Maximum (4,7 m/sn) and minimum (2,8 m/sn) horizontally wind speed values have been observed in March and September, respectively, in the study area. Maximum standard deviation values have also been observed in March (Table 3.1.c).

Maximum monthly (360 mm) and minimum (0 mm) precipitation values have been observed in December and August, respectively in the study area. Maximum standard deviation values have been observed in December (Table 3.1.d).

3.2 Time Series Analysis

Fig. 3.1.a – Fig. 3.1.d show time variations of meteorological parameters. Data was smoothed by using moving averages for lag = 7.

Monthly mean air temperature values show some increasing trend at the beginning period of the 1980s (Fig. 1a). As shown in (Fig. 1b), the first observatory (h = 1150 m above mean sea level) moved to another area (h = 997 m above mean sea level) in 1968. In recent years a slightly increasing trend has been observed on time variation of monthly mean pressure values. The decreasing trend of wind speed is related to urbanization effects in Isparta. Hence wind speed measurements are not representative; new measurements have also been recorded at two new observatories. The new systems were mounted at that airport and near the vicinity of The Suleyman Demirel University Campus. Fig. 1d shows the time variation of monthly total values of precipitation. Time series analysis and moving averages show a decreasing trend of rainfall rate values beginning from the 1960s.

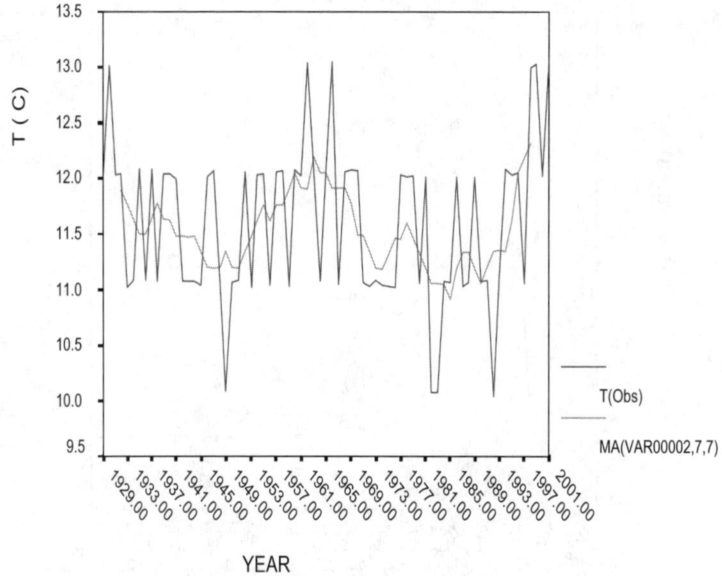

Fig.3.1.a. Time Variation of Monthly Mean Temperature (T,^{o}C) Values.

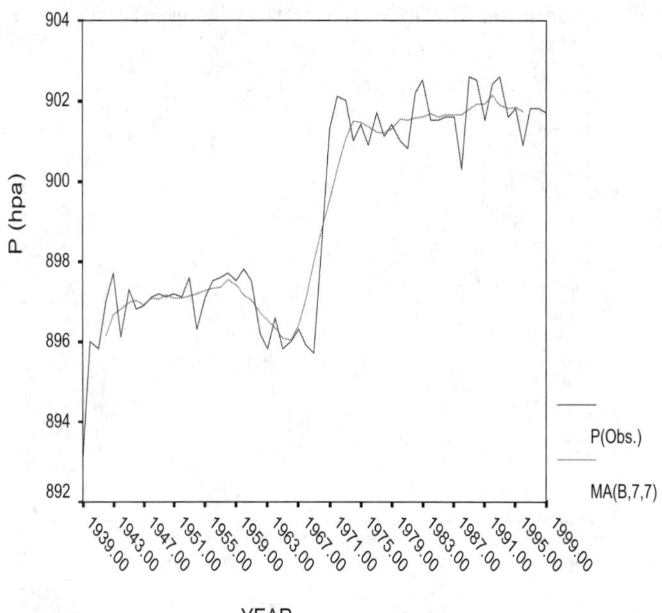

Fig.3.1.b. Time Variation of Monthly Mean Pressure (P, hPa) Values.

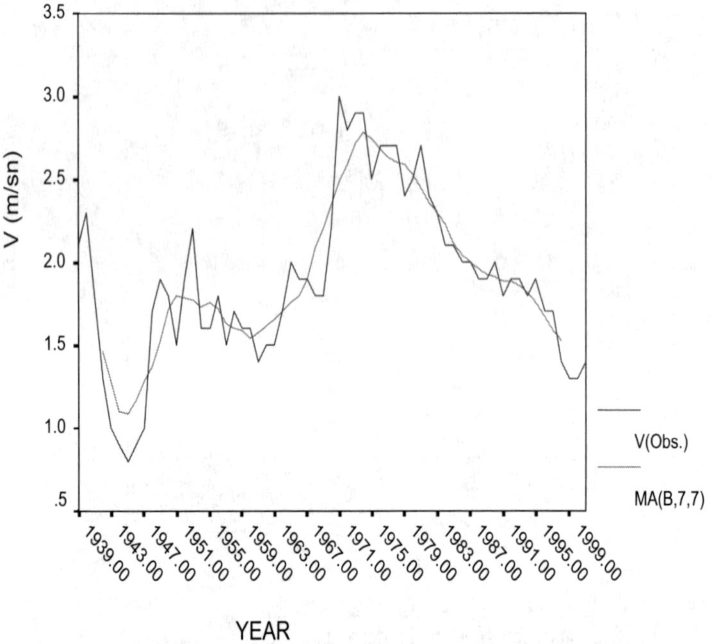

Fig.3.1.c. Time Variation of Monthly Mean Horizontal Wind Speed (V, m/s) Values.

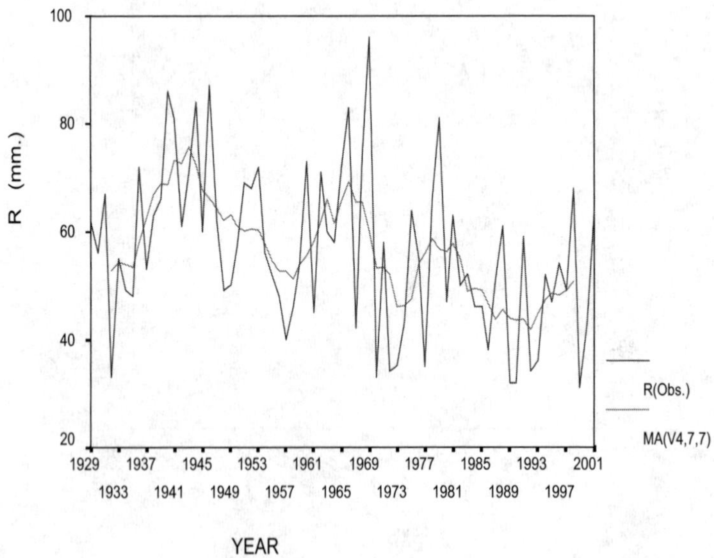

Fig.3.1.d. Time Variation of Monthly Total Precipitation (R, mm) Values.

3.3 Harmonic Analysis

Fig. 3.2.a – Fig. 3.2.d show annual variation of first, second, fourth, and sixth harmonics based on monthly mean air temperature valued observations between 1929 and 1999 in Isparta. Semi-annual and micro-scale fluctuation effects on air temperature values have increased in recent years.

Fig. 3a – Fig. 3d show annual variations of first, second, fourth and sixth harmonics based on monthly mean air pressure observations between 1929 and 1999 in Isparta. In recent years, the first, second and fourth harmonics have increased. There are no important variations of the sixth harmonics in this period.

Fig. 3.4.a – Fig. 3.4.d show annual variations of first, second, fourth, and sixth harmonics based on monthly mean wind speed observations between 1929 and 1999 in Isparta. In recent years, effects of large-, meso-, and small- scale phenomena on wind speed values have decreased.

Fig. 3.5.a – Fig. 3.5.d show annual variations of first, second, fourth, and sixth harmonics based on total monthly rain-fall rate values between 1929 and 1999 in Isparta. Annual and semi-annual fluctuations have some extreme effects between 1967 and 1970 and between 1980 and 1983. Effects of seasonal and micro-scale fluctuations on rainfall rate have decreased in recent years.

The relative contributions of the first three harmonics to the seasonal temperature variance decrease in the second term between (1931–1960) and fourth term (1990–1999). Hence, for all ratio values, $A2/A1 < 1$, it is concluded that there is no dominant semi-annual influence on temperature variations. Strong annual influence ($A2/A1 < 0.015$) on temperature variations is dominant especially in the first two terms (Table 3.2).

In the third term, semi-annual influence ($A2/A1 > 1$) is observed on pressure, wind speed, and rainfall-rate variations.

In the last two terms, increasing semi-annual influence on wind speed and precipitation is observed.

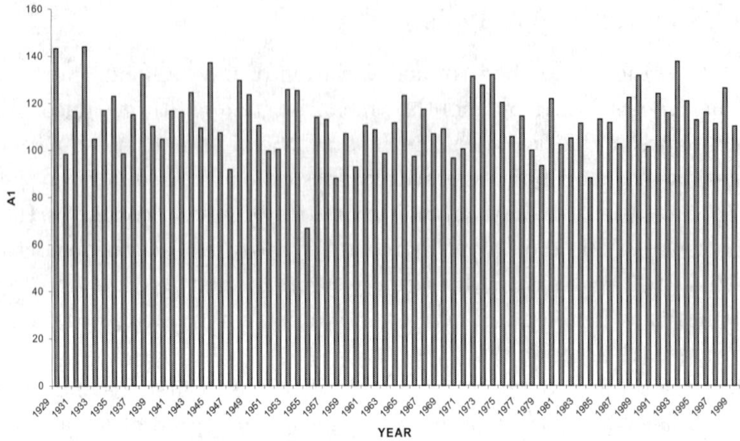

Fig.3.2.a. Time Variation of First Harmonics for Monthly Mean Air Temperature T(°C),

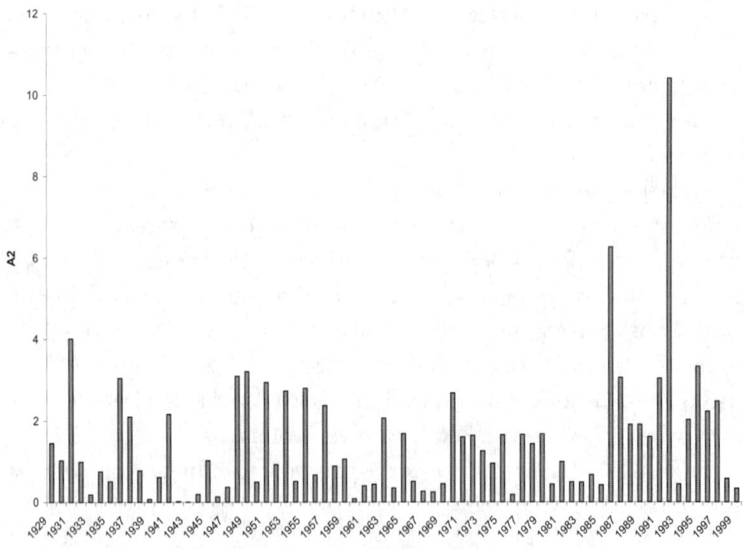

Fig.3.2.b. Time Variation of 2nd Harmonics for Monthly Mean Air Temperature T(°C),

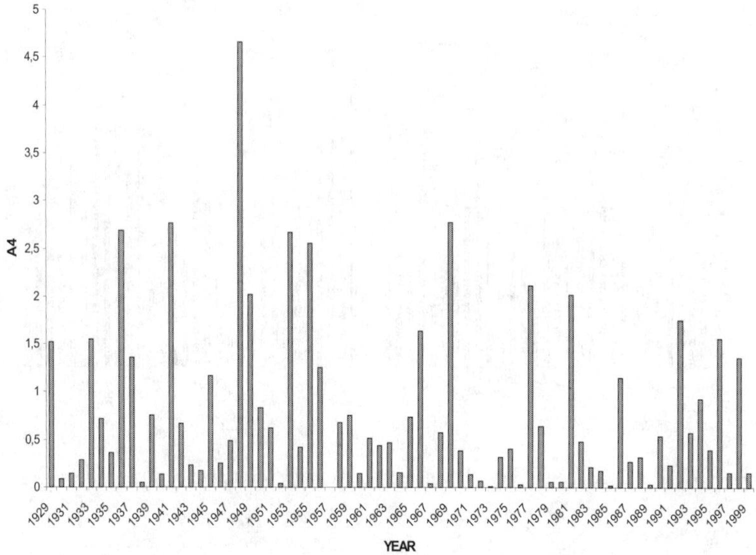

Fig.3.2.c. Time Variation of 4th Harmonics for Monthly Mean Air Temperature T(°C),

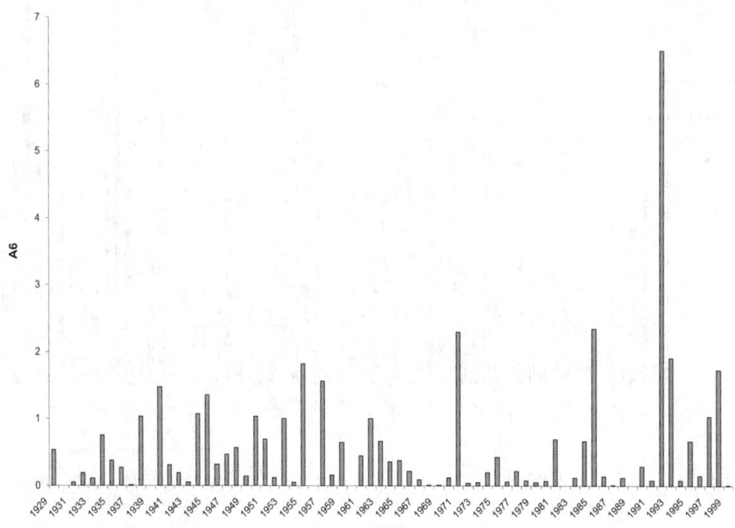

Fig.3.2.d. Time Variation of 6th Harmonics for Monthly Mean Air Temperature T(°C),

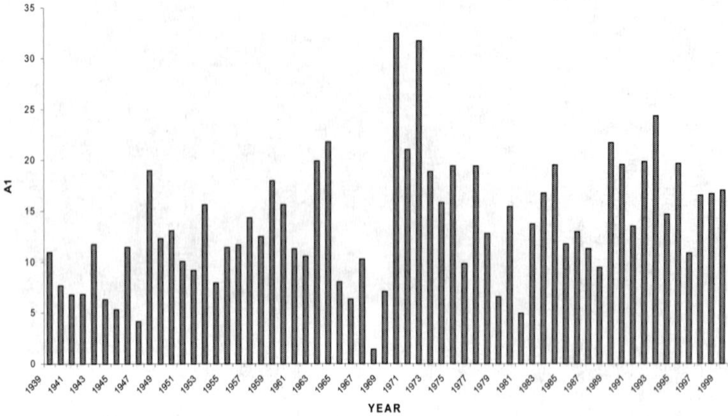

Fig.3.3.a. Time Variation of First Harmonics for Monthly Mean Pressure (P hPa),

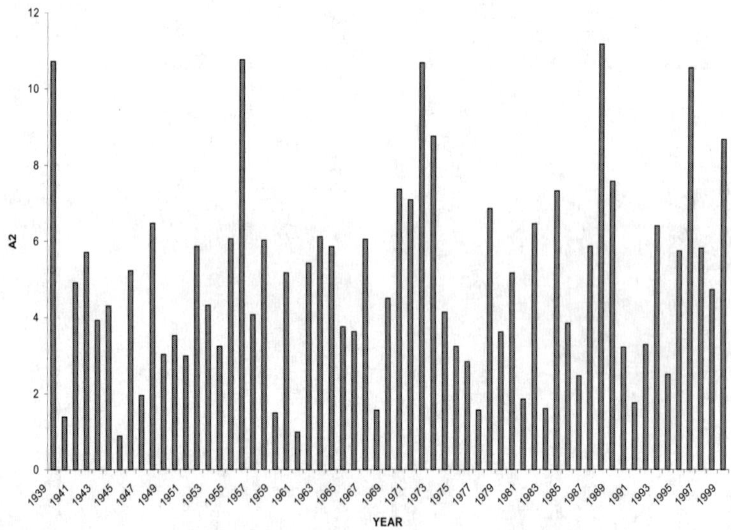

Fig.3.3.b. Time Variation of 2nd Harmonics for Monthly Mean Pressure (P hPa),

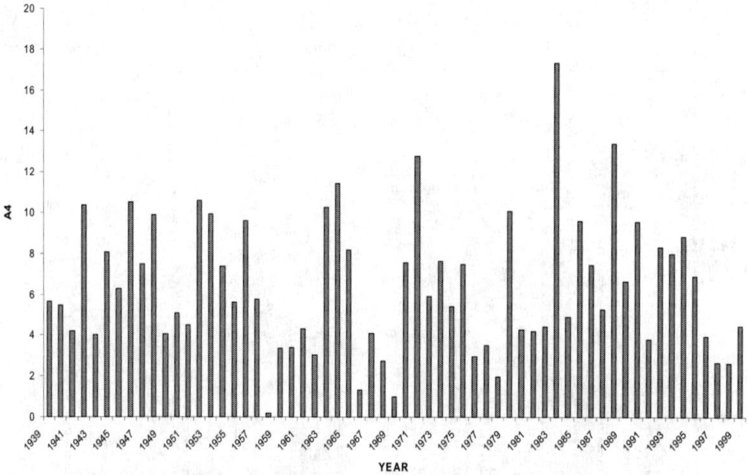

Fig.3.3.c. Time Variation of 4th Harmonics for Monthly Mean Pressure (P hPa),

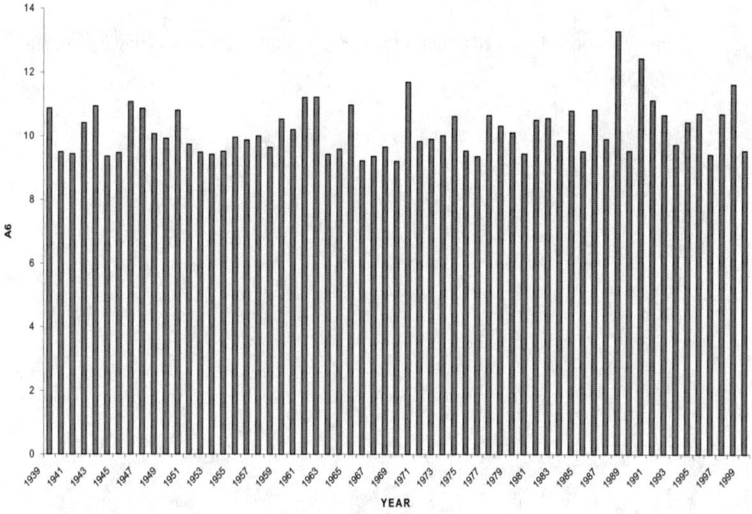

Fig.3.3.d. Time Variation of 6th Harmonics for Monthly Mean Pressure (P hPa),

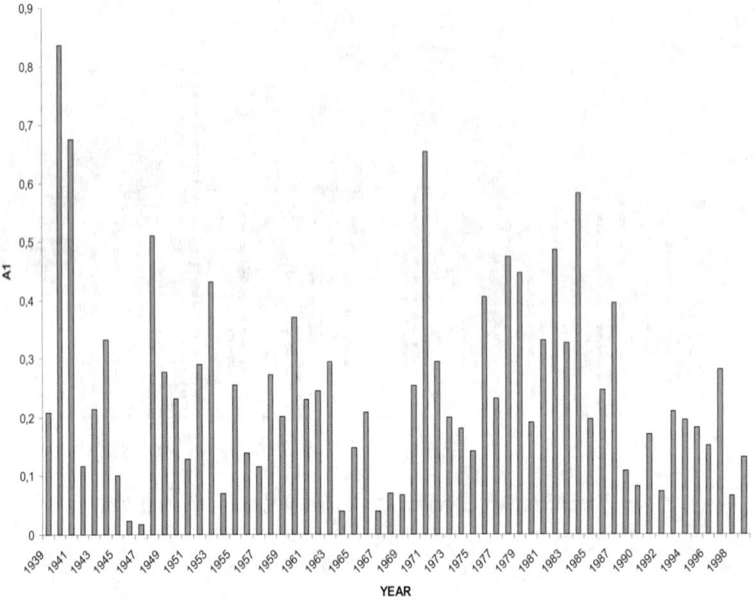

Fig.3.4.a.Time Variation of First Harmonics for Monthly Mean Horizontal Wind Speed Vm/s

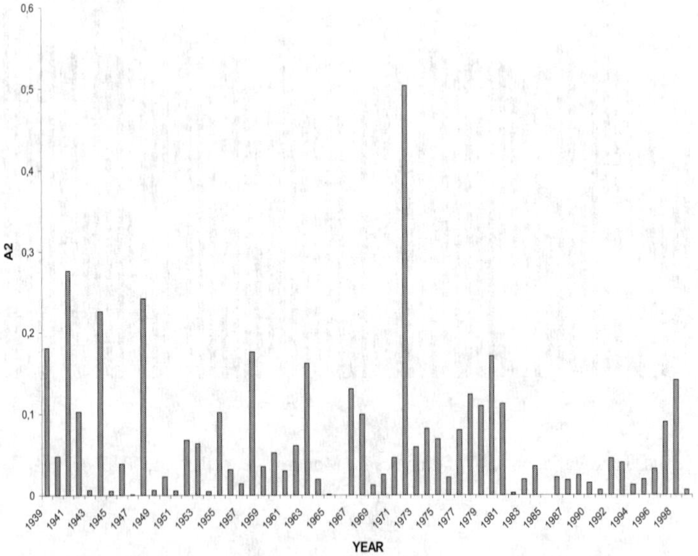

Fig.3.4.b.Time Variation of 2nd Harmonics for Monthly Mean Horizontal Wind Speed Vm/s

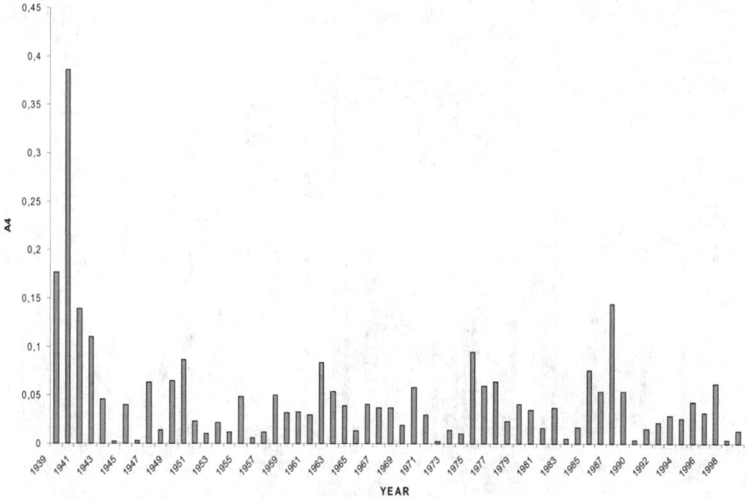

Fig.3.4.c.Time Variation of 4th Harmonics for Monthly Mean Horizontal Wind Speed Vm/s

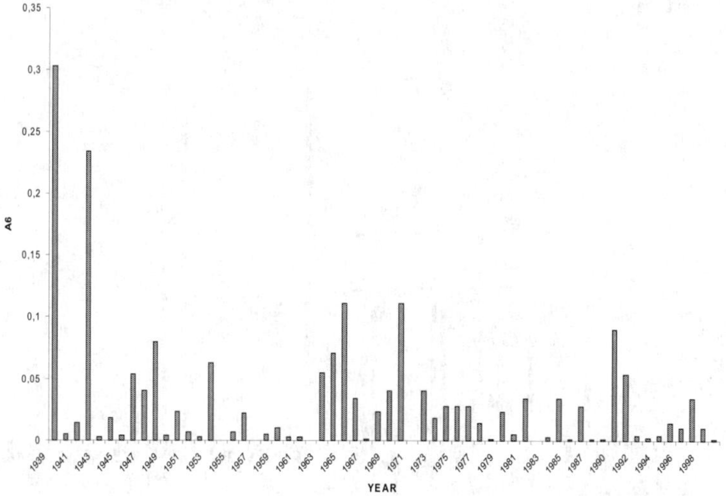

Fig.3.4.d.Time Variation of 6th Harmonics for Monthly Mean Horizontal Wind Speed Vm/s

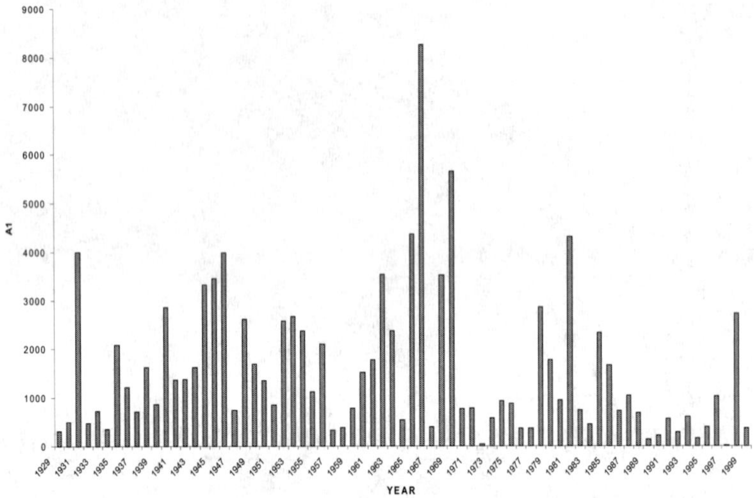

Fig.3.5.a. Time Variation of First Harmonics for Monthly Mean Precipitation (R mm),

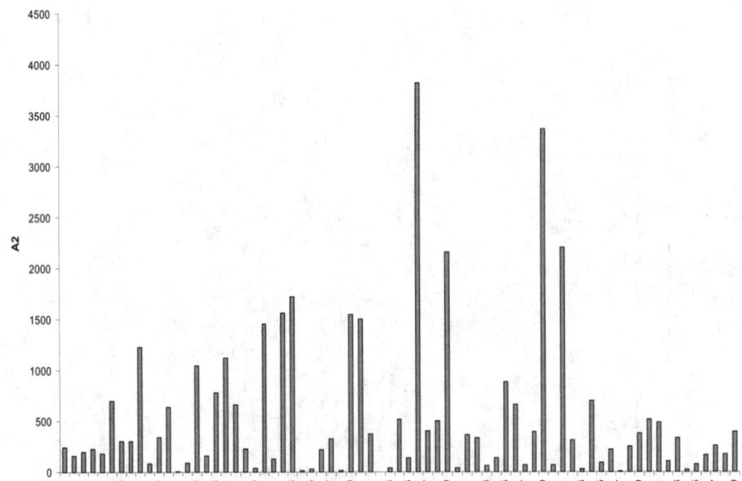

Fig.3.5.b. Time Variation of 2nd Harmonics for Monthly Mean Precipitation (R mm),

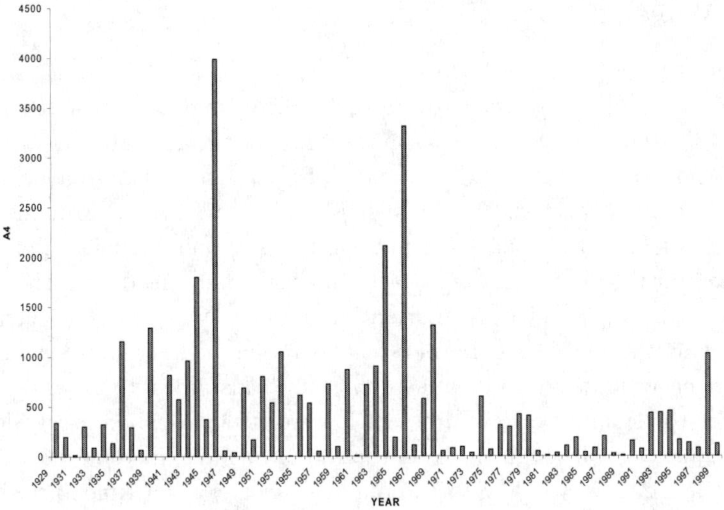

Fig.3.5.c. Time Variation of 4th Harmonics for Monthly Mean Precipitation (R mm),

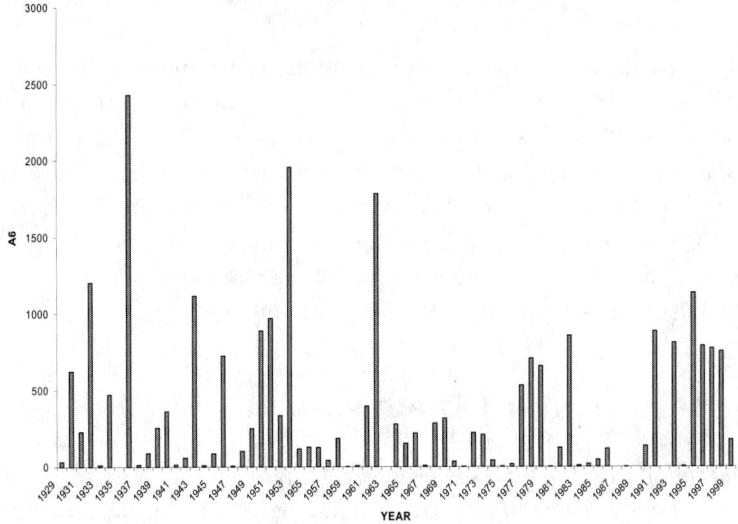

Fig.3.5.d. Time Variation of 6th Harmonics for Monthly Mean Precipitation (R mm),

3.4 Wavelet Analysis

Wavelet analysis examines how the power spectrum of a time series varies over the time of the record, see Siddiqi, Aslan, and Tokgozlu [13] and Meyer [8]. The simplest approach to studying the changing frequency response of some aspect of the climate system is to calculate a form of running Fourier transform of the available time series, see Burroughs [3]. This can be done by using a certain window size and sliding it along in time, computing the transform at each time using only the data within the window. This would give us information about the frequency spectrum, but the result will lead to an inconsistent treatment of different frequencies. For a given window width of N sampling intervals in the time series, the would be too small to resolve different low-frequency oscillations, while at high frequencies, although the resolution would be fine, it would be better to have a narrower window to examine the shorter term time variations of these oscillations. The wavelet analysis does combine both a weighted window and a defined number of oscillations of a given frequency within this window.

If we compare the wavelet series, all of them represent the raw data. Wavelet series are mathematical simulation methods to compress the data, scalogram gives the magnitudes of extreme values of the data. Based on the wavelet techniques of long-term variation for air temperature values observed in Isparta between 1929 and 1999, the maximum and minimum monthly average of temperature values are observed as 27.5°C in July and 3.5°C in December. An application of wavelets to temperature or wind speed values deals with jump detection [12–14].

4 Results and Conclusion

In real conditions, the links between various parts of the climate system will vary in strength depending on the various components of the systems (e.g. sea surface temperatures or the strength of the mid-latitude westerlies). The time of year and hence the phase of the annual cycle could also have additional control over any process. This type of behavior might explain not only why periodicities come and go, but also why the observed spectra tend to involve approximately multiples of the basic periodicity (e.g., the wavelet spectra).

In this chapter, time variations of air temperature, pressure, wind speed, and rainfall rate values were statistically analyzed. As a conclusion, some small-scale effects have a crucial role on air temperature variations. Annual influence is dominant on temperature fluctuations. Semi-annual effects on temperature values have increased in recent years. Two types of results have been studied by using wavelet analysis: (i) methodological results on the wavelet approach itself and (ii) when applied to real temperature data, results concerning the nature of temperature turbulence (small-scale fluctuations) in the real atmospheric conditions.

The contribution of the first three harmonics on precipitation values shows a decreasing trend in the fourth term (between 1990 and 1999). But the semi-annual influence increases. In the other words, a dominant semi-annual influence on precipitation variations has been observed.

Acknowledgment. The authors wish to thank Dr. Deniz Okcu for her support.

References

[1] Z. Aslan, D. Okçu, and S. Kartal, Harmonic analysis of precipition, pressure and temperature over Turkey, *Il Nuovo Cimento, Luglio*, **20**(4) (1997), 595–605.

[2] Z. Aslan and S. Topçu, Seasonal variation of surface fluxes and atmospheric interaction in Istanbul, in *International Air - Sea Interactions and Meteorology and Oceanography Congress*, AMS, Portugal, (1994), pp. 180–181.

[3] W. J. Burroughs, *Weather Cycles*, Second Edition, Cambridge University Press, Cambridge, (2003).

[4] R. G. Currie and S. Hameed, Atmospheric signals at high latitudes in a coupled ocean atmosphere general circulation model, *Geophys. Res. Lett.*, **17** (1990), 945–948.

[5] K. Kirkyla and S. Hameed, Harmonic analysis of the seasonal cycle in precipitation over the United States: A comparison between observations and general circulation model, *J. Climate*, **2** (1989), 1463–1475.

[6] E. Linacre, *Climate Data and Resources*, Routledge, Chapman and Hall Inc., London, (1991).

[7] R. S. Lindzen, B. Kirtman, D. Kirk-Davidoff, and E. K. Schneider, Seasonal surrogate for climate, *J. Climate*, **8** (1995), 1681–1684.

[8] Y. Meyer, The role of oscillations in some non-linear problems, *School on Mathematical Problems in Image Processing*, ICTP, SMR 1237/4, (2000), 4–22.

[9] D. Okçu, Variation normalised vegetation index and relation between meteorological parameters, *ITÜ. Sci. Enst.*, (1999), pp. 146, Istanbul, (In Turkish).

[10] D. Okçu, S. Kartal, and Z. Aslan, *Türkiye'de Yagis, Basinç ve Sicaklik Verilerinin Degisiminde Büyük ve Küçük Ölçekli Olaylarin Rolü*, HGK, TUMAK Kongresi, Ankara, (1995).

[11] D. Philips, *The Global Climate System Review*, WMO No. 819, Italy, (1995).

[12] J. Shukla, *Prediction of Inter-annual Climate Variations*, Springer-Verlag, New York, (1993).

[13] A. H. Siddiq, Z. Aslan, and A. Tokgozlu, Wavelet based computer simulation of some meteorological parameters: Case study in Turkey, in *Trends in Industrial and Applied Mathematics*, Edited by A.H. Siddiqi and M. Kocvara, Kluwer Academic Publishers, London, (2002).

[14] L. F. Tarleton and R. W. Katz, Statistical explanation of trends in extreme summer temperatures at Phoeniz, *Arizona J. Climate*, **8** (1994), 1704–1708.

Chapter 10

WAVELET METHODS FOR SEISMIC DATA ANALYSIS AND PROCESSING

F. M. Khène

King Fahd University of Petroleum & Minerals

Abstract

Over the past decade, wavelet transforms have been formalized into a rigorous mathematical framework and have had a large impact on solving real-world problems in diverse sciences and engineering areas. In the oil and gas prospecting industry, the involved geophysical signals are multichromatic and non-stationary and contain hidden multiscale features. The conventional *Fourier* analysis fails, for instance, to characterize the temporal evolution of the spectral content of seismic signals. Even one-dimensional time-frequency representations such as the instantaneous frequency and the group delay are not sufficient to unambiguously characterize seismic signals. Two-dimensional representations, both linear and bilinear, offer interesting joint time-frequency decompositions that yield a better structuring of the information and an improvement in the intelligibility of the representation. One linear time-frequency representation is obtained by the *short-time Fourier transform* (STFT) that can be thought of as a collection of local spectra around various analysis time points. Because the time-frequency resolution is constrained by the *Heisenberg uncertainty principle*, the *wavelet transform* yields a better linear decomposition. Indeed, the wavelet transform performs scaling rather than modulation using an adaptive window, called the wavelet, which can be used either as an integration kernel to extract information or as a basis for a more adequate decomposition. Bilinear time-frequency transforms, for their part, tend to unravel the energy distribution of signals over time and frequency. The most popular is the *Wigner-Ville transform* (WVT). It is based on the sole intrinsic properties of the analyzed signal and has numerous interesting mathematical properties. This chapter provides an introduction to the mathematical concepts of various time-frequency and time-scale representations with a particular focus on wavelet transforms and their potential in seismic data analysis and processing. After describing the various geophysical data involved in oil and gas prospecting, we will demonstrate the ability of wavelets in performing both depth-wavelength and time-frequency analyses, in filtering noise and removing unwanted interfering signals, and in compressing high-dimension seismic datasets. Numerous simulations using both synthetic and real seismic data will corroborate the theoretical concepts discussed in this chapter.

1 Introduction

Wavelet analysis represents a new branch of mathematics, which has already had a large impact on solving real-world problems in various fields [1, 5, 9]. Indeed, most of real-world signals occurring, for instance, in geophysics, engineering, and finance not only possess a frequency content varying with time, but also are scale-invariant, i.e., their temporal fluctuations cannot be described by any characteristic scale. The *Fourier* transform is not adapted for the analysis of such non-stationary signals, since it decomposes them into individual frequency components consisting of sinusoids, which give the relative intensities of each frequency component but are completely unlocalized in time. The resulting frequency-domain representation, called *spectrum*, does not provide any information either about the occurrence times of the individual frequencies or about the complexity of the signal.

To address this shortcoming, one-dimensional (1-D) techniques known as the *instantaneous frequency* and the *group delay* are used to track changes in the monochromatic spectral behavior along the time dimension. Nonetheless, in the case of multichromatic seismic signals, two-dimensional (2-D) representations, both linear and bilinear, are more suitable for joint time-frequency decompositions that yield a better structuring of the information and an improvement in the intelligibility of the representations. Linear representations can be generated either by modulation or by scaling associated with translation leading to time-frequency and time-scale representations, respectively. Linear time-frequency representations are usually produced by the *short-time Fourier transform* (STFT) that is equivalent to a sum of *Fourier* transforms of successive windowed segments of the signal. It provides a powerful tool to bring to the fore the progression with time of the frequency content of locally stationary signals. Nevertheless, since the modulating window area is constrained by a lower bound according to the *Heisenberg* uncertainty principle, a major drawback of the STFT resides in the fixed size of the window width over the entire time-frequency plane [21]. Consequently, signal components can be analyzed with either a good time resolution or a good frequency resolution.

The wavelet transforms overcome this drawback by allowing varying window widths at different scales. The adaptive analyzing window, called the wavelet, could be used either as an integration kernel to extract information from the signal or as a basis for a more adequate signal representation or decomposition. In the *continuous wavelet transform* (CWT), the time-frequency/space-scale localization is emphasized by considering analytic wavelets as an analyzing kernel. This allows the study of the features of the signal locally with a detail matched to their scale, i.e., broad features on large scales and fine features on small scales [13]. This property is especially useful for signals that are non-stationary, have short-lived transient components, have features at different scales, or have singularities. However, because the CWT is extremely redundant, it is not suited for compression and cannot be readily extended to higher dimensions. The *discrete wavelet transform* (DWT), which is conceptually a critically sampled version of the CWT [26], possesses a remarkable decorrelation power, performs compression, and can easily handle higher dimension signals such as images and volumes [17]. It also allows a simplified representation and an easy analysis of a rich class of real-world signals with increasing complexity such as seismic signals and fractals [19]. For instance, although fractional *Brownian motion* is a non-stationary and infinitely correlated process, its wavelet coefficients are stationary and practically uncorrelated. This makes the non-stationarity essentially disappear thanks to the DWT that uses wavelets as elementary building blocks in a decomposition or series expansion akin to the familiar *Fourier* series. Bilinear time-frequency representations, for their part, provide the energy distribution of signals over time and frequency. The most popular representation is provided by the *Wigner-Ville transform* (WVT) [4].

This chapter provides an introduction to the mathematical concepts of various time-frequency and time-scale representations with a particular focus on the wavelet transforms and their potential in seismic data analysis and processing. This work is motivated by the historical fact that the CWT found its origin as an analysis tool for seismic signals [10]. After explaining the particularity of the various geophysical data involved in oil and gas prospecting, we will demonstrate the ability of wavelets in performing both depth-wavelength and time-frequency analyses, in filtering noise and removing unwanted interfering signals, and in compressing high-dimension seismic datasets. Numerous simulations using both synthetic and real seismic data will corroborate the theoretical concepts discussed in this chapter.

2 Time-Frequency Representations

2.1 Time, Frequency, and Time-Frequency

The time series constitutes the most natural representation of real-world signals since these are usually obtained by sensors recording their variations in time instants. The spectrum is related to the time series by the following *Fourier* transform pair,

$$\hat{f}(\omega) \equiv \int_{-\infty}^{+\infty} f(t)e^{-i\omega t}dt \longleftrightarrow f(t) \equiv \frac{1}{2\pi}\int_{-\infty}^{+\infty} \hat{f}(\omega)e^{i\omega t}d\omega, \qquad (2.1)$$

where it is assumed that the *Fourier* integral exists for most real-world signals. The spectrum is a powerful tool for the analysis of stationary signals because the relevance of the concept of frequency is shared by many fields of science, such as geophysics, astronomy, and biology, in which periodic events occur. Moreover, the autocovariance of a stationary stochastic signal defines a convolution operator that is diagonalized by *Fourier* expansions [3]. In a number of practical situations, the information contained in a real-world signal can be found in the non-stationary property itself, e.g., start and termination of events, drifts, ruptures, and modulations. The frequency content varies with time, and the autocovariance function does not give rise to a convolution operator. Thus, the *Fourier* transform is not adapted for the analysis of non-stationary signals since it aims at projecting them on infinite length sinusoids, which are completely unlocalized in time. Consequently, the spectrum could not register frequencies varying with time. There exist both 1-D and 2-D solutions to the problem of time localization of the spectrum. 1-D techniques such as the instantaneous frequency and the group delay are only capable of handling monochromatic and non-stationary signals [4]. 2-D representations that could be either linear or bilinear are more appropriate for the analysis of such signals. The linear time-frequency representation is usually obtained with the STFT and provides a powerful tool that brings to the fore the progression with time of the frequency content of locally stationary signals. Bilinear time-frequency representations, for their part, provide the energy distribution over time and frequency. The most popular representation is provided by the WVT. It is a real-valued function and has a number of interesting properties. The WVT suffers, however, from the presence of interference terms, which could prevent a proper interpretation of some energy distributions [4].

2.2 Short-Time Fourier Transform

The STFT is a multidimensional functional that maps a given 1-D signal into a 2-D time-frequency plane [21, 26]. It introduces a certain degree of time-dependency in the spectral characteristics of a multi-chromatic and time-varying signal $f(t)$ by multiplying it by a window function $\gamma(t)$ before taking its *Fourier* transform. This will confine the spectral information to the domain of influence of the window function. The complete analysis of the spectral information in various localized time neighborhoods is obtained by using translates of the window function $\gamma(t-\tau)$ along the time axis to scan the entire time domain as follows [21],

$$\hat{f}(\omega,\tau) \equiv \int_{-\infty}^{+\infty} [f(t)\gamma^*(t-\tau)]e^{-i\omega t}dt, \tag{2.2}$$

where * stands for the complex conjugate. Theoretically, the analyzing window $\gamma(t)$ and it *Fourier* transform $\hat{\gamma}(\omega)$ could be well localized in time and in frequency, respectively. However, in practice it is impossible to achieve a perfect time-frequency localization because of the *Heisenberg* uncertainty principle. Thus, some arbitrariness is necessarily introduced into the analysis. Indeed, let the signal behavior within the STFT be illustrated using the concepts of signal *time duration* σ_t and *frequency bandwidth* σ_ω. Suppose the window $\gamma(t)$ is centered at $t=0$ and $\hat{\gamma}(\omega)$ is centered at $\omega=0$, then (2.2) gives a measure of the behavior of the signal

$f(t)$ in the vicinity of $[t - \sigma_t, t + \sigma_t]x[\omega - \sigma_\omega, \omega - \sigma_\omega]$. By virtue of the *Parseval theorem*, the time duration σ_t is tied to the frequency bandwidth σ_ω via the *Fourier* transform. Both cannot be arbitrary small and their product must satisfy the *Heisenberg* uncertainty principle, i.e.,

$$\sigma_t \sigma_\omega \succeq \frac{1}{2} \tag{2.3}$$

that imposes a theoretical lower bound on the area of the *time-frequency* window. The equality holds when $\gamma(t)$ is a *Gaussian* function, which yields the tightest time-frequency window. The STFT is then referred to as the *Gabor* transform [4].

2.3 Wigner-Ville Transform

One of the most popular bilinear (quadratic) time-frequency representations is provided by the WVT. Given a signal $f(t)$, its WVT is defined as [4]

$$W_f(t, \omega) = \int f\left(t + \frac{\tau}{2}\right) f^*\left(t - \frac{x}{2}\right) e^{-i\omega\tau} d\tau. \tag{2.4}$$

The WVT essentially amounts to considering inner products of copies $f(\tau + b)e^{-i\omega\tau}$ on the original signal shifted in the time-frequency domain with the corresponding reversed copies $f(-\tau + b)e^{-i\omega\tau}$. Simple geometrical considerations show that such a procedure provides insights on the time-frequency content of the signal [21]. The WVT is a real-valued function and has a number of interesting properties such as: i) the optimal time localization for *Dirac* signals, ii) the optimal frequency localization for pure mono-chromatic waves and for linear chirps, iii) energy conservation, and iv) translation and dilation covariance. It is also possible to extract the power spectrum and the square of the modulus of a signal from the time marginal and the frequency marginal, respectively, by [4]

$$\int W_f(t, \omega) dt = \left|\widehat{f}(\omega)\right|^2 \quad \text{and} \quad \int W_f(t, \omega) d\omega = 2\pi \left|f(t)\right|^2. \tag{2.5}$$

The WVT suffers, however, from the presence of interference terms, which could prevent a proper interpretation of some energy distributions.

3 Time-Scale Representations

3.1 Continuous Wavelet Transform

The CWT is a multidimensional mapping of a signal $f(t) \in L^2(\mathbb{R})$, depending on a continuous time (or space) parameter t (or x), to a 2-D function $\check{f}(a,b)$ depending on a scale (or dilation) parameter a and a shift (or translation) parameter b. Mathematically, the CWT is defined as the integral transform on $L^2(\mathbb{R})$ [6],

$$\check{f}(a,b) \equiv W\{f,\psi\}(a,b) = \int_{-\infty}^{+\infty} f(t)\psi_{ab}^{*}dt, \tag{3.1}$$

where the wavelet $\psi \in L^2(R)$. The wavelet kernel $\psi_{ab}(t)$ is defined as

$$\psi_{ab}(t) \equiv |a|^{-\frac{1}{2}}\,\psi\left(\frac{t-b}{a}\right), \tag{3.2}$$

with $a, b \in \mathbb{R}$, $a \neq 0$. The factor $|a|^{-\frac{1}{2}}$ is introduced to ensure energy preservation. The width of the wavelets $\psi_{ab}(t)$ is controlled by the scale parameter a such that the wavelets are compressed for $|a| < 1$, and are dilated for $|a| > 1$. The positions of the dilated or compressed wavelets are controlled by the shift parameter b. The existence of the *inverse* CWT defined by [6],

$$f(t) \equiv W^{-1}\{f,\psi\}(a,b) = C_{\psi}^{-1}\int_{-\infty}^{+\infty}\int_{-\infty}^{+\infty} \check{f}(t)\psi_{ab}(t)\frac{dbda}{a^2} \tag{3.3}$$

requires that the wavelet $\psi(t)$ fulfills the *admissibility condition* expressed by

$$0 < C_{\psi} \equiv \int_{-\infty}^{+\infty} \frac{\left|\hat{\psi}(\omega)\right|^2}{|\omega|}d\omega < \infty, \tag{3.4}$$

where $\hat{\psi}(\omega)$ is the *Fourier* transform of the wavelet $\psi(t)$. If $t\psi(t)$ decays faster than t^{-1}, then $\hat{\psi}(\omega)$ is continuous and the admissibility condition reduces to the *zero mean condition* [6],

$$\hat{\psi}(0) = 0 \Longleftrightarrow \int_{-\infty}^{+\infty} \psi(t)dt = 0. \tag{3.5}$$

The zero mean condition implies that the wavelet should oscillate and decay at infinity, whence the appellation "wavelet". The CWT shares some basic properties with the linear transform class, such as linearity, shift and scale-invariance, and energy preservation. The CWT is extremely redundant and constitutes a powerful analyzing tool as such [13]. However, the CWT is not suited for compression and cannot be readily extended to higher dimensions.

3.2 Multiresolution Analysis

The *multiresolution analysis* (MRA) provides a powerful framework to understand, formulate, and implement discrete wavelet decompositions using either the *wavelet series transform* (WST) or the DWT. Multiresolution theory consists of a nested sequence of closed linear subspaces $\{V_j\}_{j\in\mathbb{Z}}$ in $L^2(\mathbb{R})$ satisfying properties such as monotonicity, density of the union in $L^2(R)$, trivial intersection, scale and shift-invariance, and the existence of a *scaling function* $\phi_{0k}(t)$ such that the set $\{\phi(t-k)\}_{k\in\mathbb{Z}}$ forms an orthonormal basis of the central invariant subspace V_0. In other terms, the single scaling function $\phi(t)$ is the generator of the MRA. A given signal $f(t)$ can then be approximated by successive orthogonal projections $P_{V_j}f$ on the scale subspaces V_j such as [12],

$$P_{V_j}f = \sum_{k\in\mathbb{Z}} a_{jk}\phi_{jk} = \sum_{k\in\mathbb{Z}} \langle f, \phi_{jk}\rangle \phi_{jk}. \tag{3.6}$$

The coefficients a_{jk} are computed by the inner product of the signal and the set of functions

$$\phi_{jk}(t) = 2^{-\frac{j}{2}}\phi\left(2^{-j}t - k\right) \tag{3.7}$$

that form an orthonormal basis for the subspaces $V_j \subset L^2(R)$. Moreover, when going from a subspace V_{j-1} to a consecutive embedded coarser one V_j, such that $V_j \subset V_{j-1}$, the lost details lie in a complementary subspace W_j defined as the orthogonal complement of V_j in V_{j-1}, such that $V_{j-1} = V_j \oplus W_j$. From this direct sum, it appears that the subspaces W_j are mutually orthogonal, allowing the subspaces W_j to inherit a scaling property similar to the one governing scale subspaces V_j, i.e., $W_j \subset W_{j-1}$. By using this particular nested structure, the direct sum can be iterated to obtain, for any arbitrary scale index $j > J_0$,

$$V_{j-1} = V_{J_0} \oplus \bigoplus_{j=J_0}^{j} W_m, \tag{3.8}$$

where J_0 is the coarsest scale corresponding, for instance, to the subspace where the signal features are slowly changing. If an orthonormal basis $\{\psi\left(t-k\right)\}_{k\in\mathbb{Z}}$ is associated with the central detail subspace W_0, then the family of functions

$$\psi_{jk}(t) = 2^{-\frac{j}{2}}\psi\left(2^{-j}t - k\right) \tag{3.9}$$

forms an orthonormal basis for the subspaces W_j. This orthonormal basis is referred to as the *wavelet basis* and is associated to the *wavelet subspaces*

W_j. It is worthy to note that the existence of such a basis allows for direct orthogonal projections $P_{W_j} f$ of the signal $f(t)$ on wavelet subspaces W_j. This provides a way to directly characterize the details lost by going from a scale subspace V_{j-1} to a consecutive coarser one V_j using the following projection of a given signal on the wavelet subspaces,

$$P_{W_j} f = \sum_{k \in \mathbb{Z}} d_{jk} \psi_{jk} = \sum_{k \in \mathbb{Z}} \langle f, \psi_{jk} \rangle \psi_{jk}, \tag{3.10}$$

where the coefficients d_{jk} are calculated, in this case, using the inner product of the signal and the wavelet basis. In the limit case, the discrete reconstruction formula of the signal from its projections on the wavelet subspaces W_j,

$$f(t) = \sum_{j=-\infty}^{\infty} \sum_{k=-\infty}^{\infty} d_{jk} \psi_{jk}, \tag{3.11}$$

is obtained. This double indexed sum, referred to as WST, seems to have the same computation problems that an infinite *Fourier* series has. However, in practice, the summation over the scale index j and the summation over the translation index k can be made finite with little errors for a given norm. We get then the following expression for the *finite* WST [12],

$$f(t) \approx P_{VJ_1} f = \sum_{k \in \mathbb{Z}} a_{J_0 k} \phi_{J_0 k} + \sum_{j=J_0}^{J_1+1} \sum_{k \in \mathbb{Z}} d_{jk} \psi_{jk}, \tag{3.12}$$

where J_1 is the index of the sampling subspace V_{J_1} of the signal, corresponding, e.g., to the *Shannon* sampling subspace. The finite WST can be interpreted as a sequence of recursive orthogonal projections of the signal $f(t)$ onto wavelet subspaces W_j up to the coarsest scale subspace V_{J_0} such as

$$P_{V_{J1}} f = P_{V_{J_0}} f + P_{W_{J_0}} f + P_{W_{J_0 - 1}} f + ... + P_{W_{J_1+1}} f. \tag{3.13}$$

This recursive process is illustrated schematically by the diagram of Figure 1. The wavelet decomposition generates an inherent *multiresolution* or *multiscale* decomposition of a signal $f(t)$ in terms of its orthogonal projections $P_{W_j} f$ onto the wavelet subspaces W_j and the residual $P_{V_{J_0}} f$, which consists of the projection of the signal $f(t)$ on the coarsest scale subspace V_{J_0}. Due to the orthogonality of the involved subspaces, the multiresolution decomposition is unique once the subspaces V_j and W_j are selected. The wavelet reconstruction exploits a similar recursive process and is essentially the reverse of the decomposition since the original signal can be readily obtained from its projections.

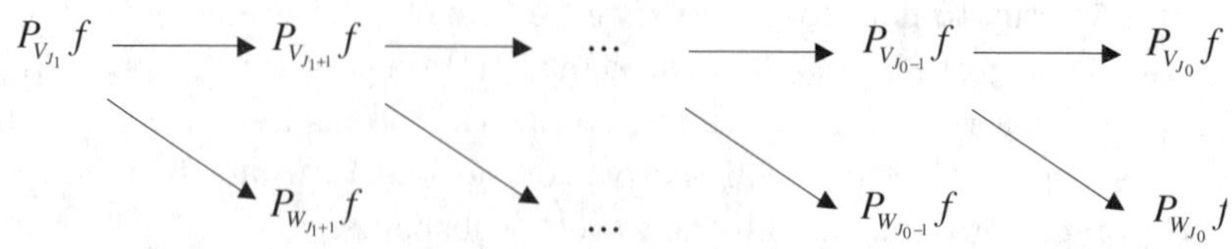

Figure 1: Wavelet multiresolution decomposition of a signal *f*.

3.3 Discrete Wavelet Transform

In order to implement the wavelet multiresolution decomposition, the projections onto the different subspaces V_j and W_j need to be computed efficiently. Recall that the two sets of functions $\{\phi_{jk}\}$ and $\{\psi_{jk}\}$ are orthogonal bases for $\{V_j, W_j\} \subset L^2(\mathbb{R})$. The set of coefficients $\{a_{jk}, d_{jk}\}_{jk\in\mathbb{Z}}$ appearing, respectively, in (3.6) and (3.10) is called the DWT of the signal $f(t)$ [26]. The coefficients a_{jk}, which convey the low-frequency trends of the signal $f(t)$, are called the *approximation* coefficients. The *detail* coefficients d_{jk}, for their part, bear the high-frequency trends. To avoid computing the DWT using inner product integrals, it take advantage of the fact that both the MRA decomposition and reconstruction only require projections between consecutive subspaces of the multiresolution ladder. Consider the scaling function $\phi \equiv \phi_{0k}(t) = \phi(t-k) \in V_0 \subset V_{-1}$, where the finer subspace is endowed with the scaling orthogonal basis $\phi_{-1} \equiv \phi_{-1k}(t) = \sqrt{2}\phi(2t-k)$. The scaling function ϕ can be expressed by the following linear combination,

$$\phi(t) = \sum_{k\in\mathbb{Z}} c_k \phi_{-1k} = \sqrt{2}\sum_{k\in\mathbb{Z}} c_k \phi(2t-k), \tag{3.14}$$

where the *scaling filter coefficients* c_k, that unambiguously characterize the scaling function $\phi(t)$, are given by [12]

$$c_k = \langle \phi_{0k}, \phi_{-1k} \rangle. \tag{3.15}$$

Equation (3.14) is referred to as the *two-scale difference equation* and plays a crucial role in MRA [24]. Similarly, given a wavelet $\psi \equiv \psi_{0k}(t) = \psi(t-k) \in W_0 \subset V_{-1}$ and some real coefficients b_k, we can write

$$\psi(t) = \sum_{k\in\mathbb{Z}} b_k \phi_{-1k} = \sqrt{2}\sum_{k\in\mathbb{Z}} b_k \phi(2t-k), \tag{3.16}$$

where the *wavelet filter coefficients* are given by

$$b_k = \langle \psi_{0k}, \phi_{-1k} \rangle. \tag{3.17}$$

By using the two-scale difference equations (3.14) and (3.16), we can express the scaling and wavelet bases, defined by (3.7) and (3.9), respectively, at scale j, in terms of the scaling basis functions of the scale subspace V_{j-1} at finer subsequent scale $j-1$. The substitution of (3.14) into (3.7) yields, after an appropriate change of variables [12],

$$\phi_k^j(t) = \sum_{n\in\mathbb{Z}} c_{n-2k}\phi_n^{j-1}(t). \tag{3.18}$$

By taking the inner product of the signal $f(t)$ with both sides of (3.18), the following recursive equation is obtained

$$a_k^j = \sum_{n\in\mathbb{Z}} c_{n-2k}^* a_k^{j-1} \equiv H_0 a_k^{j-1} \tag{3.19}$$

that expresses the approximation coefficients at a given scale j in terms of the ones at the next finer scale $j-1$. Similarly, the following recursion, relating the details coefficients at scale j and the approximation coefficients at the next finer scale $j-1$,

$$d_k^j = \sum_{n\in\mathbb{Z}} b_{n-2k}^* a_k^{j-1} \equiv H_1 a_k^{j-1}, \tag{3.20}$$

is obtained. Note that H_0 and H_1 stand for the low-pass and the high-pass analysis wavelet (discrete) filters, respectively. Equations (3.19) and (3.20) represent the *decomposition (or analysis) Mallat* algorithm [18], which is behind the so-called *fast DWT*. The forward transform converts the samples of a signal $f(t)$ into a set of wavelet coefficients. The reconstruction part of the *Mallat* algorithm generates the coefficients of the scale representation a^{j-1} from the coefficients of the wavelet representation a^j and d^j such as [18],

$$\begin{aligned} a_n^{j-1} &= \sum_{k\in\mathbb{Z}} c_{n-2k} a_k^j + \sum_{k\in\mathbb{Z}} b_{n-2k} d_k^j \\ &\equiv G_0 a_k^j + G_1 d_k^j, \end{aligned} \tag{3.21}$$

where G_0 and G_1 represent the *synthesis* wavelet filters.

The success of the fast DWT is incontestably due to the implementation of the Mallat algorithm using multirate *filter banks* [26]. A generic 1-D, two-channel filter bank is shown in Figure 2. The input sequence $x(m)$ is first passed through the *analysis* filters pair that comprises the low-pass filter $H_0(\omega)$ and the high-pass filter $H_1(\omega)$. The filtered versions of the input signal, i.e., $y_0(m)$ and $y_1(m)$, have a bandwidth that is half the

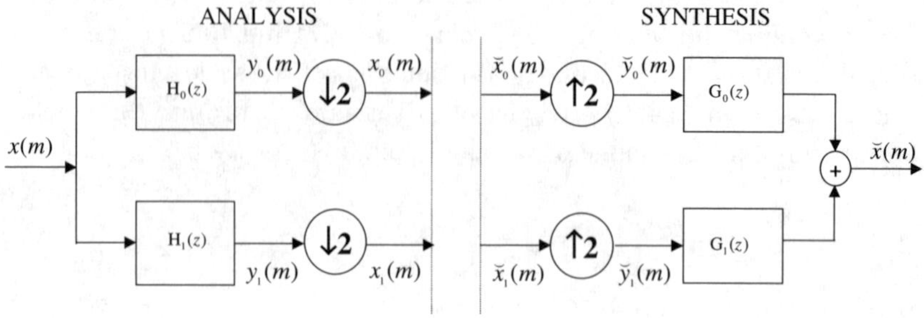

Figure 2: Generic two-channel filter bank.

bandwidth of the original input signal. Consequently, both are subsampled by a factor of two, resulting in the approximation and details wavelet coefficients $x_0(m)$ and $x_1(m)$ with half the number of samples of the input signal $x(m)$. The input signal is reconstructed from the processed wavelet coefficients $\breve{x}_0(m)$ and $\breve{x}_1(m)$ by first upsampling them to full bandwidth signals $\breve{y}_0(m)$ and $\breve{y}_1(m)$ and then filtering them, respectively, by the *synthesis* low-pass and high-pass filters $G_0(\omega)$ and $G_1(\omega)$, which could be different from the analysis filters. Finally, the filters outputs are merged to form the reconstructed signal $\breve{x}(m)$. The fast DWT is executed in linear time $O(N)$, which is faster than the *fast Fourier transform* (FFT) that is performed in logarithmic time $N\log(N)$ [27].

3.4 Wavelets Characteristics

In any analysis, the choice of the wavelet determines the nature of information that can be extracted or represented about a given real-world signal $f(t)$. The best wavelet representation requires a good understanding of the properties of wavelets and how, for instance, each property could be used for the analysis of the frequency content, irregular structures, or singularities. Besides the admissibility condition (3.5), it is often desirable to impose that more higher order moments vanish, i.e., $\int_{-\infty}^{\infty} t^m \psi(t)dt = 0, \quad m = 0, 1, ..., p-1$. The higher the number of *vanishing moments* is, the better smooth signals can be approximated with fewer wavelet coefficients. However, because of the wavelets orthogonality constraint, the adequate selection of a particular wavelet is subject to a trade-off between the number of vanishing moments and the size of the support of the wavelets. As a rule of thumb, if the signal is smooth with a few isolated singularities, a wavelet with many vanishing moments and thus of large support size is desirable to produce a large number of negligible wavelet coefficients. On the contrary, if the density of the singularities increases, it might be better to decrease the size of the support of the wavelet at the cost of reducing the number of vanishing moments, because wavelets that overlap the singularities create high amplitude coefficients, which may increase the dynamic range of the wavelet coefficients. The importance of the regularity or smoothness* of wavelets appears in many respects. First, the regularity has mostly a cosmetic effect on the error introduced by wavelet coefficients processing. Second, wavelets are good tools to study either the local or the global regularity of signals. Last, the regularity of discrete wavelets is important to assess the convergence of the iterated filter bank that is used to implement the fast DWT. Usually, with symmetric wavelet and scaling functions, fewer artifacts appear in the reconstructed data, and the handling of boundaries is simplified. For computer implementation, it is preferable that the scaling and wavelet filters coefficients are rationals or, even better, dyadic rationals because a binary multiplication by a power of two translates to bits shifts. In practice, wavelets can be classified in three broad categories that can guide the selection of an appropriate wavelet for a specific task. The analytic wavelets are most useful for the CWT, as they usually have well-defined time-frequency characterization. However, the scaling function does not exist, and often no fast wavelet transform is available. The two most widely used analytic wavelets are the *Mexican hat* and the *Morlet* wavelets. The Mexican hat wavelet is defined as the second derivative of the zero mean, unit variance *Gaussian* density function. The *Morlet wavelet (Modulated Gaussian)* was first used in geophysical explorations by *Grossman* and *Morlet* in 1985 [10] and is at the origin of the development of the wavelet analysis. The *Morlet* wavelet used in this work is a locally periodic wave train. It is complex and is obtained by shifting a *Gaussian* function in the *Fourier* domain or equivalently multiplying it by an exponential in the time domain. Let ω_0 be the *internal frequency*, then the normalized *Morlet* wavelet is approximated by [25]

$$\psi(t) \approx \pi^{-\frac{1}{4}} e^{-i\omega_0 t} e^{-\frac{t^2}{2}} \quad \text{for} \quad \omega_0 \approx 5. \tag{3.22}$$

*Strictly speaking, the regularity concerns the wavelet filter and the smoothness concerns the wavelet function.

Orthogonal wavelets are mostly characterized by either *finite impulse response* (FIR) or *infinite impulse response* (IIR) discrete filters. They are most often used for MRA, and the associated filter bank uses the same analysis and synthesis filters. The FIR wavelets have the advantage of speeding up the fast DWT since they are compactly supported, i.e., have only a finite number of taps, while IIR wavelets cannot. However, the IIR wavelets are symmetric and more regular than FIR wavelets. Haar, Daubechies, symmlets, and coiflets wavelets are FIR wavelets. *Meyer* and *Battle-Lemarié* wavelets are IIR wavelets. Finally, the biorthogonal wavelets use a pair of dual biorthogonal FIR filters, one for analysis and the other for synthesis. They are based on B-splines and are symmetric. They are useful for data compression because the human eye is quite sensitive to errors introduced by asymmetric FIR wavelets. For all wavelets, there is a one-to-one relationship between the scale and the period. The scale can be defined as the distance between oscillations in the wavelet (e.g., for the *Morlet*), or it can be some average width of the entire wavelet (e.g., for Mexican hat). The period (or inverse frequency) is the approximate *Fourier* period that corresponds to the oscillations within the wavelet. The relationship can be derived by finding the wavelet transform of a pure cosine wave with a known *Fourier* period and then computing the scale at which the wavelet power spectrum reaches its maximum. For some wavelets the period has more meaning than others. For the wavelets, which have irregularly-spaced oscillations (e.g., for FIR wavelets), the period has less meaning and should probably be ignored. For more information about the choice of wavelets, see [13, 25, 15].

4 Wavelet Analysis of Seismic Data

4.1 Geophysical Data Description

Seismic waves excited by a source of energy are extensively used in the oil and gas industry to probe the earth subsurface in order to retrieve useful information about the location of hydrocarbons reservoirs and their characteristics. Seismic waves generated at the surface tend to reflect at subsurface regions where the earth material properties show rapid variations, i.e., where singularities in the medium properties occur. They convey information not only about the locations, but also about the nature of the dominant singularities. Practically, the seismic wave field could be measured either by an array of sensors called geophones or by a probe sliding

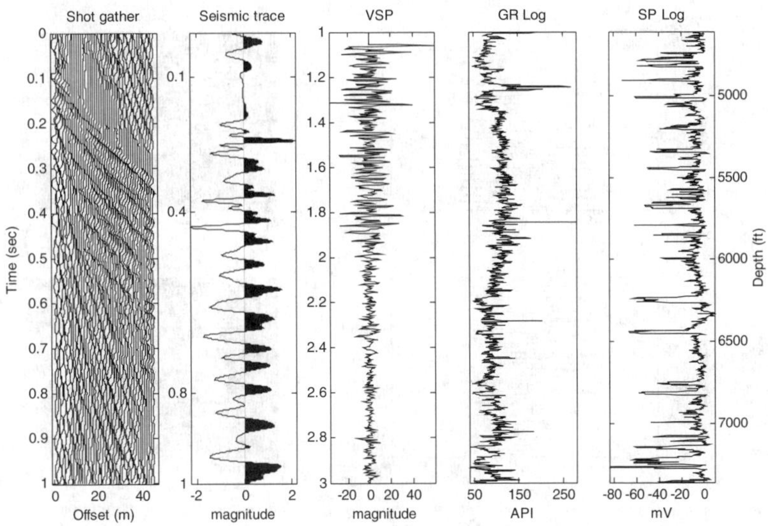

Figure 3: Geophysical data used in oil and gas prospecting.

inside a borehole. Measurements taken along the surface are referred to as surface seismic data and are collected in seismic shot records or in stacked sections after processing. Measurements taken inside the borehole yield a *vertical seismic profile* (VSP) that records higher frequency data than surface seismic data but can be used to enhance their bandwidth [28]. Another type of borehole measurements consists of geophysical logs that provide a detailed measure of physical quantities as a function of depth. Changes in lithology, porosity, bulk density, electrical conductivity, temperature, and hole diameter can be measured using various logging probes. For example, the *gamma ray* (GR) logs record the amount of natural gamma radiation emitted by the rocks surrounding the borehole, whereas the *spontaneous-potential* (SP) logs record voltages developed between the borehole fluid and the surrounding rock and fluids. Well logs are used to identify and correlate underground rocks and to determine the mineralogy and physical properties of potential reservoir rocks and the nature of the fluids they contain [2]. As can be inferred from the visual inspection of the various signals displayed in Figure 3, geophysical signals, as many real-world signals, are non-linear and non-stationary and contain multiscale features. This fact gives an *a priori* justification of the introduction of the wavelet trans-

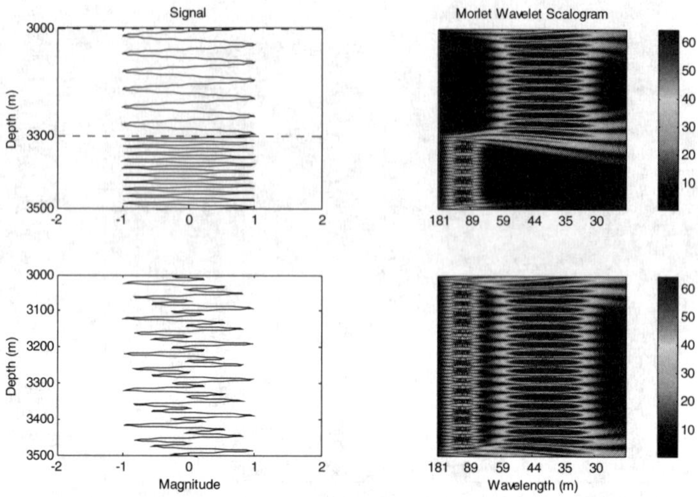

Figure 4: Morlet wavelet analysis of two synthetic logs modeling a para-sequence boundary (top) and a superimposed sedimentary sequence (bottom).

forms as an alternative to the classical spectral analysis and geo-statistical techniques for the analysis and processing of such signals. The results presented in the sequel aim at corroborating these assumptions.

4.2 Depth-Wavelength Analysis

To demonstrate the suitability of wavelet analysis for cyclo-stratigraphic interpretation, two models that may occur in strata are considered [8]. The first instance is a synthetic signal with an abrupt change in wavelength, which may represent a para-sequence boundary. This signal has two components, one from 3000 to 3300 meters with a wavelength of 40 meters and the second from 3300 to 3500 meters with a wavelength of 100 meters. The second example consists of another synthetic signal with two superimposed sedimentary sequences with wavelengths of 40 and 100 meters. The results of the CWT can be represented by either the modulus or the phase of the wavelet coefficients arranged in a 2-D display, know as the *scalogram*, that relates the scales or wavelengths to the translations. Because interpretation of the phase is more complex, only modulus scalograms will be considered. The *Morlet* scalograms depicted in Figure 4 clearly confirm the ability of the CWT to identify the two different events by quantifying their wavelengths and by detecting either their exact locations or the presence of superimposed cycles. To test the power of the CWT in identifying cyclicity in sedimentary sequences, real GR and SP logs from a well in the *Stratton* field (*South Texas*) are processed using the CWT. The stratigraphic interval studied is the *Oligocene Frio Formation*, a tick fluvially deposited sand-shale sequence that has been a prolific gas producer [11]. The cyclicities observed in Figure 5 correlate well with sedimentary features of the formation, e.g., channels and channel sets. The hierarchical scales of systematic variation are evident from the scalogram. The small wavelengths heterogeneities may be associated with thinning-up and fining-up channel fill (multiple curved bands of the coefficients), while the organization at larger scales may be related to the increasing-upwards separation of the channel bodies (inclined band of high amplitude coefficients). When applied to several hypothetical signals similar to those observed in petro-physical measurements in hydrocarbon reservoirs, the wavelet scalogram provides a clear indication of when the cyclic element is present and where wavelength changes occur in the logs. The advantage of the wavelet analysis over geo-statistical techniques consists of preserving the true character of the data by avoiding the removal of trends that may constitute a valuable part of the log data and does not require *a priori* zoning of the data.

4.3 Time-Frequency Analysis

The time-frequency analysis that can provide higher frequency resolution at lower frequencies and higher time resolution at higher frequencies is desirable for analyzing seismic data [28]. This is due to the fact that the hydrocarbons in the reservoir are diagnosed at lower frequencies and thin beds can be resolved with enhanced time resolution at higher frequencies. The time-frequency spectrum is commonly used to compute various frequency attributes of the seismic signal like single frequency, dominant frequency, center frequency, and so forth. The conventional approach is to use either the STFT or the WVT. The STFT time-frequency resolution is limited by the choice of a window of fixed length whereas the WVT suffers from quadratic interferences. The CWT time-frequency analysis, for its part, has the ability to adapt with the frequency content of the signal. The objective of this section is to demonstrate the capabilities of the CWT in

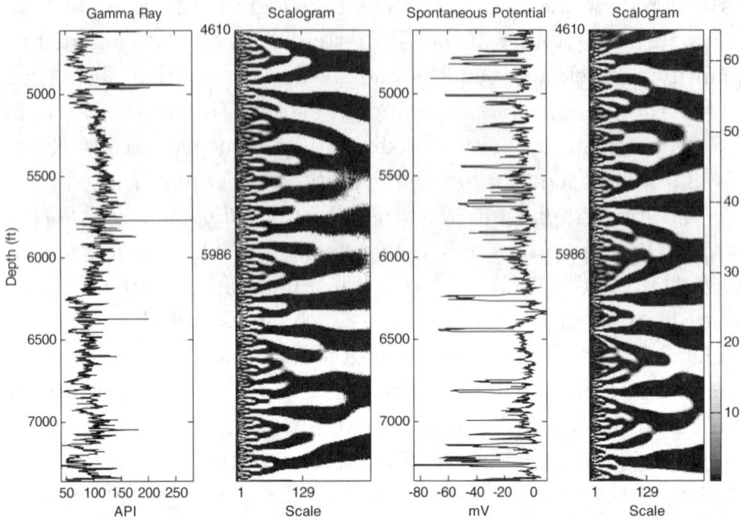

Figure 5: Morlet wavelet analysis of real well logs.

analyzing seismic data and to compare its performance with those of the STFT and WVT. We first start with a synthetic seismic signal consisting of five pure frequency components, namely, 50, 70, 20, 40, and 30 Hz, that are superimposed to a sweep signal varying from 0 to 100 Hz. Figure 6 compares the *Morlet* scalogram with the spectrograms of the STFT and the WVT. The best frequency resolution for the mono-frequency components is provided by the STFT because the sine and cosine bases are optimally localized in frequency. However, the time localization introduced by the STFT *Hamming* window of 80 taps lacks the necessary time resolution to neatly identify the transitions times of the different frequency components. The performance for the WVT is worse around the transition times, but nonetheless, the WVT exhibits a very good time-frequency localization for the linear chirp. Note also the presence of quadratic interferences that may compromise the proper interpretation of more intricate frequency components. As expected, only the scalogram ensures a fairly good compromise between the time and frequency resolutions where it can be noticed that both the frequency components and their transition instants are well resolved. The hyperbolic pattern of the linear chirp is due to the discrete algorithm used to implement the CWT [27].

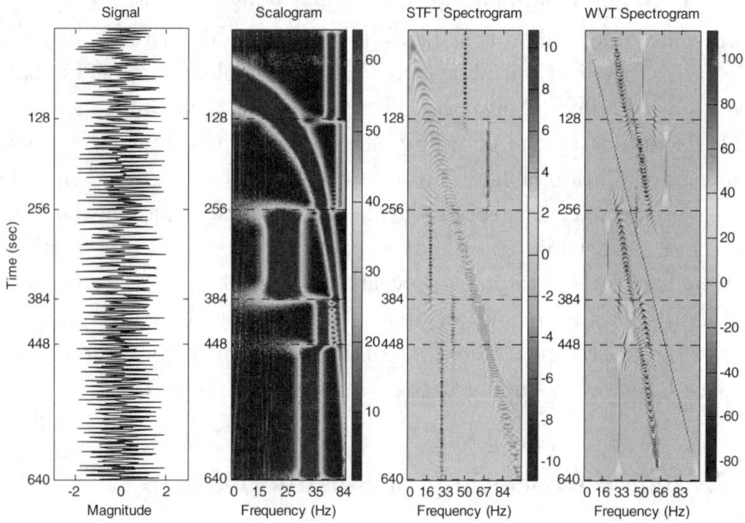

Figure 6: Time-frequency analysis of a time-varying, multicomponent synthetic seismic trace.

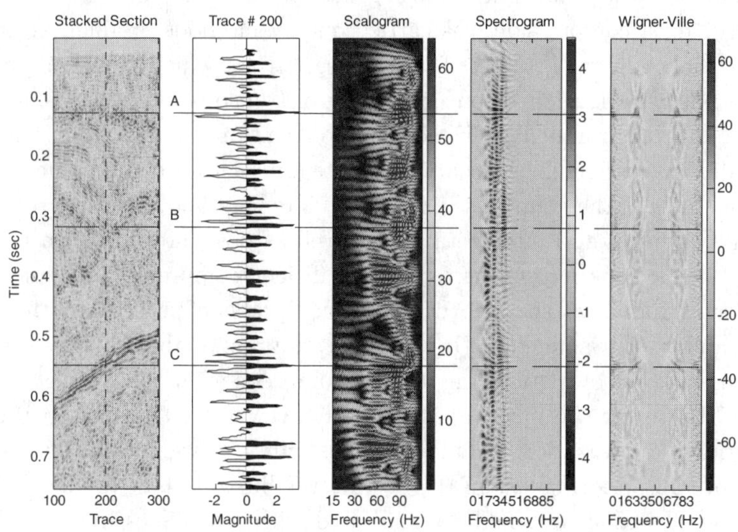

Figure 7: Time-frequency analysis of a real seismic trace taken arbitrarily from a stacked section.

In the second example, the three time-frequency tools are applied to a real seismic trace taken arbitrarily from a stacked seismic section. From the energy distributions shown in Figure 7, it clearly appears that both spectrograms failed to properly characterize the significantly time-varying spectral content of the seismic trace. On the contrary, the best time-frequency resolution is again given by the *Morlet* scalogram. This allows a very clear characterization of the time-varying attenuation of the frequency content of the seismic trace that varies from 100 to 20 Hz. Furthermore, the time locations of important events such as the three reflections labeled A, B, and C can be readily evidenced by tracking the converging maxima lines of the scalogram. These maxima lines usually converge to the time or space location of singularities present in the signal [18].

5 Wavelet Processing of Seismic Data

5.1 Seismic Compression

It is widely recognized that seismic compression is a key technology for managing seismic data in a world of ever-increasing datasets to maintain productivity without compromising interpretation results [16]. Until recently, data compression was carried out by methods originally designed for signal coding [14]. Most of the applications for which those compression methods were developed concern storage or transmission of the compressed data, but they rarely, if ever, require significant processing of the signals after decompression. The situation is quite different for compressing seismic data. Indeed, the compressed dataset is intended for visual inspection for quality control and interpretation, as well as post-interpretation processing. Thus, it is important to design efficient compression techniques in order to minimize the impact of the compression process on the interpretability of seismic data. In this respect, a remarkable improvement has been achieved by performing seismic compression in the wavelet domain [22]. However, the mere association of the wavelet transform with conventional quantization techniques yields only a marginal improvement of the overall performance. In this section we illustrate how the flexibility of the wavelet analysis could be associated with techniques from non-linear theory to design an elaborated wavelet-based compression scheme [16]. A multiresolution decomposition is first generated using a biorthogonal 2-D DWT [18]. The separable 2-D DWT transform consists of alternating the 1-D DWT on rows and columns of the input matrix data to generate three detail subbands and a unique approximation subband [18, 23]. Three *anisotropic* biorthogonal wavelets $\{\psi^H, \psi^V, \psi^D\}$, emphasizing details at different scales and orientations, i.e., horizontally, vertically, and diagonally, and a scaling function producing a smoothed low-resolution version of the input data are applied to the seismic data. This yields a low-resolution residual at the coarsest scale and a number of detail subbands. For each detail subband an adaptive *threshold uniform scalar quantizer (TUSQ)* is designed. The thresholds are designed using the *wavelet shrinkage* technique, originally proposed by *Donoho* and *Johnstone* [7] for near-optimal non-parametric regression. The heuristic of this approach is that for spatially inhomogeneous signals (e.g., seismic signals) most of the information is concentrated in a small subset of the wavelet space. This is mainly due to the inherent space and frequency localization properties of the wavelet transform. So under the *additive white Gaussian noise (AWGN)* assumption, noise contaminates all the coefficients equally (i.e., in all the directions), whereas only a very few wavelet coefficients contribute a signal. Wavelet shrinkage exploits those spectral and structural differences of the underlying signal and noise across scales to efficiently separate their respective components. The proposed compression strategy is illustrated with real seismic signals representing a seismic stacked section from the *Midyan Basin (Saudi Arabia)* [16, 20]. The 2-D biorthogonal wavelets and scaling functions of Figure 8b are used to generate the three-level MRA decomposition depicted in Figure 8a. The compression results showing the original, the reconstructed and the residual seismic data are displayed in Figure 8c. The visual inspection of the compression results clearly indicates that most of the geological structure present in original seismic data is very well preserved for a relatively high compression ratio, i.e., 93:1. This is due to the good decorrelation power of the DWT, which by concentrating most of the data energy, i.e., 94%, in a few detail wavelet coefficients, allows a large number of them to be safely discarded . The compression error is small as confirmed by a high *peak signal-to-noise ratio* (PSNR) of 27.021 dB and the well peaked *autocorrelation function* (ACF) of the L^1- and L^2-norms of the error shown in Figure 8d.

5.2 Coherent Noise Filtering

Ground roll is a surface wave energy that propagates along and near the surface with relatively low velocity, often with low frequency and usually with high amplitudes relative to other events of interest in land seismic sur-

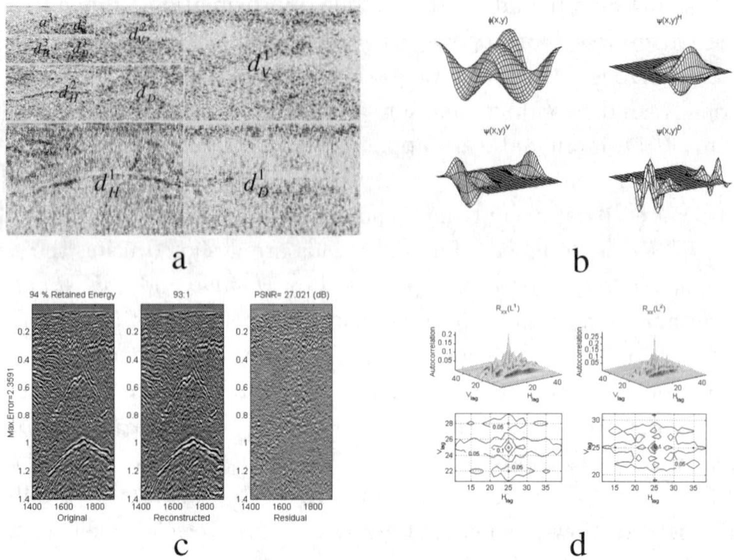

Figure 8: Wavelet-based seismic compression results.

veys [28]. The conventional wave-number $(f - k)$ filtering, which is based on the 2-D *Fourier* transform, is not suitable to remove such a time- and space-variant coherent seismic noise. An adaptive 2-D, wavelet-based filtering technique is proposed. The rationale of the denoising strategy is to first generate a multiresolution decomposition of the data using a biorthogonal 2-D DWT. Then, the characteristic patterns of the mapping of the ground roll into the horizontal, vertical, and diagonal detail subbands are analyzed. Next, the level-dependent thresholds are determined using the wavelet shrinkage technique. These are applied to all detail subbands but not the vertical subbands. The denoised data is then reconstructed by the inverse wavelet transform.

The proposed technique is illustrated with a seismic time-offset shot gather from Alaska [28] depicted in Figure 9a. Notice that this seismic dataset exhibits a series of reflections and refractions and is corrupted by random noise and ground roll. A spline biorthogonal wavelet filter pair is used, and a three-level multiresolution decomposition is generated. The visual inspection of the reconstructed vertical subband at level 1 of Figure 9b suggests that most of the ground roll energy is confined to that subband.

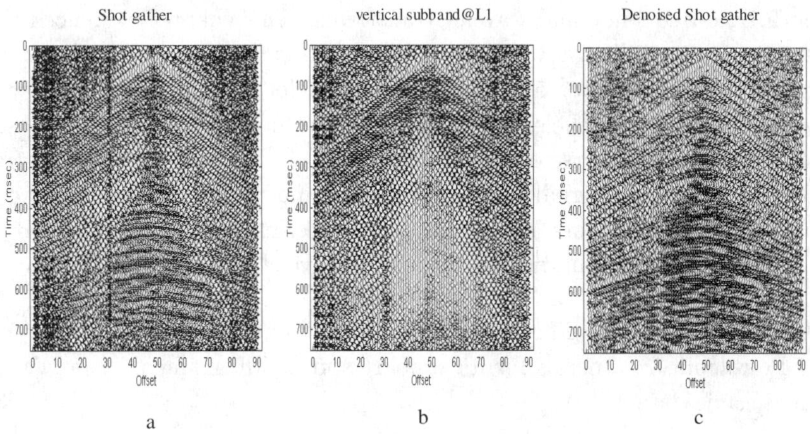

Figure 9: Ground roll and random noise filtering in the wavelet domain using the wavelet shrinkage technique.

An explanation to this could be that the slope of the envelope containing the ground roll lies below the diagonal, whereas the reflections lie above it. However, because the refracted events share the same vertical subbands, the vertical thresholds are tuned to remove the former without seriously affecting the refractions. Finally, it is clear from the reconstructed denoised data of Figure 9c that the undesirable ground roll is significantly filtered with a little effect on the rest of the record, while the random noise is significantly attenuated.

6 Conclusions

Mathematical rigorous analysis and processing tools that can highlight important features of a real-world process and reveal structure not apparent from direct observations are a key component of process analysis and processing. For geophysical data used in oil and gas prospecting, tools that offer the ability to examine the variability of the geological structure of reservoirs at different wavelengths and frequencies are of paramount importance for the success of any hydrocarbons exploration enterprise. Wavelet transforms offer such tools and have already proven useful in the study of many real-world processes in diverse areas of sciences and engineering. In this chapter, we focused on presenting the most important mathematical concepts together with features and properties relevant to both time-frequency and wavelet transforms. This chapter evidenced the power and flexibility of wavelet methods by discussing several applications of both the continuous and the discrete wavelet transforms. We showed examples of depth-wavelength analysis for the detection of cyclic and transient geological events, of spectral analysis of the time-varying spectral content of seismic signals, of compression, and of random and coherent noise filtering. Readers are encouraged to consult the reference list for a more detailed mathematical treatment of wavelets theory.

Acknowledgment. The author would like thank King Fahd University of Petroleum & Minerals for its support and Saudi Aramco for providing the seismic data.

References

[1] J.J. Benedetto and M. Frazier (Eds.), *Wavelets: Mathematics and Applications*, CRC Press, Boca Raton, FL, 1994.

[2] H. Blatt, W.B.N. Berry, and S. Brande, *Principles of Stratigraphic Analysis,* Blackwell Scientific Publications: Boston,1991.

[3] R. N. Bracewell, *The Fourier Transform & Its Applications,* McGraw-Hill, New York, 1986.

[4] R. Carmona , W.-L. Hwang, and B.Torresani, *Practical Time-Frequency Analysis,* Academic Press, New York, 1998.

[5] C. K. Chui, L. Montefusco, and L. Puccio (Eds.), *Wavelets: Theory, Algorithms, and Applications,* Academic Press, New York, 1994.

[6] I. Daubechies, *Ten Lectures on Wavelets,* SIAM, Philadelphia, 1992.

[7] D. Donoho and I. Johnstone, Ideal Spatial Adaptation by Wavelet Shrinkage, *Biometrika*, **81**, (1994), 425-455.

[8] D. Emery and K.J. Myers, *Sequence Stratigraphy,* Blackwell Science, Oxford, 1996.

[9] E. Foufoula-Georgiou and P. Kumar (Eds.), *Wavelets in Geophysics,*. Academic Press, New York, 1994.

[10] P. Goupillaud, A. Grossmann, and J. Morlet, Cycle Octave and Related Transforms in Seismic Signal Analysis, *Geoexploration*, **23**, (1985), 85-102.

[11] B. Hardage et al., A 3-D Seismic History Evaluation Fluvially Deposited Thin-bed Reservoirs in a Gas-producing Property, *Geophysics,* **59**(11), (1994), 1650-1665.

[12] E. Hernández and G. Weiss, *A First Course on Wavelets*, CRC Press, Boca Raton, FL, 1996.

[13] M. Holschneider, *Wavelets: An Analysis Tool,* Clarendon Press, Oxford, 1995.

[14] N. Jayant and P. Noll, *Digital Coding of Waveforms- Principles and Applications to Speech and Video,* Prentice-Hall, Englewood Cliffs, NJ, 1984.

[15] G. Kaiser, *A Friendly Guide to Wavelets,* Birkhäuser, Boston, 1994.

[16] M. F. Khène, *Fast Wavelet Transforms and Seismic Compression,* PhD. Diss., KFUPM 2001.

[17] J. Laurent et al. (Eds.), *Wavelets, Images, and Surface Fitting,* Peters Wellesley, MA, 1994.

[18] S. Mallat, *A Wavelet Tour of Signal Processing,* Academic Press, New York, 1998.

[19] P. R. Massopust, Fractal Functions, *Fractal Surfaces, and Wavelets,* Academic Press, New York, 1994.

[20] D. Mougenot and A. Al-Shakhis, Depth Imaging a Pre-Salt Faulted Block: A Case Study from the Midyan Basin (Red Sea), *Saudi Aramco Jour. of Tech.*, **Fall**, (1998), 2-10.

[21] S. Qian and D. Chen, *Joint Time-Frequency Analysis,* Prentice-Hall, Englewood Cliffs, NJ, 1996.

[22] E. Reiter and P. Heller, Wavelet Transform-based Compression of NMO-corrected CDP Gathers, *64th Ann. Int. Mtg SEG*, *Expanded Abstracts,* (1994), 731-734.

[23] E. Stollnitz, T. DeRose, and D. Salesin, *Wavelets for Computer Graphics: Theory and Applications,* Morgan Kaufmann Publishers, Inc., San Francisco, 1996.

[24] G. Strang, Wavelets and Dilation Equations: A Brief Introduction, *SIAM Rev.*, **31**, (1989), 614-627.

[25] A. Teolis, *Computational Signal Processing with Wavelets,* Birkhäuser, Boston, 1997.

[26] M. Vetterli and J. Kovacevic, *Wavelets and Subband Coding,* Prentice-Hall PTR, New Jersey, 1995.

[27] M. Wickerhauser, *Adapted Wavelet Analysis from Theory to Software,* Peters Wellesley, 1994.

[28] O. Yilmaz , *Seismic Data Analysis,* SEG, Tulsa, OK, 2001.

Part III

Classical and Fractal Methods for Physical Problems

Chapter 11

GRADIENT CATASTROPHE IN HEAT PROPAGATION WITH SECOND SOUND

S. A. Messaoudi

King Fahd University of Petroleum & Minerals

A. S. Al Shehri

Dammam Girl's College

Abstract

In this chapter we consider a hyperbolic nonlinear system describing heat propagation with second sound in an inhomogeneous material. We establish a blow up result for the classical solution with large-gradient initial data.

1 Introduction

In the absence of deformation and external sources, the equation of balance of energy in the one-dimensional heat propagation is

$$\mathcal{E}(\theta)_t + q_x = 0, \tag{1.1}$$

where $\theta > 0$ is the difference temperature, q is the heat flux, and $\mathcal{E}$ is a positive strictly increasing function. In the classical theory, the flux q is given by Fourier' s law

$$q + \kappa(\theta)\theta_x = 0,$$

where κ is a strictly positive function characterizing the material in consideration. In the case where $c = \mathcal{E}'$ and κ are independent of θ, we get the familiar linear heat equation

$$\theta_t = k\theta_{xx}, \qquad k = \frac{\kappa}{c}.$$

This equation provides a useful description of heat conduction under a large range of conditions and predicts an infinite speed of propagation; that is, any thermal disturbance at one point has an instantaneous effect elsewhere in the body. This is not always the case. In fact, experiments showed that heat conduction in some dielectric crystals at low temperatures is free of this paradox and that disturbances which are almost entirely thermal propagate in a finite speed. This phenomenon in dielectric crystals is called second sound. To overcome this contradictory paradox, many theories have merged. One theory suggests that we should account for memory effects (see [7], [12], [13], [14]). For this purpose, an internal parameter p has been introduced as

$$q = -a(\theta)p. \tag{1.2}$$

If the memory effect is considered as a functional of a history of temperature gradient, then

$$p(x,t) = \int_{-\infty}^{t} e^{-b(t-s)}\theta_x(x,s)ds, \qquad b > 0. \tag{1.3}$$

A differentiation of (1.3) with respect to time gives

$$p_t = -bp + \theta_x. \tag{1.4}$$

If $a(\theta)$ is constant, then (1.2) and (1.4) yield

$$q_t + bq = -a\theta_x.$$

This is a linear equation and does not fully describe the heat propagation in solids (see [7], [13]). In fact, this is a special case of Cattaneo' s law [1], which has the form

$$\tau(\theta)q_t + q = -\kappa(\theta)\theta_x.$$

Here τ and κ are strictly positive functions depending on the absolute temperature and characterizing the material in consideration. In this case the system governing the evolution of θ and q becomes

$$\begin{aligned} c(\theta)\theta_t + q_x &= 0 \\ \tau(\theta)q_t + q + \kappa(\theta)\theta_x &= 0. \end{aligned} \tag{1.5}$$

Global existence and decay of classical solutions, for smooth and small initial data, to the Cauchy problem, as well as to some initial boundary value problems, have been established by Coleman, Hrusa, and Owen [2]. In their work, the authors considered a system, which satisfies the requirements imposed by the second law of thermodynamics discussed in [3], and showed that (θ, q) tends to the equilibrium state; however, no rate of decay has been discussed. Messaoudi [11] showed that if the initial data are small enough, then the solution decays exponentially to the rest state.

Concerning formation of singularities, Messaoudi [9], [10] showed, under the same restrictions on τ, c, and κ, that classical solutions to the Cauchy problem of (1.5) break down in finite time if the initial data are chosen small in the L^∞ norm with large enough derivatives.

In this chapter, we consider the situation when a in (1.2) is a function of x only. This may be regarded as inhomogeneity in the material in consideration. Therefore, the system we study takes the form

$$\begin{gathered} c(\theta(x,t))\theta_t(x,t) + q_x(x,t) = 0 \\ q_t(x,t) + bq(x,t) = -a(x)\theta_x(x,t), \qquad x \in I = (0,1), \qquad t \geq 0. \end{gathered} \tag{1.6}$$

This is a hyperbolic system for (θ, q), and it will take care of the paradox of infinite speed propagation known in the classical theory of heat propagation.

We associate with (1.6) the initial and the boundary conditions

$$\begin{gathered} \theta(x,0) = \theta_0(x), \;\; q(x,0) = q_0(x), \qquad x \in I = [0,1] \\ \theta(0,t) = \theta(1,t) = 0, \qquad t \geq 0 \end{gathered} \tag{1.7}$$

and prove a finite time blow up result similar to one in [9]. We should note here that hyperbolic systems similar to (1.6) have been discussed by many mathematicians [6], [8], [14], and various results concerning global existence and blow up have been established.

In order to make this chapter self-contained we state, without proof, a local existence result. The proof can be established by either a classical energy argument [4] or by using the nonlinear semigroup theory [5]. We first start with the hypotheses on the functions a, c, and the initial data.

(**H1**) $a \in C^2([0,1])$ such that $a \geq a_0 > 0$.
(**H2**) $c \in C^2(\mathbb{R})$ such that $c \geq c_0 > 0$.
(**H3**) $\theta_0 \in H^2(I) \cap H_0^1(I)$ and $q_0 \in H^2(I)$.

Proposition 1.1 *Assume that* $(H1), (H2)$, *and* $(H3)$ *hold. Then problem (1.6)-(1.7) has a unique local solution* (θ, q) *on a maximal time interval*

$[0, T)$ *satisfying*

$$\begin{aligned} \theta \in C([0,T), \quad H^2(I) \cap H_0^1(I)) \cap C^1([0,T), \quad H_0^1(I)) \\ q \in C([0,T), \quad H^2(I)) \cap C^1([0,T), \quad H^1(I)). \end{aligned} \tag{1.8}$$

Remark 1.2 θ, q are in $C^1([0,1] \times [0,T))$ by the Sobolev embedding theorem.

2 Formation of Singularities

In this section, we state and prove our main result. We first begin with a result, which gives uniform bounds on the solution in terms of the initial data.

Theorem 2.1 *Assume that* $(H1), (H2)$, *and* $(H3)$ *hold. Then the solution (1.8) satisfies*

$$\max_{(x,t)\in[0,1]\times\{0,T)} \{|\theta(x,t)| + |q(x,t)|\} \le \Gamma \max_{x\in[0,1]} \{|\theta_0(x)| + |q_0(x)|,\} \tag{2.1}$$

where Γ *is a constant independent of* θ, q, *and* t.

Proof. We introduce the quantities

$$r(x,t) := \frac{q(x,t)}{\sqrt{a(x)}} - A(\theta(x,t)), \qquad s(x,t) := \frac{q(x,t)}{\sqrt{a(x)}} + A(\theta(x,t)) \tag{2.2}$$

and the differential operators

$$\partial_t^- := \frac{1}{\rho}\frac{\partial}{\partial t} - \frac{\partial}{\partial x}, \quad \partial_t^+ := \frac{1}{\rho}\frac{\partial}{\partial t} + \frac{\partial}{\partial x},$$

where

$$\rho(x,t) = \sqrt{a(x)/c(\theta(x,t))}, \qquad A(\theta) = \int_0^\theta \sqrt{c(\xi)}d\xi.$$

We then compute

$$\begin{aligned} \partial_t^- r &= \frac{1}{\rho} r_t - r_x \qquad (2.3) \\ &= \sqrt{\frac{c(\theta)}{a}}(\frac{q_t}{\sqrt{a}} - \sqrt{c(\theta)}\theta_t) - (\frac{q_x}{\sqrt{a}} - \sqrt{c(\theta)}\theta_x) + \frac{1}{2}a'a^{-3/2}q \\ &= -\frac{1}{\sqrt{a}}(c\theta_t + q_x) + \frac{\sqrt{c}}{a}(q_t + a\theta_x) + \frac{1}{2}a'a^{-3/2}q \\ &= (-b\frac{\sqrt{c}}{a} + \frac{1}{2}a'a^{-3/2})q(x,t) = (-b\sqrt{\frac{c}{a}} + \frac{a'}{2a})\frac{r+s}{2}, \end{aligned}$$

by virtue of the system (1.6). Similar computations also yield

$$\partial_t^+ s = (-b\sqrt{\frac{c}{a}} - \frac{a'}{2a})\frac{r+s}{2}. \tag{2.4}$$

We then define the nonnegative Lipschitz functions

$$R(t) := \max_{x\in[0,1]} |r(x,t)|, \quad S(x,t) := \max_{x\in[0,1]} |s(x,t)|. \tag{2.5}$$

For each fixed $t > 0$, we pick x_1 and x_2 in $[0,1]$ so that

$$R(t) = |r(x_1,t)|, \qquad S(t) = |s(x_2,t)|; \tag{2.6}$$

therefore, for so small $h \in (0,t)$, we have

$$\begin{aligned} R(t-h) &\geq |r(x_1 + h\rho(x_1,t), t-h)| \\ S(t-h) &\geq |s(x_1 - h\rho(x_2,t), t-h)|. \end{aligned} \tag{2.7}$$

By subtracting (2.7) from (2.6), dividing by h, and then letting h go to zero, we get

$$\begin{aligned} R'(t) &\leq \rho(x_1,t)|\partial_t^- r(x_1,t)| \leq \sqrt{\frac{a}{c}}|-b\sqrt{\frac{c}{a}} + \frac{a'}{2a}|\frac{r+s}{2} \\ &\leq |-b + \frac{a'}{2\sqrt{ac}}|\frac{r+s}{2} \leq \frac{\gamma}{2}(R(t)+S(t)) \end{aligned}$$

and

$$S'(t) \leq \frac{\gamma}{2}(R(t)+S(t)),$$

where $\gamma = b + (\max a')/a_0c_0$. Therefore, we have

$$\frac{d}{dt}[R(t)+S(t)] \leq \gamma[R(t)+S(t)].$$

A simple integration leads to

$$[R(t)+S(t)] \leq [R(0)+S(0)]e^{\gamma T}. \tag{2.8}$$

By using (2.2) and (2.5), the assertion of Theorem 2.3 is established. ■

Theorem 2.2 *Assume that $(H1), (H2)$, and $(H3)$ hold. Assume further that $c'(0) > 0$. Then there exist initial data θ_0 and q_0, for which the solution (1.8) blows up in finite time.*

Proof. We take a t-partial derivative of (2.3) to have

$$(\partial_t^- r)_t = [(-b\sqrt{\frac{c}{a}} + \frac{a'}{2a})\frac{r+s}{2}]_t \tag{2.9}$$

which, in turn, implies

$$\begin{aligned} \partial_t^- r_t &= \frac{\rho_t}{\rho^2} r_t + [(-b\sqrt{\frac{c}{a}} + \frac{a'}{2a})\frac{r+s}{2}]_t \\ &= \frac{-c'}{2\sqrt{ac}}\theta_t r_t + (-b\sqrt{\frac{c}{a}} + \frac{a'}{2a})\frac{r_t+s_t}{2} - \frac{bc'}{4\sqrt{ac}}(r+s)\theta_t. \end{aligned} \tag{2.10}$$

By using (2.2), it is easy to see that

$$\theta_t = \frac{s_t - r_t}{2c(\theta)};$$

thus substituting in (2.10), we obtain

$$\begin{aligned} \partial_t^- r_t &= \frac{-c'}{2\sqrt{ac}}\frac{s_t - r_t}{2c} r_t + (-b\sqrt{\frac{c}{a}} + \frac{a'}{2a})\frac{r_t+s_t}{2} \\ &\quad - \frac{bc'}{4\sqrt{ac}}(r+s)\frac{s_t - r_t}{2c} \\ &= \frac{c'}{4c\sqrt{ac}} r_t^2 - \frac{c'}{4c\sqrt{ac}} r_t s_t + (-b\sqrt{\frac{c}{a}} + \frac{a'}{2a})\frac{r_t+s_t}{2} \\ &\quad - \frac{bc'}{4\sqrt{ac}}(r+s)\frac{s_t - r_t}{2c}. \end{aligned} \tag{2.11}$$

In order to eliminate the second term in the RHS of (2.11), we set

$$W := c^{1/4} r_t.$$

Consequently, we get

$$\begin{aligned} \partial_t^- W &= c^{1/4}\partial_t^- r_x + \frac{1}{4}c^{-3/4} r_t \partial_t^- c \\ &= \frac{c^{-7/4}c'}{4\sqrt{a}} W^2 - \frac{c^{-3/4}c'}{4\sqrt{a}} r_t s_t + c^{1/4}[(-b\sqrt{\frac{c}{a}} + \frac{a'}{2a})\frac{r_t+s_t}{2} \\ &\quad - \frac{bc'}{4\sqrt{ac}}(r+s)\frac{s_t - r_t}{2c}] + \frac{1}{4}c^{-3/4} r_t c'(\sqrt{\frac{c}{a}}\theta_t - \theta_x). \end{aligned} \tag{2.12}$$

At this point we should note that, by (1.6) and (2.2), we have

$$-\theta_x = \frac{1}{a}(q_t + bq) = \frac{s_t}{\sqrt{a}} - \frac{\sqrt{c}}{\sqrt{a}}\theta_t + \frac{b}{a}q.$$

Therefore, (2.12) becomes

$$\begin{aligned}\partial_t^- W &= \frac{c^{-7/4}c'}{4\sqrt{a}}W^2 - \frac{c^{-3/4}c'}{4\sqrt{a}}r_t s_t + c^{1/4}[(-b\sqrt{\frac{c}{a}}+\frac{a'}{2a})\frac{r_t+s_t}{2} \qquad (2.13)\\ &-\frac{bc'}{4\sqrt{ac}}(r+s)\frac{s_t-r_t}{2c}] + \frac{1}{4}c^{-3/4}r_t c'(\frac{s_t}{\sqrt{a}}+\frac{b}{a}q)\\ &= \frac{c^{-7/4}c'}{4\sqrt{a}}W^2 + \frac{1}{2}[\frac{a'}{2a} - b\sqrt{\frac{c}{a}} + \frac{bc'}{8c\sqrt{ac}}(r+s) + \frac{c'}{4c}(r+s)]W\\ &+\frac{1}{2}[\frac{a'}{2a} - b\sqrt{\frac{c}{a}} - \frac{bc'}{8c\sqrt{ac}}(r+s)]s_t.\end{aligned}$$

A direct computation, using (1.6) and (2.2) again, gives

$$s_t = \sqrt{a}\partial_t^-\theta - \frac{b}{2\sqrt{a}}(s+r), \qquad r+s = 2r+2A(\theta).$$

So substituting in (2.13) yields

$$\begin{aligned}\partial_t^- W &= \frac{c^{-7/4}c'}{4\sqrt{a}}W^2 + \frac{1}{2}[\frac{a'}{2a} - b\sqrt{\frac{c}{a}} + \frac{bc'}{8c\sqrt{ac}}(r+s) + \frac{c'}{4c}(r+s)]W\\ &+[\frac{a'}{2a} - b\sqrt{\frac{c}{a}} - \frac{bc'}{8c\sqrt{ac}}(r+A)]\sqrt{a}\partial_t^-\theta \qquad (2.14)\\ &-\frac{b}{4\sqrt{a}}(s+r)[\frac{a'}{2a} - b\sqrt{\frac{c}{a}} - \frac{bc'}{8c\sqrt{ac}}(r+s)].\end{aligned}$$

We now estimate the third term of (2.14) as follows:

$$\begin{aligned}\frac{a'}{2\sqrt{a}}\partial_t^-\theta &= \partial_t^-(\frac{a'}{2\sqrt{a}}\theta) + \theta(\frac{a'}{2\sqrt{a}})'\\ -b\sqrt{c(\theta)}\partial_t^-\theta &= -\partial_t^- bA\\ -\frac{bc'}{8c\sqrt{c}}r\partial_t^-\theta &= -\frac{b}{8}rc^{-3/2}(\theta)c'(\theta)\partial_t^-\theta \qquad (2.15)\\ &= \frac{b}{4}r\partial_t^- c^{-1/2}(\theta) = \frac{b}{4}\partial_t^-[rc^{-1/2}(\theta)] - \frac{b}{4}c^{-1/2}(\theta)\partial_t^- r\\ &= \frac{b}{4}\partial_t^-[rc^{-1/2}(\theta)] - \frac{b}{4}c^{-1/2}(-b\sqrt{\frac{c}{a}}+\frac{a'}{2a})\frac{r+s}{2}\\ -\frac{bc'}{8c\sqrt{c}}A\partial_t^-\theta &= -\frac{b}{8}c^{-3/2}(\theta)c'(\theta)A(\theta)\partial_t^-\theta = -\frac{b}{8}\partial_t^-(\int_0^\theta c^{-3/2}c'A(\xi)d\xi).\end{aligned}$$

By setting

$$f(x,t) = (\frac{a'}{2\sqrt{a}}\theta) - bA + \frac{b}{4}rc^{-1/2}(\theta) - \frac{b}{8}(\int_0^\theta c^{-3/2}c'A(\xi)d\xi),$$

the estimate (2.14) takes the form

$$\partial_t^- W = \frac{c^{-7/4}c'}{4\sqrt{a}} W^2 + \partial_t^- f + MW + N,$$

where

$$\begin{aligned} M &= \frac{1}{2}[\frac{a'}{2a} - b\sqrt{\frac{c}{a}} + \frac{bc'}{8c\sqrt{ac}}(r+s) + \frac{c'}{4c}(r+s)] \\ N &= \theta(\frac{a'}{2\sqrt{a}})' - \frac{b}{4}c^{-1/2}(-b\sqrt{\frac{c}{a}} + \frac{a'}{2a})\frac{r+s}{2} \\ &\quad - \frac{b}{4\sqrt{a}}(s+r)[\frac{a'}{2a} - b\sqrt{\frac{c}{a}} - \frac{bc'}{8c\sqrt{ac}}(r+s)]. \end{aligned}$$

We then set $F = W - f$ to obtain, from (2.16),

$$\partial_t^- F = \frac{c^{-7/4}c'}{4\sqrt{a}} F^2 + BF + C, \tag{2.16}$$

where $B = (M - 2\frac{c^{-7/4}c'}{4\sqrt{a}} f)$ and $C = N + \frac{c^{-7/4}c'}{4\sqrt{a}} f^2 - Mf$ are functions depending on a, b, θ, and q only. Therefore, by choosing the initial data small enough in L^∞ norm, with sufficiently large derivatives (hence F is large enough), it is standard to deduce that F blows up in finite time. ■

Remark 2.3 The same result holds for $c'(0) < 0$. In this case consider the evolution of s_t on the forward characteristics.

Remark 2.4 Similar results can be obtained for the Cauchy problem, as well as other types of boundary conditions.

Acknowledgment. The authors would like to thank KFUPM for its continuous support.

References

[1] C. Cattaneo, Sulla conduzione del calore, *Atti Sem. Math. Fis Univ. Modena*, **3** (1948), 83–101.

[2] B.D. Coleman, W.J. Hrusa, and D.R. Owen, Stability of equilibrium for a nonlinear hyperbolic system describing heat propagation by second sound in solids, *Arch. Rational Mech. Anal.*, **94** (1986), 267–289.

[3] B.D. Coleman, M. Fabrizio, and D.R. Owen, On the thermodynamics of second sound in dielectric crystals, *Arch. Rational Mech. Anal.*, **80** (1982), 135–158.

[4] C.M. Dafermos and W.J. Hrusa, Energy methods for quasilinear hyperbolic initial-boundary value problems. Applications to elastodynamics, *Arch. Rational Mech. Anal.*, **87** (1985), 267–292.

[5] T.J.R. Hughes, T. Kato, and J.E. Marsden, Well-posed quasilinear second order hyperbolic systems with applications to nonlinear elastodynamic and general relativity, *Arch. Rational Mech. Anal.*, **63** (1977), 273–294.

[6] P.D. Lax, Development of singularities in solutions of nonlinear hyperbolic partial differential equations, *J. Math. Physics*, **5** (1964), 611–613.

[7] H. Li and K. Saxton, Asymptotic behavior of solutions to quazilinear hyperbolic equations with nonlinear damping, *Quarterly Appl. Math.*, **2** (2003), 295–313.

[8] R.C. MacCamy and V.J. Mizel, Existence and nonexistence in the large solutions of quasilinear wave equations, *Arch. Rational Mech. Anal.*, **25** (1967), 299–320.

[9] S.A. Messaoudi, Formation of singularities in heat propagation guided by second sound, *J. Diff. Eqns.*, **130** (1996), 92–99.

[10] S.A. Messaoudi, On the existence and nonexistence of solutions of a nonlinear hyperbolic system describing heat propagation by second sound, *Appl. Anal.*, **73** (1999), 485–496.

[11] S.A. Messaoudi, Decay of solutions of a nonlinear hyperbolic system describing heat propagation by second sound, *Appl. Anal.*, **81**(2) (2002), 201–209.

[12] K. Saxton, R. Saxton, and W. Kosinsky, On second sound at the critical temperature, *Quarterly Appl. Math.*, **57**(4) (1999), 723–740.

[13] K. Saxton and R. Saxton, Nonlinearity and memory effects in low temperature heat propagation, *Arch. Rational Mech. Anal.*, **52** (2000), 127–142.

[14] M. Slemrod, Instability of steady shearing flows in nonlinear viscoelastic fluid, *Arch. Rational Mech. Anal.*, **3** (1978), 211–225.

Chapter 12

ACOUSTIC WAVES IN A PERTURBED LAYERED OCEAN

F. D. Zaman

King Fahd University of Petroleum & Minerals

A. M. Al-Marzoug

Saudi Aramco

Abstract

The study of acoustic wave propagation in an ocean is of interest due to our need to understand naval detection and marine seismology. If the ocean is assumed to be homogeneous with a plane seabed, normal mode analysis can be employed to the depth equation obtained by separation of variables from the acoustic wave equation. The analysis is simpler if a rigid seabed is assumed. In practical situations, however, the ocean may have depth-dependent properties due to an increase in density due to depth, salinity, or a change in temperature. This change can often be modeled by considering a layered model of ocean. Moreover, the seabed may not be rigid, but may satisfy reflecting-type boundary conditions. One interesting situation can arise if the seabed undergoes undulation so that separation of variables is no longer feasible. We use the layered model of an ocean and employ the perturbation method to discuss the solution of the depth equation arising from these situations.

1 Introduction

The study of acoustic wave propagation in ocean waveguides has attracted considerable attention due to its applications in geophysics, underground acoustics, and marine detection problems. There have been several studies using the homogeneous model of the ocean in which the density and hence the acoustic speed are assumed to be uniform. In such a case, the acoustic pressure, which is the exact solution of the acoustic wave equation, can be obtained using separation of variables resulting in the Helmholtz equation (Ahluwalia and Keller, [1]). While the speed of the sound propagation in the homogeneous ocean is nearly uniform, small variations in speed may occur due to a change in temperature, salinity, depth or other inhomogeneities. The change of density with depth of the ocean water leads to a depth-dependent velocity model.

Duston, Verma, and Wood [3] and Ricardo [5] used the perturbation theory to study the problem of variable speed with depth. They used normal modes, involving the eigenvalues and eigenfuctions of a depth-dependent Sturm-Liouville's problem, and obtained corrections to the solution of the perturbed problem arising from depth-dependent properties.

In many cases, it has been observed that the variation in the physical characteristics is not continuous as depth changes, but change occurs in a discontinuous way. These properties remain piecewise constant within layers and only change across the interface [4]. Boyles [2] has studied a model consisting of two or more homogeneous layers to account for such situations. Zaman and Al-Muhiameed [6] used the perturbation method to study variation in density with depth in a layered model of the ocean.

In addition to nonhomogeneous models of the ocean, we sometimes need to study the acoustic field due to a source. The presence of the source makes the wave equation inhomogeneous, and the separation of variables to obtain a range equation and a depth equation fails. In addition to failure of separation of variables in the presence of a source, we observe that it can only be applied to the wave equation if the boundary surface of the problem coincides with the coordinate surface. We focus our attention to this case in a layered model of the ocean and use the Fourier transform, Green's function, and perturbation methods to study the variations in the seabed. The seabed is assumed to have a small smooth undulation which introduces a perturbed boundary condition at the bottom of the sea. The boundary conditions at the seabed are taken to be rigid or of reflecting type to allow a more realistic situation in which part of the energy is reflected back by the seabed. Some examples to demonstrate the effect of parabolic, quadratic, and linear perturbations in the sea bottom on the acoustic pressure have also been considered.

2 Formulation of the Problem

With an irregular seabed, we consider the problem of an ocean consisting of two layers: the top layer is assumed to be flat with density ρ_1 and velocity c_1 and the second layer is assumed to be non–horizontal with density ρ_2 and velocity c_2. A point source is assumed to be situated at $x = x_s = 0$ and $z = z_s$. We will assume that the source lies in layer 1, but the problem with the source in the second layer can be handled similarly. Figure 1 shows the geometry of the problem.

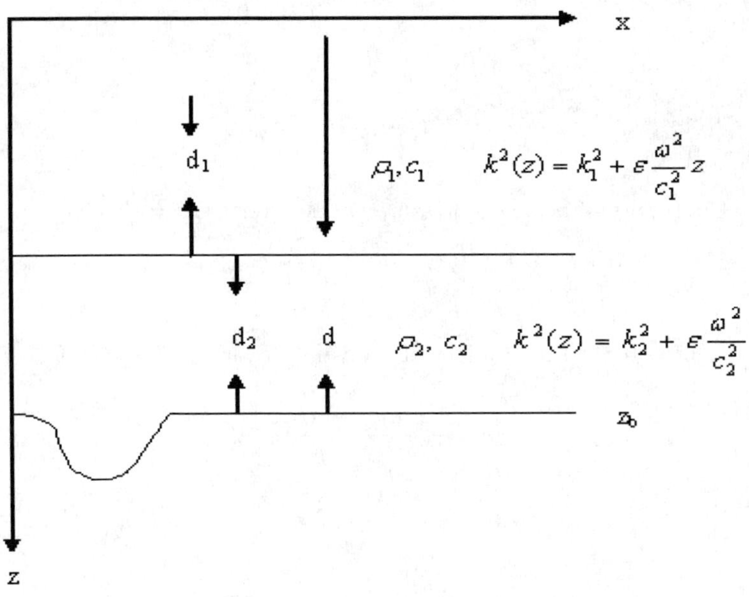

Figure 1: Geometry of the problem

The top layer has depth $z = d_1$, and the seabed has depth

$$z_b = \begin{cases} d_2 + \epsilon h(x) & 0 \leq x \leq L, \text{ where } \quad 0 < \epsilon < 1. \\ d_2 & \text{otherwise} \end{cases} \tag{2.1}$$

The acoustic pressure $p^{(i)}$, density $\rho^{(i)}$, velocity c_i, and wavenumber k_i refer to the quantities in the i-th layer, $i = 1, 2$. The Helmholtz equation in layer 1 is

$$\frac{\partial^2 p^{(1)}}{\partial x^2} + \frac{\partial^2 p^{(1)}}{\partial z^2} + k_1^2(z)p^{(1)} = -2\pi\delta(z - z_s), \tag{2.2}$$

where $k_1 = \frac{\omega}{c_1}$ is the wavenumber, ω is the angular frequency, and z_s is the source depth.

The Helmholtz equation in layer 2 is

$$\frac{\partial^2 p^{(2)}}{\partial x^2} + \frac{\partial^2 p^{(2)}}{\partial z^2} + k_2^2(z)p^{(2)} = 0, \tag{2.3}$$

where $k_2 = \frac{\omega}{c_2}$ is the wavenumber.

In case the seabed is assumed to be rigid, we have the following boundary conditions.

1. Free surface at $z = 0$ gives

$$p^{(1)}(x, 0) = 0. \tag{2.4}$$

2. Continuity of acoustic pressure at the interface $z = d_1$ gives

$$p^{(1)}(x, d_1) = p^{(2)}(x, d_1). \tag{2.5}$$

3. Continuity of the gradient of acoustic pressure at the interface gives

$$\frac{1}{\rho_1}\frac{\partial p^{(1)}(x, d_1)}{\partial z} = \frac{1}{\rho_2}\frac{\partial p^{(2)}(x, d_1)}{\partial z}. \tag{2.6}$$

4. Rigid bottom at seabed $z = z_b$ gives

$$\frac{\partial p^{(2)}}{\partial n}(x, z_b) = 0.$$

Here $\dfrac{\partial}{\partial n}$ denotes the derivative in the normal direction to the seabed. Using (2.1), we can re–write it as

$$\frac{\partial p^{(2)}(x, z_b)}{\partial z} - \epsilon\frac{dh}{dx}\frac{\partial p^{(2)}(x, z_b)}{\partial x} = 0. \tag{2.7}$$

3 Solution of the Problem

Let us introduce the one–dimensional Fourier transform pair

$$\hat{f}(\xi, z) = \frac{1}{2\pi}\int_{-\infty}^{\infty} f(x, z)e^{i\xi x}dx,$$

$$f(x, z) = \int_{-\infty}^{\infty} \hat{f}(\xi, z)e^{-i\xi x}d\xi. \tag{3.1}$$

Taking the Fourier transform of the Helmholtz equations (2.2) and (2.3), we get

$$\frac{d^2\hat{p}^{(1)}(\xi, z)}{dz^2} + \alpha_1^2\hat{p}^{(1)}(\xi, z) = -\delta(z - z_s), \tag{3.2}$$

and

$$\frac{d^2\hat{p}^{(2)}(\xi, z)}{dz^2} + \alpha_2^2\hat{p}^{(2)}(\xi, z) = 0, \tag{3.3}$$

where $\alpha_i^2 = k_i^2 - \xi^2, \quad i = 1, 2$, and the boundary conditions (2.4)-(2.7), in the Fourier transform domain become the following boundary conditions.

1F. Free surface at $z = 0$ gives

$$\hat{p}^{(1)}(\xi, 0) = 0. \tag{3.4}$$

2F. Continuity of acoustic pressure at the interface $z = d_1$ gives

$$\hat{p}^{(1)}(\xi, d_1) = \hat{p}^{(2)}(\xi, d_1). \tag{3.5}$$

3F. Continuity of the derivative of acoustic pressure at the interface gives

$$\frac{1}{\rho_1}\frac{d\hat{p}^{(1)}(\xi, d_1)}{dz} = \frac{1}{\rho_2}\frac{d\hat{p}^{(2)}(\xi, d_1)}{dz}. \tag{3.6}$$

4F. Rigid bottom at seabed at $z = z_b$ gives

$$\frac{d\hat{p}^{(2)}(\xi, z)}{dz} - \epsilon[i\xi\hat{h}(\xi) * i\xi\hat{p}^{(2)}(\xi, z)] = 0. \tag{3.7}$$

Here $\hat{h}(\xi)$ is the Fourier transform of $h(x)$ and * denotes the convolution operation defined as

$$g(t) = f_1(\xi) * f_2(\xi) = \int_{-\infty}^{\infty} f_1(\xi)f_2(t-\xi)d\xi = \int_{-\infty}^{\infty} f_1(t-\xi)f_2(\xi)dx. \tag{3.8}$$

We shall use Green's function to solve our transformed Helmholtz equations (3.2) and (3.3). Green's functions associated with these equation satisfy

$$\frac{d^2G^{(1)}(\xi, z_s, z)}{dz^2} + \alpha_1^2 G^{(1)}(\xi, z_s, z) = -\delta(z - z_s), \tag{3.9}$$

and

$$\frac{d^2G^{(2)}(\xi, z_s, z)}{dz^2} + \alpha_2^2 G^{(2)}(\xi, z_s, z) = 0, \tag{3.10}$$

together with the following boundary conditions.

1G. Free surface boundary conditions give

$$G^{(1)}(\xi, z_s, 0) = 0. \tag{3.11}$$

2G. Continuity of the Green's function at the source location gives

$$G^{(1)}(\xi, z_s, z_s^+) = G^{(1)}(\xi, z_s, z_s^-), \tag{3.12}$$

where z_s^- and z_s^+ are above and below the source depth, respectively.

3G. Jump discontinuity in the derivative of Green's function at the source depth gives

$$\frac{dG^{(1)}(\xi, z_s, z_s^+)}{dz} - \frac{dG^{(1)}(\xi, z_s, z_s^-)}{dz} = -1. \tag{3.13}$$

4G. Continuity of Green's functions at the interface $z = d_1$gives

$$G^{(1)}(\xi, z_s, d_1) = G^{(2)}(\xi, z_s, d_1). \tag{3.14}$$

5G. Continuity of the derivative of Green's function at the interface $z = d_1$gives

$$\frac{1}{\rho_1}\frac{dG^{(1)}}{dz}(\xi, z_s, d_1) = \frac{1}{\rho 2}\frac{dG^{(2)}}{dz}(\xi, z_s, d_1). \tag{3.15}$$

6G. Condition at rigid seabed gives

$$\frac{dG^{(2)}(\xi, z_s, z_b)}{dn} = 0. \tag{3.16}$$

Green's function solution for the two layers is

$$G^{(1)}(\xi, z_s, z) = \begin{cases} A\sin\alpha_1 z + B\cos\alpha_1 z & 0 \le z \le z_s \\ C\sin\alpha_1 z + D\cos\alpha_1 z & z_s \le z \le d_1 \end{cases} \tag{3.17}$$

for layer 1 and

$$G^{(2)}(\xi, z_s, z) = E\sin\alpha_2\, z + F\cos\alpha_2 z \quad d_1 \le z \le z_b \tag{3.18}$$

for layer 2. The conditions (3.11)–(3.16) stated above lead to

$$A\sin\alpha_1 z_s - C\sin\alpha_1 z_s - D\cos\alpha_1 z_s = 0. \tag{3.19}$$

$$-\alpha_1 A\cos\alpha_1 z_s\alpha_1 + C\cos\alpha_1 z_s - \alpha_1 D\sin\alpha_1 z_s = 1. \tag{3.20}$$

$$\frac{1}{\rho_1}[\alpha_1 C\cos\alpha_1 d_1 - \alpha_1 D\sin\alpha_1 d_1] - \frac{1}{\rho_2}[\alpha_2 E\cos\alpha_2 d_2 - \alpha_2 F\sin\alpha_2 d_2] = 0. \tag{3.21}$$

$$C\sin\alpha_1 d_1 + D\cos\alpha_1 d_1 - E\sin\alpha_2 d_1 - F\cos\alpha_2 d_1 = 0. \tag{3.22}$$

$$EM_1 - FM_2 = 0, \text{ where} \tag{3.23}$$

$$M_1 = \alpha_2\cos\alpha_2 z_b - \epsilon R_1(\xi, z_b) \tag{3.24}$$

and

$$M_2 = \alpha_2\sin\alpha_2 z_b + \epsilon R_2(\xi, z_b), \tag{3.25}$$

where

$$R_1(\xi, z_b) = -(i\xi\hat{h}(\xi)) * (i\xi\sin\alpha_2 z_b) \tag{3.26}$$

$$R_2(\xi, z_b) = -(i\xi\hat{h}(\xi)) * (i\xi\cos\alpha_2 z_b). \tag{3.27}$$

In order that these equations have a nontrivial solution, the determinant of the coefficients of the unknown quantities $A, C, D, E,$ and F must vanish. Denoting the determinant by Det, we have

$$Det = \left(\frac{\alpha_1^2}{\rho_1}\right) M_2\cos\alpha_1 d_1\sin\alpha_2 d_1 + \left(\frac{\alpha_1^2}{\rho_1}\right) M_1\cos\alpha_1 d_1\cos\alpha_2 d_1$$
$$-\left(\frac{\alpha_1\alpha_2}{\rho_2}\right) M_2\cos\alpha_2 d_1\sin\alpha_1 d_1 + \left(\frac{\alpha_1\alpha_2}{\rho_2}\right) M_1\sin\alpha_2 D_1\sin\alpha_1 D_1 = 0. \tag{3.28}$$

This is the characteristic equation, and the zeros $\xi^2 = \xi_n^2$ are the eigenvalues of the problem. Solving for these constants, we obtain

$$A = \left(\frac{1}{Det}\right)\Bigg[\left(\frac{\alpha_1}{\rho_1}\right) M_2\sin\alpha_2 d_1\cos\alpha_1(z_s - d_1)$$
$$+\left(\frac{\alpha_1}{\rho_1}\right) M_1\cos\alpha_2 d_1\cos\alpha_1(z_s - d_1)$$
$$+\left(\frac{\alpha_2}{\rho_2}\right) M_2\cos\alpha_2 d_1\sin\alpha_1(z_s - d_1)$$

$$+\left(\frac{\alpha_2}{\rho_2}\right) M_1 \sin\alpha_2 d_1 \sin\alpha_1 (d_1 - z_s), \tag{3.29}$$

$$C = \left(\frac{1}{Det}\right)\left[-\left(\frac{\alpha_1}{\rho_1}\right)\sin\alpha_1 z_s \sin\alpha_1 d_1 (M_2 \sin\alpha_2 d_1 + M_1\cos\alpha_2 d_1)\right], \tag{3.30}$$

$$D = \left(\frac{1}{Det}\right)\left[\left(\frac{\alpha_1}{\rho_1}\right)\sin\alpha_1 z_s \cos\alpha_1 d_1 (M_2\sin\alpha_2 d_1 + M_1 \cos\alpha_2 d_1) \right. \tag{3.31}$$

$$\left. -\left(\frac{\alpha^2}{\rho_2}\right) M_2 \cos\alpha_2 d_1 \sin\alpha_1 d_1 + \left(\frac{\alpha_2}{\rho_2}\right) M_1 \sin\alpha_2 d_1 \sin\alpha_1 d_1\right],$$

$$E = \left(-\frac{1}{Det}\right)\left(\frac{\alpha_1}{\rho_1}\right) M_2 \sin\alpha z_s, \tag{3.32}$$

and

$$F = \left(\frac{1}{Det}\right)\left(\frac{\alpha_1}{\rho_1}\right) M_1 \sin\alpha_1 z_s. \tag{3.33}$$

Substituting the values for $A, C, D, E,$ and F in (3.17) and (3.18), we can write the solution for (3.2), and (3.3) which, upon taking the inverse Fourier transform and the Cauchy residue theorem, we obtain for $z_s \leq z \leq d$.

$$p^{(1)}(x,z) = 2\pi i \sum_{n=1}^{\infty} \left.\frac{\sin\alpha_{1,n} z}{\frac{\partial Det}{\partial\xi^2}}\right|_{\xi^2=\xi_n^2}$$
$$\times \left[\left(\frac{\alpha_{1,r}}{\rho_1}\right) M_{2,n} \sin\alpha_{2,n} d_1 \cos\alpha_{1,n}(z_s - d_1)\right.$$
$$+\left(\frac{\alpha_{1,r}}{\rho_1}\right) M_{1,n}\cos\alpha_{2,n} d_1 \cos\alpha_{1,n}(z_s - d_1)$$
$$+\left(\frac{\alpha_{2,n}}{\rho_2}\right) M_{2,n}\cos\alpha_{2,n} d_1 \sin\alpha_{1,n}(z_s - d_1)$$
$$\left. +\left(\frac{\alpha_{2,n}}{\rho_2}\right) M_{1,n}\sin\alpha_{2,n} d_1 \sin\alpha_{1,n}(d_1 - z_s)\right] e^{-i\xi_n x}. \tag{3.34}$$

For $z_s \leq z \leq d_1$, we can obtain in a similar way

$$p^{(2)}(x,z) = 2\pi i \sum_{n=1}^{\infty} \left.\frac{1}{\frac{\partial Det}{\partial\xi^2}}\right|_{\xi^2=\xi_n^2}$$

$$\times \left[-\left(\frac{\alpha_1}{\rho_1}\right)\sin\alpha_{1,n} z \sin\alpha_{1,n} z_s \sin\alpha_{1,n} d_1\right.$$

$$\times(M_{2,n}\sin\alpha_{2,n}d_1 + M_{1,n}\cos\alpha_{2,n}d_1)$$
$$+\left\{\left(\frac{\alpha_{1,n}}{\rho_1}\right)\sin\alpha_{1,n}z_s\cos\alpha_{1,n}d_1(M_{2,n}\sin\alpha_{2,n}d_1 + M_{1,n}\cos\alpha_{2,n}d_1)\right)$$
$$-\left(\frac{\alpha_{2,n}}{\rho_2}\right)M_{2,n}\cos\alpha_{2,n}d_1\sin\alpha_{1,n}d_1$$
$$+\left(\frac{\alpha_{2,n}}{\rho_2}\right)(M_{1,n}\sin\alpha_{2,n}d_1\sin\alpha_{1,n}d_1)\}\cos\alpha_{1,n}z\Big]e^{-i\xi_n x}. \tag{3.35}$$

For $d_1 \leq z \leq z_b$, we have

$$p^{(3)}(x,z) = 2\pi i\sum_{n=1}^{\infty}\frac{1}{\frac{\partial Det}{\partial\xi^2}}\Bigg|_{\xi^2=\xi_n^2}\left[-\left(\frac{\alpha_{1,n}}{\rho_1}\right)M_{2,n}\sin\alpha_{2,n}z\sin\alpha_{1,n}z_s\right.$$
$$\left.+M_{1,n}\cos\alpha_{2,n}z\sin\alpha_{1,n}z_s\right]e^{-i\xi_n x}, \tag{3.36}$$

where $\xi^2 = \xi_n^2$ are the zeros of the determinant.

Now,

$$\frac{\partial Det}{\partial\xi^2} = (M_2\sin\alpha_2 d_1 + M_1\cos\alpha_2 d_1)\left(\frac{-1}{\rho_1}\right)\cos\alpha_1 d_1$$
$$+\left(\frac{\alpha_1}{2\rho_1}\right)d_1\sin\alpha_1 d_1 + \left(\frac{\alpha_1^2}{\rho_1}\right)\cos\alpha_1 d_1\left(\frac{\partial M_2}{\partial\xi^2}\right)\sin\alpha_2 d_1$$
$$- M_2\left(\frac{d_1}{2\alpha_2}\right)(\cos\alpha_2 d_1) + \left(\frac{\partial M_1}{\partial\xi^2}\right)\cos\alpha_2 d_1$$
$$+ M_1\left(\frac{d_1}{\alpha_2}\right)(\sin\alpha_2 d_1) + (M_1\sin\alpha_2 d_1 - M_2\cos\alpha_2 d_1)$$
$$-\left(\frac{\alpha_2}{2\alpha_1\rho_2}\right)\sin\alpha_1 d_1 - \left(\frac{\alpha_1}{2\alpha_2\rho_2}\right)\sin\alpha_1 d_1 \tag{3.37}$$
$$-\left(\frac{\alpha_2 d_1}{2\rho_2}\right)\cos\alpha_1 d_1 + \left(\frac{\alpha_1\alpha_2}{\rho_2}\right)\sin\alpha_1 d_1\left(\frac{\partial M_1}{\partial\xi^2}\right)\sin\alpha_2 d_1$$
$$- M_1\left(\frac{d_1}{2\alpha_2}\right)(\cos\alpha_2 d_1) - \left(\frac{\partial M_2}{\partial\xi^2}\right)\cos\alpha_2 d_1$$
$$- M_2\left(\frac{d_1}{2\alpha_2}\right)(\sin\alpha_2 d_1),$$

where $\frac{\partial M_1}{\partial\xi^2}$ and $\frac{\partial M_2}{\partial\xi^2}$ can be found from equations (3.24)–(3.27) as follows:

$$\frac{\partial M_1}{\partial\xi^2} = -\left(\frac{1}{2\alpha_2}\right)\cos\alpha_2 z_b + z_b\sin\alpha_2\frac{z_b}{2}$$
$$+\epsilon\left[(\xi\hat{h}(\xi) * \left(\cos\frac{\alpha_2 z_b}{2\xi} + \left(\frac{z_b\xi^2}{2\alpha_2}\right)\sin\alpha_2 z_b\right)\right] \tag{3.38}$$

and

$$\frac{\partial M_2}{\partial \xi^2} = -\left(\frac{\xi}{\alpha_2}\right)\sin\alpha_2 z_b - z_b\xi\cos\alpha_2 z_b \tag{3.39}$$

$$-\epsilon\left[(\xi\hat{h}(\xi)) * \left(\cos\frac{\alpha_2 z_b}{2\xi} + \left(\frac{z_b\xi}{2\alpha_2}\right)\sin\alpha_2 z_b\right)\right].$$

In Section 4, we study the acoustic pressure for the model in Section 3, but the rigid bottom condition is replaced by the reflecting–type condition, as this gives a more realistic ocean model.

4 Reflecting–Type Boundary Condition

In this section, we assume that the sea bottom is of a reflecting type as this is a more realistic case. The two–layer model discussed in Section 3 is studied with the rigid boundary condition replaced by the reflecting or impedance boundary condition.

To avoid undue repetitions, we do not repeat the boundary conditions (2.4)–(2.6), their corresponding Fourier transformed boundary conditions (3.4)–(3.6) and their Green's functions (3.11)–(3.16) which are the same in this case. However, the rigid boundary condition at the seabed,

$$\frac{\partial p^{(2)}(x, z_b)}{\partial n} = 0, \tag{4.1}$$

is replaced by the reflecting condition

$$\frac{\partial p^{(2)}(x, z_b)}{\partial n} = \alpha p^{(2)}(x, z_b), \tag{4.2}$$

where α is the reflection coefficient $-1 \leq \alpha \leq 1$.

Equation (4.2) can be rewritten as

$$\frac{\partial p^{(2)}(x, z_b)}{\partial z} - \epsilon\frac{dh}{dz}\frac{\partial p^{(2)}(x, z_b)}{\partial x} = \alpha p^{(2)}(x, z_b). \tag{4.3}$$

Taking the Fourier transform of (4.3) gives

$$\frac{d\hat{p}^{(2)}(\xi, z_b)}{dz} + \epsilon\left[(\xi\hat{h}(\xi)) * (\xi\hat{p}^{(2)}(\xi, z_b))\right] = \alpha\hat{p}^{(2)}(\xi, z_b). \tag{4.4}$$

Equation (4.4) can be written as

$$E(\alpha_2\cos\alpha_2 z_b - \alpha\sin\alpha_2 z_b) - F(\alpha_2\sin\alpha_2 z_b + \alpha\cos\alpha_2 z_b) \tag{4.5}$$

$$= -\epsilon \left[(\xi \hat{h}(\xi)) * (\xi \{ E \sin \alpha_2 z_b + F \cos \alpha_2 z_b \}) \right].$$

Equation (4.5) can be written as

$$EM_3 - FM_4 = 0, \tag{4.6}$$

where

$$M_3 = \alpha_2 \cos \alpha_2 z_b - \alpha \sin \alpha_2 z_b - \epsilon R_1(\xi, z_b) \tag{4.7}$$

$$M_4 = \alpha_2 \sin \alpha_2 z_b + \alpha \cos \alpha_2 z_b + \epsilon R_2(\xi, z_b), \tag{4.8}$$

and $R_1(\xi, z_b)$ and $R_2(\xi, z_b)$ are as defined in (3.26) and (3.27), respectively. The constants A, C, D, E and F and the determinant Det are the same as determined in Section 3, but with the terms M_1 and M_2 replaced by M_3 and M_4, respectively as derived in (4.7) and (4.8).

The acoustic pressure $p^{(1)}, p^{(2)}$ and $p^{(3)}$ will be the same expressions as determined in (3.34) to (3.36) with the terms $\frac{\partial M_1}{\partial \xi^2}$ and $\frac{\partial M_2}{\partial \xi^2}$ replaced by $\frac{\partial M_3}{\partial \xi^2}$ and $\frac{\partial M_4}{\partial \xi^2}$, respectively, which can be derived from (4.7) and (4.8) as follows:

$$\frac{\partial M_3}{\partial \xi^2} = -\left(\frac{1}{2\alpha_2}\right) \cos \alpha_2 z_b + \frac{z_b}{2} \sin \alpha_2 z_b + \frac{\alpha}{2\alpha_2} z_b \sin \alpha_2 z_b \tag{4.9}$$

$$+\epsilon \left[(\xi \hat{h}(\xi)) * \left(\cos \frac{\alpha_2 z_b}{2\xi} + \left(\frac{z_b \xi^2}{2\alpha_2}\right) \sin \alpha_2 z_b \right) \right],$$

and

$$\frac{\partial M_4}{\partial \xi^2} = -\left(\frac{\xi}{\alpha_2}\right) \sin \alpha_2 z_b - z_b \xi \cos \alpha_2 z_b + \frac{\alpha z_b}{2\alpha_2} \sin \alpha_2 z_b$$

$$-\epsilon \left[(\xi \hat{h}(\xi)) * \left(\cos \frac{\alpha_2 z_b}{2\xi} + \left(\frac{z_b \xi}{2\alpha_2}\right) \sin \alpha_2 z_b \right) \right]. \tag{4.10}$$

If we take the reflection coefficient to be a random function of $x, \alpha = \alpha(x, \gamma)$, where γ is a random parameter, the boundary condition (4.4) is replaced by

$$\frac{d\hat{p}^{(2)}}{dz} + \epsilon \left[(\xi \hat{h}(\xi)) * (\xi \hat{p}^{(2)}(\xi, z_b)) \right] = \hat{\alpha}(\xi, \gamma) * \hat{p}^{(2)}(\xi, z_b), \tag{4.11}$$

where $\hat{\alpha}$ is the Fourier transform of α.

Equation (4.6) can be written as

$$EM_5 - FM_6 = 0, \tag{4.12}$$

where

$$M_5 = \alpha_2 \cos \alpha_2 z_b - \hat{\alpha} * \sin \alpha_2 z_b - \epsilon\, R_1(\xi, z_b) \tag{4.13}$$

$$M_6 = \alpha_2 \sin \alpha_2 z_b + \hat{\alpha} * \cos \alpha_2 z_b + \epsilon R_2(\xi, z_b), \tag{4.14}$$

and $\dfrac{\partial M_5}{\partial \xi^2}$ and $\dfrac{\partial M_6}{\partial \xi^2}$ appearing in $\dfrac{\partial Det}{\partial \xi^2}$ can be derived similarly as

$$\frac{\partial M_5}{\partial \xi^2} = -\left(\frac{1}{2\alpha_2}\right) \cos \alpha_2 z_b + \frac{z_b}{2} \sin \alpha_2 z_b + \hat{\alpha} * \left(\frac{z_b}{2\alpha_2} \sin \alpha_2 z_b\right)$$

$$+\epsilon \left[(\xi \hat{h}(\xi)) * \left(\cos \frac{\alpha_2 z_b}{2\xi} + \left(\frac{z_b \xi^2}{2\alpha_2}\right) \sin \alpha_2 z_b\right)\right], \tag{4.15}$$

and

$$\frac{\partial M_6}{\partial \xi^2} = -\left(\frac{\xi}{\alpha_2}\right) \sin \alpha_2 z_b - z_b \xi \cos \alpha_2 z_b + \hat{\alpha} * \left(\frac{z_b}{2\alpha_2}\right) \sin \alpha_2 z_b$$

$$-\epsilon \left[(\xi \hat{h}(\xi)) * \left(\cos \frac{\alpha_2 z_b}{2\xi} + \left(\frac{z_b \xi}{2\alpha_2}\right) \sin \alpha_2 z_b\right)\right]. \tag{4.16}$$

If $\alpha = 0$ (rigid case), then all the equations derived in this section become the same equations as those derived in Section 3. In Section 5, we apply he acoustic pressure derived in Sections 3 and 4 for some particular smooth undulation functions. We take the perturbations as sine, linear, and quadratic functions. Also, we implement the reflecting–type condition for the same model discussed in this section.

5 Numerical Results

In this section, we implement the acoustic pressure equations (3.34–3.36) derived in Section 3 for a two-layered model with smooth undulations in the sea bottom. The seabed perturbation shape is taken to be sine, quadratic, and linear functions, as these are structurally realistic shapes of ocean layers. The perturbation functions are assumed to be smooth and differentiable possessing Fourier transforms. We show the effect of these perturbations on the acoustic pressure and compare the results with the case of a horizontal seabed.

5.1 Parabolic Perturbation Sea Bottom

If we take the sea bottom as $z_b = d + \epsilon h(x)$, ϵ is a small parameter $0 \leq \epsilon < 1$, and if

$$h(x) = \begin{cases} \alpha \sin\left(\frac{x\pi}{L}\right) & 0 \leq x \leq L \\ 0 & \text{otherwise} \end{cases},$$

then the Fourier transform of $h(x)$ is

$$\hat{h}(\xi) = \frac{\alpha}{2\pi^2} L \left[\frac{\xi(e^{i\xi L} + 1)}{1 - \left(\frac{\xi L}{\pi}\right)^2} \right], \quad \frac{\xi L}{\pi} \neq 1. \tag{5.1}$$

Figure 2 shows the pressure plots for the two-layered model with and without sine undulations in the seabed with range x. We take $L = 20$ meters, $\alpha = 25$, and $\epsilon = 1$ in Figure 2. There are obvious differences between the sound pressure for the sine perturbed and unperturbed sea bottom. The accuracy of the calculated pressure relies principally on the precision of the calculated eigenvalues.

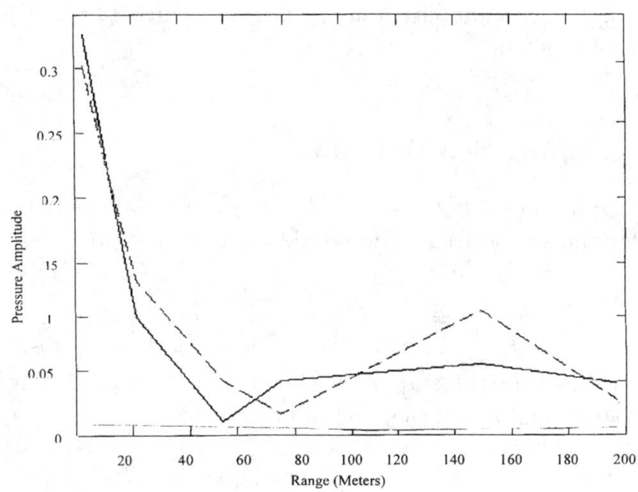

Figure 2: Pressure amplitude for horizontal (solid) and sine perturbed (dashed)

Figure 3 shows the effect of ϵ on the pressure amplitude when the range $x = 10$ meters. It is obvious that the pressure amplitude decreases with increasing ϵ, as expected.

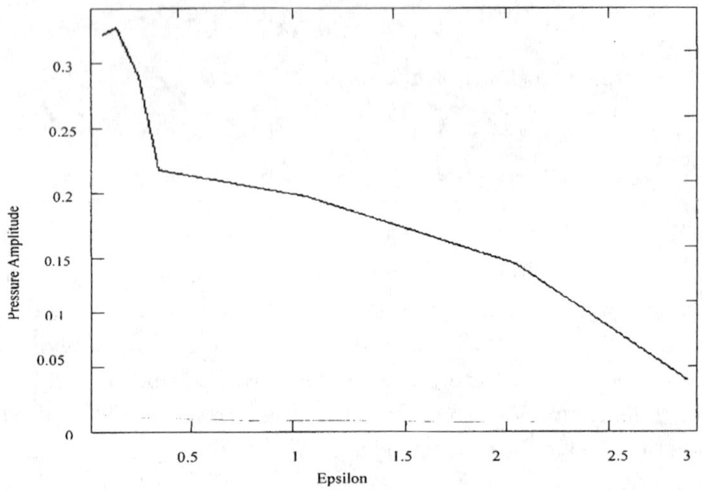

Figure 3: Pressure amplitude for different perturbation parameter values at range 10 meters

5.2 Dipping Sea Bottom

In this case, we take $h(x) = ax, \quad 0 \leq x \leq L, L = 200$ meters, and α is the slope for the sea bottom. The Fourier transform of $h(x)$ defined by (3.1) is

$$\hat{h}(\xi) = \frac{\alpha}{2\pi}\left[L\frac{e^{i\xi L}}{i\xi} + \frac{e^{i\xi L}}{\xi^2} - \frac{1}{\xi^2}\right] \xi \neq 0. \qquad (5.2)$$

Figure 4 shows the pressure plots when $\alpha = 1$ (bottom sloping at an angle of 45 degrees and $\alpha = 0$ (horizontal bottom)).

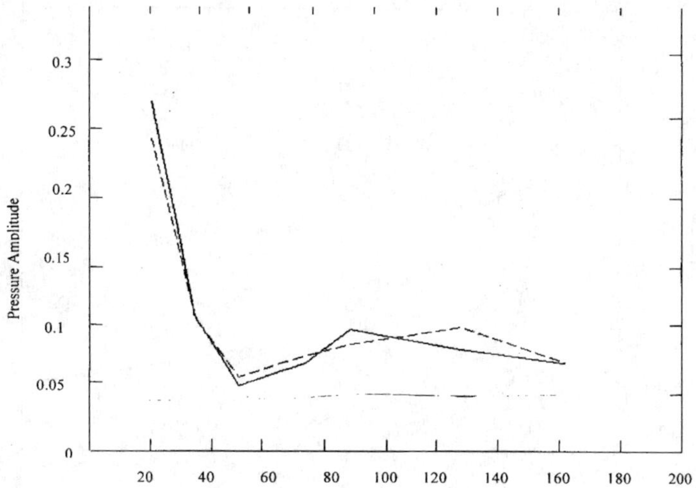

Figure 4: Pressure amplitude for horizontal (solid) and linear perturbed (dashed)

5.3 Quadratic Perturbation Sea Bottom

If we have

$$h(x) = \begin{cases} x(L-x) & 0 \le x \le L \\ 0 & \text{otherwise}, \end{cases} \tag{5.3}$$

then $h(x)$ has the Fourier transform

$$\hat{h}(\xi) = \frac{1}{2\pi}\left[-L\frac{e^{i\xi L}}{\xi^2} + \frac{2e^{i\xi L}}{i\xi^2} - \frac{L}{\xi^2} - \frac{2}{i\xi^3}\right] \xi \neq 0.$$

Figure 5 shows the plot for the acoustic pressure for the same model with the seabed having the quadratic shape defined above. We take $\epsilon = 0$ (horizontal), $\epsilon = .01$, and $\epsilon = .1$. The three plots coincide with each other beyond $L = 20$ meters where the effect of undulation does not exist.

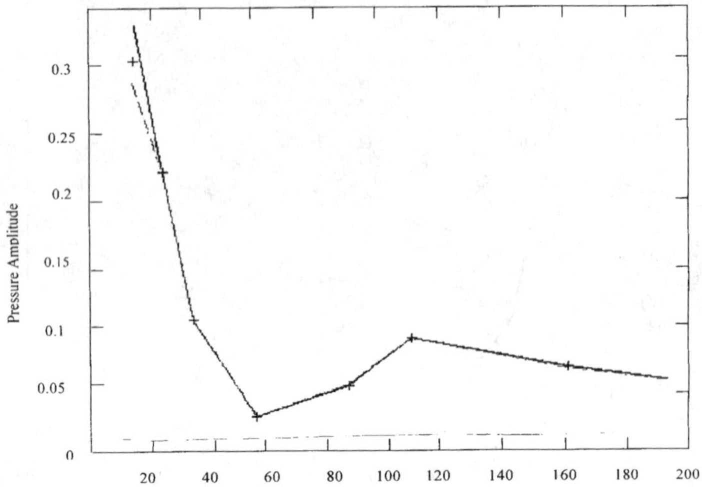

Figure 5: Pressure amplitude for horizontal (solid) and quadratic perturbed (dashed)

Acknowledgment. The first author wishes to acknowledge the support provided by the King Fahd University of Petroleum & Minerals, Dhahran, Saudi Arabia.

References

[1] D.J. Ahluwalia and J.B. Keller, Exact and asymptotic representations of the sound field in a stratified ocean, In *Wave Propagation and Underwater Acoustic*, Edited by J.B. Keller and J.S. Papadakis, Lecture Notes in Physics, Vol. **70** , Springer-Verlag, Heidelberg, 1977.

[2] C. A. Boyles, *Acoustic Waveguide, Applications to Oceanic Science*, John Wiley & Sons, New York, 1984.

[3] M.D. Duston, C. Verma, and D. Wood, *Changes in Eigenvalues to Bottom Interaction Using Perturbation Theory, Numerical Mathematics and Applications,*, Elsevier Science Publishers, Amsterdam; New York, 1986.

[4] P.C. Etter, *Underwater Acoustic Modeling Principles, Techniques and Applications*, Elsevier Applied Sciences, New York, 1986.

[5] W. Ricardo, *Spectral Theory and Scattering Theory for Wave Propagation in Perturbed Stratified Media*, Applied Mathematical Sciences, Springer-Verlag, Heidelberg, Vol. **87**, 1991.

[6] F.D. Zaman and Z. Al-Muhiameed, Acoustic waves in a layered inhomogeneous ocean, *Applied Acoustics*, **61** (2000), 427–440.

Chapter 13

NON-LINEAR PLANAR OSCILLATION OF A SATELLITE LEADING TO CHAOS UNDER THE INFLUENCE OF THIRD-BODY TORQUE

R. Bhardwaj and R. Tuli

Guru Gobind Singh Indraprastha University

Abstract

For thousand of years humans have noticed that small causes could have large effects and that it was hard to predict anything for certain. What had caused a stir among scientists was that in some systems small changes of initial conditions could lead to predictions so different that prediction itself became useless. Henri Poincare explained the chaos and showed the sensitive dependence on initial conditions. The non-linear planar oscillation of a satellite in an elliptic orbit under the influence of third-body torque is being studied in this chapter. By using Melnikov's method, we have shown that the equations of motion are non-integrable. Using the BKM method, it is observed that the amplitude of the oscillation remains constant up to the second order of approximation. The main and parametric resonances have been shown to exist and have been studied by the BKM method. The analysis regarding the stability of the stationary planar oscillation of a satellite near the resonance frequency shows that discontinuity occurs in the amplitude of oscillation at a frequency of the external periodic force which is less than the frequency of natural oscillation. The half-width of the chaotic separatrix has been estimated by Chrikov's criterion. Through Poincare's surface of section method, it has been observed that the third-body torque parameter (ε), the eccentricity of the orbit (e), and the mass ratio (n) play an important role in changing the regular motion into a chaotic one. The theory is applied to the rotational motion of Hyperion, a satellite of Saturn. It is observed that Hyperion tumbles chaotically, and as the third-body torque parameter increases, it tumbles more chaotically.

1 Introduction

Planar oscillations of a satellite in a elliptic orbit have been studied by Beletskii [2], Zltanstov et al. [36], Singh [31], Bhatnagar et. al. [3], Maciejewski [26]. None of them have taken the third body effect. Bhatnagar and Bharadwaj [4,6], have taken the effect of the third-body torue, but the satellite is assumed to move in an almost circular orbit. Bharadwaj, Tuli [7], have modified the problem by taking the orbit of the satellite as elliptic. We have determined hyperbolic equilibrium solution and double asymptotic solutions corresponding to unperturbed Hamiltonian H_0. The non-integrability of the system has been shown through Melnikov's integral [25]. Poincare [29], has discovered that transversal crossing of asymptotic surfaces of unstable periodic solution leads to complex structure of phase curves. Melnikov's result has allowed us to formulate theorems about non-integrability of systems with transversal homoclinic (hetrolinic) orbits. With the help of the theory developed in Smale [30], Wisdom et al. [33], Wiggins [34,35] and Henon [18] we have also made a study for the Earth-Moon-Artificial Satellite (1958 B2 Vanguard 1) system. Resonance is a relationship in which the orbital period of one body is related to another by a simple integer fraction. Novak [28], Chiang [10] have studied the resonance effects in the solar system. Kne ˆ zevic, Milani, Farnella [19] have studied the 5:2 mean motion resonance, Varadi [32] has studied about periodic orbits in 3:2 orbitalresonance and their stability and Nauenberg [27] has studied the stability and eccentricity of period orbits for two planets in 1:1 resonance. Konaki, Maciejewcki, Wolszczan [20], Marcy et al. [23], Beauge et al. [1] have studied resonance in extrasolar planets. Bruce [8] has shown that the 3:2 spin orbit resonance of Mercury is a potential contributer to the global thermal budget. Goldreich and

Peale [16] have studied the resonant rotation of Venus. Cynthia Phillips [12] have studied resonance in the satellites and asteroids of Jupiter. Dermott et al. [15], Cuk et al. [11], have studied resonance in the Uranium and Saturnium satellite systems. Lotko and Strelstov [22] have studied the magnetospheric resonance. Haghighipour [17], have discussed resonance in a planar circular restricted three-body system. Michtchencko [24] have studied the resonant structure of the outer solar system in the neighbourhood of the planets. Bharadwaj and Bhatnagar [6] have studied the resonance in the nonlinear planar oscillation of a satellite moving under the influence of a third-body torque in an almost circular orbit using BKM method by Bogoliubov [9,21] and Cayley [13]. The width of the chaotic zone has been estimated by Chirikov [14].

2 Equation of Motion

Let us consider a rigid satellite moving in an elliptic orbit (semi-major axis a, eccentricity e) under the influence of a central body of mass M and its moon of mass m whose orbit is assumed circular and coplanar with the orbit of the satellite (fig 1). The satellite is assumed to be a triaxial ellipsoid with principal moments of inertia $A < B < C$, and C is the moment of inertia about the spin axis which is perpendicular to the orbital plane. We approximate the influence of the moon by resolving the potential of the torque is with respect to the r/R ratio, where r is the radius of the satellite and R is the radius of moon's orbit.

Let the true anomaly be v, and the orientation of the satellite's long axis be θ. Then $\theta - v = \delta/2$ measures the orientation of the satellite's long axis relative to the satellite's radius vector. Let e be the eccentricity of the satellite's elliptical orbit, m be the mass of the moon $\approx 7.348 \times 10^{22}$ kg, M be the mass of the earth $\approx 5.9742 \times 10^{24}$ kg, Ω be the angular velocity of the moon $= \frac{2\pi}{30\times24\times60\times60}$ rad/s, and l be the length of semi-latus rectum of the satellite's orbit.

The equation of motion, which we have finally obtained, is

$$[1 + e\cos v]\frac{d^2q}{dv^2} - 2e\sin v\frac{dq}{dv} - 4e\sin v + n^2\sin q - n^2\varepsilon(1 - 3e\cos v)$$
$$[\sin(cv - q) + de\sin v\cos(cv - q)] = 0, \qquad (2.1)$$

where ε is the parameter due to the third-body torque $n^2 = 3\frac{B-A}{C}$, $c = 2\left[1 - \Omega\sqrt{\frac{l^3}{\mu}}\right]$, $d = 4\Omega\sqrt{\frac{l^3}{\mu}}, \in = \frac{m}{M}\Omega^2\frac{l^3}{\mu}$, $q = \delta$, and μ = gravitational constant of the earth $\cong GM \approx 3.986005 \times 10^5$ km^3/s^2.

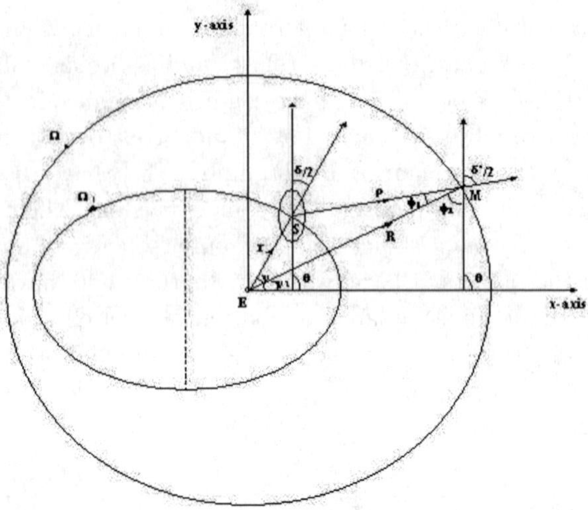

Figure 1: Satellite's motion in elliptic orbit with third-body perturbation.

2.1 Hamilton's Equation

Equation (2.1) is equivalent to Hamilton's equations

$$\frac{dq}{dv} = \frac{\partial H}{\partial p}, \frac{dp}{dv} = -\frac{\partial H}{\partial p}, \tag{2.2}$$

where p is the generalized momenta. Taking e of the order of ϵ (parameter due to the third-body torque), $e = e_1 \in$, $(0 < e_1 << 1, 0 <\in<< 1.)$ The Hamiltonian function H can be written as $H = H_0 + \in H_1 + 0\left(\in^2\right)$.

In our problem

$$H = -2p + \frac{p^2}{2} - n^2 \cos q + \in \left[-p^2 e_1 \cos v - n^2 \cos(cv - q) - \right.$$
$$\left. -n^2 e_1 \cos v \cos q\right],$$

where

$$\begin{aligned} H_0 &= -2p + \frac{p^2}{2} - n^2 \cos q \\ H_1 &= b - p^2 e_1 \cos v - n^2 \cos(cv - q) + n^2 e_1 \cos v \cos q. \end{aligned}$$

With $\in$ being small, $0\left(\in^2\right)$ and higher are neglected.

2.2 Equilibrium and Double Asymptotic Solution

The equilibrium solutions corresponding to H_0 are given by

$$\frac{dq}{dv} = 0, \quad \frac{dp}{dv} = 0.$$

Corresponding to H_0, we get from (2.2)

$$\frac{dq}{dv} = p - 2, \quad \frac{dp}{dv} = -n^2 \sin q$$

which gives us $p = 2$ and $q = 0, \pi$. This in the phase space $[q, p]$; and $[0, 2]$ and $[\pi, 2]$ are the equilibrium points. We have further calculated that $[0, 2]$ is stable, whereas $[\pi, 2]$ is unstable. Also $\frac{dp}{dq} = \pm\frac{n^2 \sin q}{p-2}$, which on integration gives

$$\frac{p^2}{2} - 2p = n^2 \cos q + n^2 - 2, \text{but} p = 2 + \frac{dq}{dv}.$$

Hence, the unperturbed double asymptotic solutions at $[\pi, 2]$ are obtained as

$$\begin{aligned} p^{\pm}(v) &= 2 \pm \frac{2n}{\cosh(nv)} \\ \sin\left[q^{\pm}(v)\right] &= \pm\frac{\sinh(nv)}{\cosh^2(nv)} \\ \cos\left[q^{\pm}(v)\right] &= \frac{2}{\cosh^2(nv)} - 1. \end{aligned}$$

2.3 Melnikov's Function

Melnikov's function is obtained as

$$M^{\pm}(v_0) = \pm 2\pi e_1 \sin v_0 \left[4 \sec\mathrm{h}\frac{\pi}{2n} + 3 \cos\mathrm{ech}\frac{\pi}{2n}\right] \pm 2\pi c^2 \sin cv_0$$

$$\left[\sec\mathrm{h}\frac{\pi c}{2n} + \cos\mathrm{ech}\frac{\pi c}{2n}\right]. \qquad (2.3)$$

It is easy to observe that for any value of the mass parameter $n > 0$ and the third-body torque parameter e_1 $(0 < e_1 << 1)$ the above function has a simple zero. Thus, the equations in (2.2) are non-integrable.

2.4 Graphical Representation of Melnikov's Function

We have studied the graphical representation of Melnikov's function in the Earth-Moon-Artificial Satellite (1958 B2 Vanguard 1) system. For the Artificial Satellite (1958 B2 Vanguard 1) the data is eccentricity $e = 0.190$, semi-major axes $a = 8676$ km. In our problem $c = 2\left[1 - \Omega\sqrt{\frac{l^3}{\mu}}\right]$. For the Earth-Moon-Artifical Satellite (1958 B2 Vanguard 1) system $c = 1.9999$.

For this fixed value of c we have studied the graphs of Melnikov's function.

1. Figures 2, 3, 4, 5, 6, 7, 8 and 9 illustrate the graphs of $M^+ (v_0, e_1, n)$ and $M^- (v_0, e_1, n)$ for $0 \leq v_0 \leq 10$.

   We observe from the figures that as V_0 changes from 0 to 10, the Melnikov's functions $M^+ (v_0, n, e_1)$ and $M^- (v_0, n, e_1)$ behave almost like sine functions. They have simple zeroes. Also, we observe that in each graph when $3 \leq v_0 \leq 3.5$, $M^+ (v_0, n, e_1)$ changes sign from positive to negative and when $6 \leq v_0 \leq 6.5$, $M^+ (v_0, n, e_1)$ changes sign from negative to positive. Thus, in all these abscissae remain almost the same. We also observe from Table 1 that as e_1 and n both vary from $e_1 = 0.2, 0.4, 0.6, 0.8$, $n = 0.1, 0.3, 0.5, 0.7$, $M^+ (v_0, n, e_1)$ and $M^- (v_0, n, e_1)$ elongate along the ordinate, with the abscissa remaining almost the same.

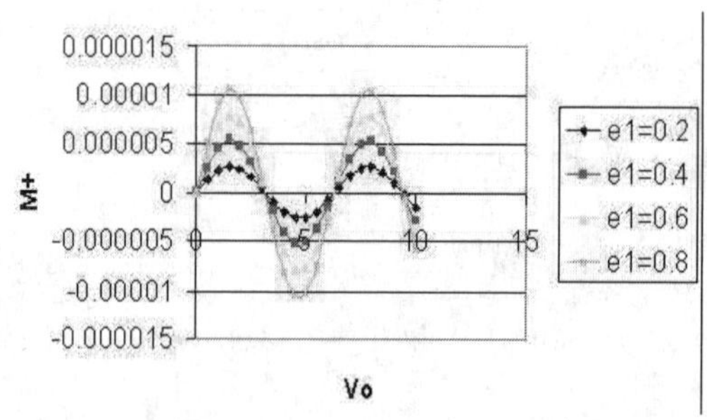

Figure 2: Melnikov's function $M^+(v_0, n, e_1)$ for $c = 1.9999$, $0 \leq v_0 \leq 10$, $n = 0.1$, $e_1 = 0.2,\ 0.4,\ 0.6,\ 0.8$.

Table 1: Figure 2–5 illustrating $M^{+}(R_0, e_1, x)$ and $M^{-}(r_0, e_1, n)$ for $0 \le r_0 \le 10$.

Fig. no.	n	e_1	$M^{+}(v_0, n, e_1)$		$M^{-}(v_0, n, e_1)$	
			Max. value	Min. value	Max. value	Min. value
2	0.1	0.2	2.66E-06	-2.61E-06	2.61E-06	-2.66E-06
	0.1	0.4	5.32E-06	-5.22E-06	5.22E-06	-5.32E-06
	0.1	0.6	7.99E-06	-7.83E-06	7.83E-06	-7.99E-06
3	0.3	0.2	0.094	-0.091	0.091	-0.-94
	0.3	0.4	0.187	-0.182	0.182	-0.187
	0.3	0.6	0.281	-0.274	0.274	-0.281
	0.3	0.8	0.375	-0.366	0.366	-0.375
4	0.5	0.2	0.836	-0.832	0.832	-0.836
	0.5	0.4	1.550	-1.562	1.562	-1.550
	0.5	0.6	2.303	-2.291	2.291	-2.303
	0.5	0.8	3.062	-3.021	3.021	-3.062
5	0.7	0.2	2.598	-2.447	2.447	-2.598
	0.7	0.4	4.234	-4.190	4.190	-4.234
	0.7	0.6	5.982	-5.977	5.977	-5.982
	0.7	0.8	7.730	-7.765	7.765	-7.730

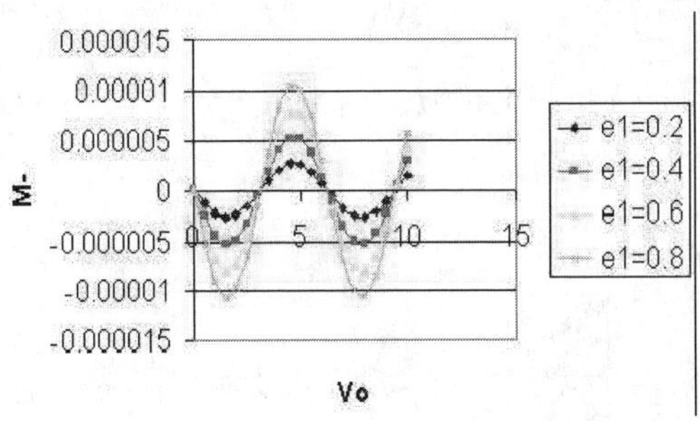

Figure 3: Melnikov's function $M^{-}(v_0, n, e_1)$ for $c = 1.9999$, $0 \le v_0 \le 10$, $n = 0.1$, $e_1 = 0.2,\ 0.4,\ 0.6,\ 0.8$.

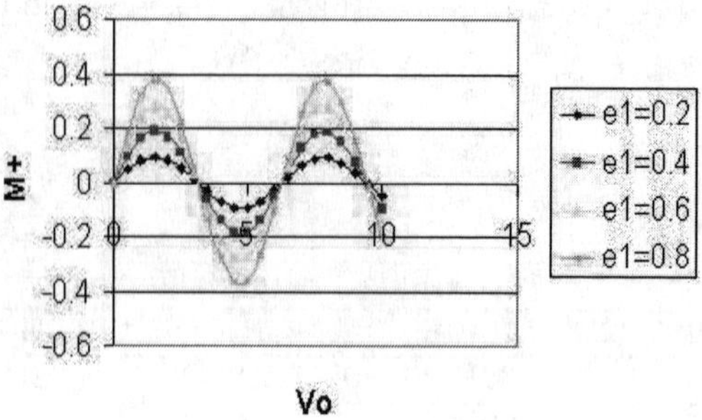

Figure 4: Melnikov's function $M^+(v_0, n, e_1)$ for $c = 1.9999$, $0 \le v_0 \le 10$, $n = 0.3$, $e_1 = 0.2, 0.4, 0.6, 0.8$.

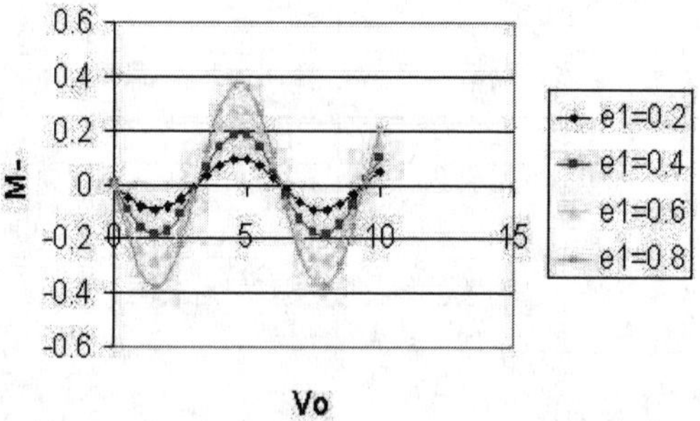

Figure 5: Melnikov's function $M^-(v_0, n, e_1)$ for $c = 1.9999$, $0 \le v_0 \le 10$, $n = 0.3$, $e_1 = 0.2, 0.4, 0.6, 0.8$.

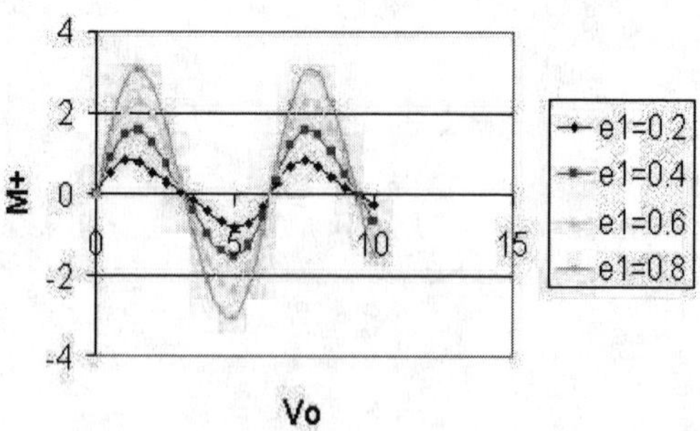

Figure 6: Melnikov's function $M^+(v_0, n, e_1)$ for $c = 1.9999$, $0 \le v_0 \le 10$, $n = 0.5$, $e_1 = 0.2, 0.4, 0.6, 0.8$.

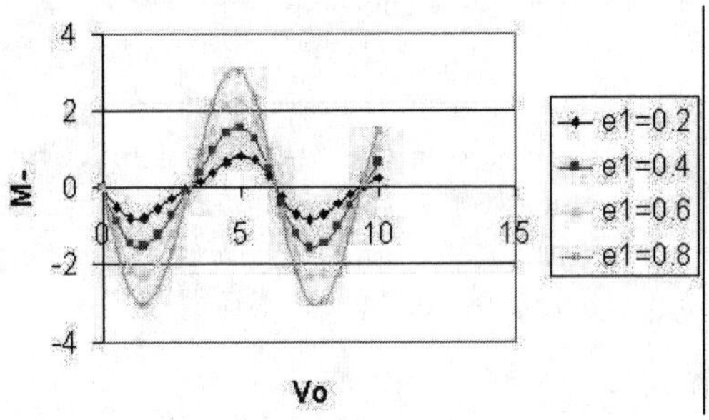

Figure 7: Melnikov's function $M^{-}(v_0, n, e_1)$ for $c = 1.9999$, $0 \leq v_0 \leq 10$, $n = 0.5$, $e_1 = 0.2, 0.4, 0.6, 0.8$.

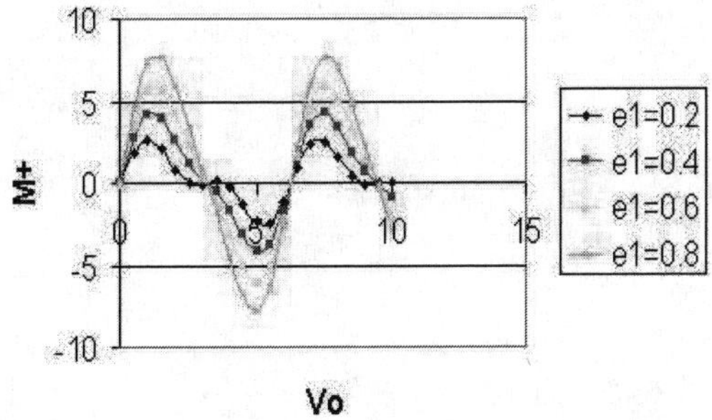

Figure 8: Melnikov's function $M^{+}(v_0, n, e_1)$ for $c = 1.9999$, $0 \leq v_0 \leq 10$, $n = 0.7$, $e_1 = 0.2, 0.4, 0.6, 0.8$.

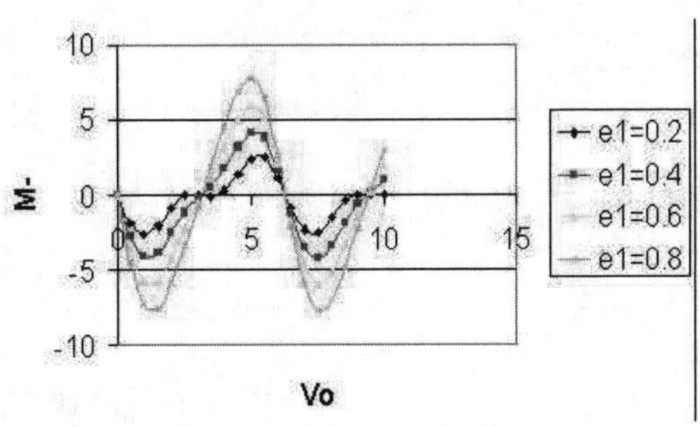

Figure 9: Melnikov's function $M^{-}(v_0, n, e_1)$ for $c = 1.9999$, $0 \leq v_0 \leq 10$, $n = 0.7$, $e_1 = 0.2, 0.4, 0.6, 0.8$.

2. Figures 10 and 11 illustrate the graphs of $M^{\pm}(v_0, e_1, n)$ for $0 \leq e_1 \leq 0.9$, $n = 0.1$, and $v_0 = 0.2, 0.4, 0.6, 0.8$. It has been observed that as the parameter due to the third-body effect $'e_1'$ changes from 0 to 0.9 $(0 \leq e_1 \leq 0.9)$, the value of the Melnikov's function $M^{+}(v_0, e_1, n)$ increases monotonically for each value of v_0. Also, as we vary both v_0 and e_1, $v_0 = 0.2, 0.4, 0.6$ and 0.8 and $0 \leq e_1 \leq 0.9$, the graphs elongate along the ordinate.

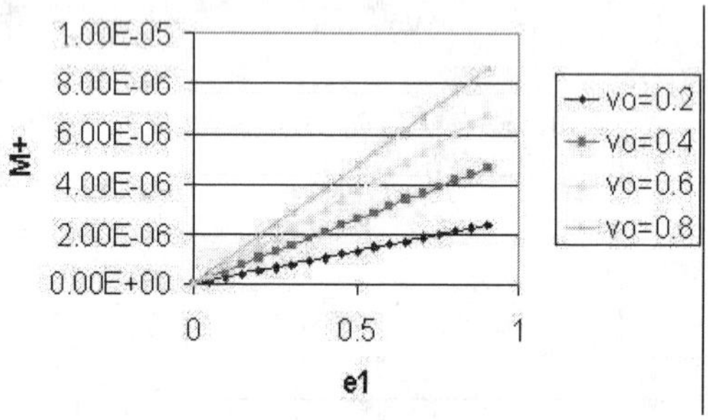

Figure 10: Melnikov's function $M^{+}(v_0, n, e_1)$ for $c = 1.9999$, $0 \leq e_1 \leq 0.9$, $n = 0.1$, $v_0 = 0.2, 0.4, 0.6, 0.8$.

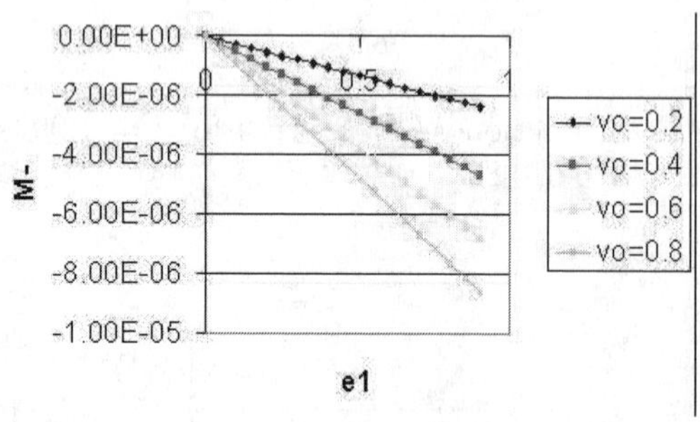

Figure 11: Melnikov's function $M^{-}(v_0, n, e_1)$ for $c = 1.9999$, $0 \leq e_1 \leq 0.9$, $n = 0.1$, $v_0 = 0.2, 0.4, 0.6, 0.8$.

3. Figures 12 and 13 illustrate the graphs of $M^{\pm}(v_0, e_1, n)$ for $0.1 \leq n \leq 0.9$, $v_0 = 0.1$, and $e_1 = 0.2, 0.4, 0.6, 0.8$. We have observed that as the mass distribution parameter $'n'$ of the satellite changes from 0.1 to 0.9 ($0.1 \leq n \leq 0.9$), the value of $M^{+}(v_0, e_1, n)$ for each e_1 initially increases very slowly and then exponential increases to $+\infty$. Also, the value of $M^{-}(v_0, e_1, n)$ for each e_1 initially decreases very slowly and then exponentially decreases to $-\infty$. Also, as we vary both e_1 and n, $e_1 = 0.2, 0.4, 0.6, 0.8$ and $0.1 \leq n \leq 0.9$, the graphs elongate along the ordinate.

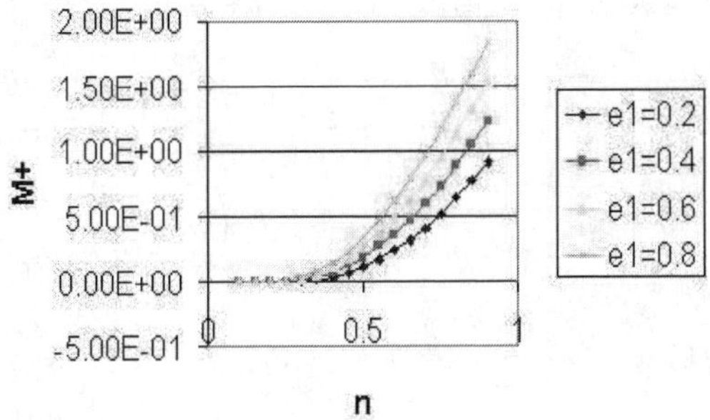

Figure 12: Melnikov's function $M^{+}(v_0, n, e_1)$ for $c = 1.9999$, $0.1 \leq n \leq 0.9$, $v_0 = 0.1$, $e_1 = 0.2, 0.4, 0.6, 0.8$.

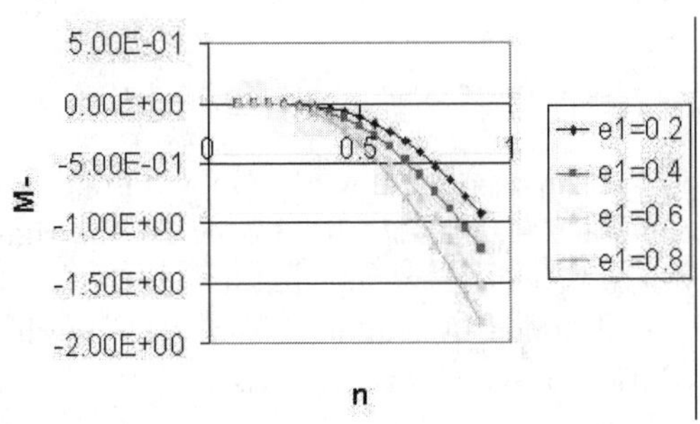

Figure 13: Melnikov's function $M^{+}(v_0, n, e_1)$ for $c = 1.9999$, $0.1 \leq n \leq 0.9$, $v_0 = 0.1$, $e_1 = 0.2, 0.4, 0.6, 0.8$.

3 Non-Resonant Planar Oscillations of A Satellite

Equation (2.1) can be rewritten as

$$\frac{d^2q}{dv^2} + n^2 q = 4e\sin v + 2e\sin v\frac{dq}{dv} - e\cos v\frac{d^2q}{dv^2} + n^2\left[q - \sin q\right] + n^2 \in$$
$$\{1 - 3e\cos v\}\sin(cv - q) + n^2 \in de\,(1 - 3e\cos v)\sin v\cos(cv - q)\,. \quad (3.1)$$

In (3.1) the non-linearity $n^2\left[q - \sin q\right]$ is sufficiently weak, and therefore, it can be taken to be of the order of $\in$ (the third-body torque parameter). So by taking $n^2 = \alpha \in$ and also $e = e_1 \in$, we get

$$\frac{d^2q}{dv^2} + n^2 q = \in f_1\left(v,\, q,\, \frac{dq}{dv},\, \frac{d^2q}{dv^2}\right) + \in^2 f_2\left(v,\, q,\, \frac{dq}{dv},\, \frac{d^2q}{dv^2}\right)$$
$$+ \in^2 f_3\left(v,\, q,\, \frac{dq}{dv},\, \frac{d^2q}{dv^2}\right) + \in^4 f_4\left(v,\, q,\, \frac{dq}{dv},\, \frac{d^2q}{dv^2}\right), \quad (3.2)$$

where

$$f_1\left(v,\, q,\, \frac{dq}{dv},\, \frac{d^2q}{dv^2}\right) = 4e_1\sin v + 2e_1\sin v\frac{dq}{dv} - e_1\cos v\frac{d^2q}{dv^2} + \alpha q - \alpha\sin q$$

$$f_2\left(v,\, q,\, \frac{dq}{dv},\, \frac{d^2q}{dv^2}\right) = \alpha\sin(cv - q)$$

$$f_3\left(v,\, q,\, \frac{dq}{dv},\, \frac{d^2q}{dv^2}\right) = -3e_1\alpha\cos v\sin(cv - q) + de_1\alpha\sin v\cos(cv - q)$$

$$f_4\left(v,\, q,\, \frac{dq}{dv},\, \frac{d^2q}{dv^2}\right) = -3\alpha de_1^2\cos v\sin v\cos(cv - q)\,.$$

The dynamical system described by (3.2) moves under forced vibrations due to the presence of the periodic sine forces on the right-hand side of the equation. We are benefited by the smallness of $\in$ in (3.2), and hence the solution can be obtained by exploting the BKM method. For, $\in = 0$, the generating solution of the zeroth order is

$$q = a\cos\Psi, \text{ where } \Psi = nv_1 + \Psi^*,$$

where the amplitude $'a'$ and phase Ψ^* are constants, which can be determined by the initial conditions. Proceeding exactly as in [6], the solution of (3.2) is obtained in the form

$$q = a\cos\Psi + \in u_1(a,\, \Psi,\, v) + \in^2 u_2(a,\, \Psi,\, v)\,, + \in^3 u_3(a,\, \Psi,\, v) + \ldots, \quad (3.3)$$

where $'a'$ and Ψ are determined by the differential equations

$$\frac{da}{dv} = \in A_1(a) + \in^2 A_2(a) + \in^3 A_3(a) + \dots \tag{3.4}$$

$$\frac{d\Psi}{dv} = n + \in B_1(a) + \in^2 B_2(a) + \dots \tag{3.5}$$

and where $A_1(a)$, $B_1(a)$, $u_1(a, \Psi, v)$, $A_2(a)$, $B_2(a)$ etc. are found to be

$A_1(a) = 0,$

$B_1(a) = \frac{\alpha}{2an}[2J_1(a) - a],$

$u_1(a, \Psi, v) = \frac{1}{n^2-1}4e_1 \sin v - \frac{ane_1}{2}\frac{n+2}{2n+1}\cos(v+\Psi) + \frac{ane_1}{2}\frac{n-2}{2n-1}$

$\cos(v-\Psi) + \frac{\alpha}{2n^2}\sum(-1)^k \frac{J_{2k+1}(a)}{k(k+1)}\cos(2k+1)\Psi,$

$A_2(a) = 0,$

$B_2(a) = -\frac{\alpha^2}{8a^2n^3}[2J_1(a) - a]^2 + \frac{3ne_1^2}{4(4n^2-1)}(n^2-1) + \frac{\alpha^2}{2an^3}\sum\frac{1}{k(k+1)}$

$J_{2k+1}(a)J'_{2K+1}(a)$

$u_2(a, \Psi, v) = \cos(v+\Psi)\frac{\alpha e_1}{2n(2n+1)^2}\left[(2J_1(a) - a)(n+1)(n^2-1)\right.$

$\left.+ an^2(n+2)(1 - J_0(a))\right]$

$+\cos(v-\Psi)\frac{\alpha e_1}{2n(2n-1)^2}\left[(2J_1(a) - a)(n-1)(n^2-1) + an^2(n-2)\right.$

$\left.[1 - J_0(a)]\right] +$

$\sin 2v\frac{6e_1^2}{(n^2-4)(n^2-1)} + \sin v\frac{4\alpha e_1(1 - J_0(a))}{(n^2-1)^2} + \sin cv\frac{2J_0(a)}{n^2-c^2}$

$-\cos(2v+\Psi)\frac{ane_1^2(n+2)(n+3)}{16(2n+1)} + \cos(2v-\Psi)\,ane_1^2\frac{(n-2)(n-3)}{16(2n-1)}$

$-\sum_{k=1}^{\infty}(-1)^k J_{2k+1}(a)\cos(2k+1)\Psi$

$\frac{\alpha^2[4k(k+1)(2J_1(a) - a) + 2J_1(a) - aJ_0(a)]}{8an^4k^2(k+1)^2}$

$-\sum_{k=1}^{\infty}(-1)^k J_{2k+1}(a)\cos(v-(2k+1)\Psi)\frac{1}{(2nk-1)[2n(1+k)-1]}$

$\left[\frac{e_1\alpha(2k+1)(2nk+n-2)}{4nk(k+1)}\right]$

$+\sum_{k=1}^{\infty}(-1)^k J_{2k+1}(a)\cos(v-(2k+1)\Psi)\frac{1}{(2nk-1)[2n(1+k)-1]}$

$\left[\frac{ane_1}{2}\alpha\frac{n-2}{2n-1}\right]$

$$-\sum_{k=1}^{\infty}(-1)^k J_{2k+1}(a)\cos(v+(2k+1)\Psi)\frac{1}{(2nk+1)[2n(1+k)+1]}$$
$$\left[\frac{e_1\alpha(2k+1)(2nk+n+2)}{4nk(k+1)}\right]$$
$$-\sum_{k=1}^{\infty}(-1)^k J_{2k+1}(a)\cos(v+(2k+1)\Psi)\frac{1}{(2nk+1)[2n(1+k)+1]}$$
$$\left[\tfrac{ane_1}{2}\alpha\tfrac{n+2}{2n+1}\right]$$
$$-\frac{4e_1\alpha}{n^2-1}k=\sum_{k=1}^{\infty}(-1)^k J_{2k}(a)\sin(v+2k\Psi)\frac{1}{[n(1+2k)+1][n(1-2k)-1]}$$
$$-\frac{4e_1\alpha}{n^2-1}k=\sum_{k=1}^{\infty}(-1)^k J_{2k}(a)\sin(v-2k\Psi)\frac{1}{[n(1+2k)-1][n(1-2k)+1]}$$
$$+\alpha\frac{ane_1}{2}\frac{n+2}{2n+1}k=\sum_{k=1}^{\infty}(-1)^k J_{2k}(a)\cos(v-(2k-1)\Psi)$$
$$\tfrac{1}{(2nk-1)[2n(1-k)+1]}$$
$$-\alpha\frac{ane_1}{2}\frac{n-2}{2n-1}k=\sum_{k=1}^{\infty}(-1)^k J_{2k}(a)\cos(v+(2k-1)\Psi)\frac{1}{(2nk+1)[2n(1-k)+1}$$
$$+\frac{\alpha^2}{2n^2}=\sum_{k,m=1}^{\infty}(-1)^{k+m}\cos(2k+2m+1)\Psi\frac{J_{2k+1}(a)J_{2m}(a)}{k(k+1)4n^2(k+m)(k+m+1)}$$
$$+\frac{\alpha^2}{2n^2}\sum_{k,\,m=1}^{\infty}(-1)^{k+m}\cos(2k-2m+1)\Psi\frac{J_{2k+1}(a)J_{2m}(a)}{k(k+1)4n^2(k-m)(k-m+1)}$$
$$+\alpha k=\sum_{k=1}^{\infty}(-1)^k J_{2k}(a)\sin(cv+2k\Psi)\frac{1}{[n(1+2k)+c][n(1-2k)-c]}$$
$$+\alpha=\sum_{k=1}^{\infty}(-1)^k J_{2k}(a)\sin(cv-2k\Psi)\frac{1}{[n(1-2k)+c][n(1+2k)-c]}$$
$$+\alpha k=\sum_{k=1}^{\infty}(-1)^k J_{2k+1}(a)\cos(cv+(2k+1)\Psi)\frac{1}{(2kn+c)[2n(1+k)+c]}$$
$$+\alpha k=\sum_{k=0}^{\infty}(-1)^k J_{2k+1}(a)\cos(cv-(2k+1)\Psi)\frac{1}{(2kn-c)[2n(1+k)-c]}$$
$$A_3(a)\ =0$$
$$B_3(a)=\frac{\alpha^3}{16a^3n^5}[2J_1(a)-a]^3-\frac{\alpha e_1^2}{2}\frac{n^2-1}{4n^2-1}J_2(a)-\frac{\alpha}{8}a\,ne_1^2\frac{n^2-2}{4n^2-1}J_2'(a)$$
$$+\frac{\alpha e_1^2}{8an(4n^2-1)^2}\left[(2J_1(a)-a)\left[8n^6+6n^4-39n^2+19\right]\right]$$
$$+\frac{\alpha e_1^2}{4(4n^2-1)^2}(n^2-1)(n-2)\left(4n^2+1\right)(1-J_0(a))$$

$$+\frac{\alpha^3}{8a^2n^5}k=\sum_1^\infty \frac{J_{k+1}(a)J'_{2k+1}(a)}{k^2(k+1)^2}\left[6k\left(k+1\right)\left(2J_1\left(a\right)-a\right)+2J_1\left(a\right)-aJ_0\left(a\right)\right]$$
$$-\frac{\alpha^3}{8an^5}k,m=\sum_1^\infty \frac{J_{2k+1}(a)J_{2m(a)}J'_{2k+2m+1}(a)}{k(k+1)(k+m)(k+m+1)}$$
$$-\frac{\alpha^3}{32an^5}k,m=\sum_1^\infty \frac{J_{2k+1}\left(a\right)J_{2m+1}\left(a\right)}{k\left(k+1\right)m\left(m+1\right)}\left[J_{2k+2m+1}\left(a\right)-J_{2k+2m+3}\left(a\right)\right]$$
$$-\frac{\alpha^3}{16an^5}k=\sum_1^\infty \frac{J_{2k+1}\left(a\right)J_{2m}\left(a\right)}{k\left(k+1\right)\left(k-m\right)\left(k-m+1\right)}\left[J_{2m-2k}\left(a\right)+J_{2k-2m}\left(a\right)\right]$$
$$-\frac{\alpha^3}{16an^5}=\sum_1^\infty \frac{J_{2k+1}\left(a\right)J_{2m}\left(a\right)}{k\left(k+1\right)\left(k-m\right)\left(k-m+1\right)}\left[-J_{2m-2k-2}\left(a\right)-J_{2k-2m+2}\left(a\right)\right]$$
(3.6)

$J'_{2k+1}\left(a\right)=\frac{dJ_{2k+1}(a)}{da}$, and $J_k\left(a\right)$ is the Bessel's function of the kth order.

Thus, in the first approximation, the solution is given by

$$q=a\cos\Psi, \tag{3.7}$$

where amplitude $'a'$ phase $'\Psi'$ are given by $\frac{da}{dv}=0\Longrightarrow a=$ constant,

$$\frac{d\Psi}{dv}=n+\in B_1\left(a\right). \tag{3.8}$$

In the second approximation, the solution is obtained as

$$q=a\cos\Psi+\in u_1\left(a,\,\Psi,\,v\right), \tag{3.9}$$

where amplitude $'a'$ phase $'\Psi'$ are given by $\frac{da}{dv}=0\Longrightarrow a=$ constant,

$$\frac{d\Psi}{dv}=n+\in B_1\left(a\right)+\in^2 B_2\left(a\right). \tag{3.10}$$

And in the third approximation, the solution is obtained as

$$q=a\cos\Psi+\in u_1\left(a,\,\Psi,\,v\right)+\in^2 u_2\left(a,\,\Psi,\,v\right), \tag{3.11}$$

where amplitude $'a'$ phase $'\Psi'$ are given by $\frac{da}{dv}=0\Longrightarrow a=$ constant,

$$\frac{d\Psi}{dv}=n+\in B_1\left(a\right)+\in^2 B_2\left(a\right)+\in^3 B_3\left(a\right), \tag{3.12}$$

where the values of $A_1\left(a\right)$, $A_2\left(a\right)$, $A_3\left(a\right)$, $B_1\left(a\right)$, $B_2\left(a\right)$, $B_3\left(a\right)$, $u_1\left(a,\Psi,v\right)$, and $u_2\left(a,\,\Psi,\,v\right)$ are substituted from (3.6). It may be observed from (3.8), (3.10), and (3.12) that the amplitude of the oscillation remains constant even up to the third order of approximation. Moreover, it is observed that the main resonance occurs at $n=\pm1,\pm2,\pm c$. Up to the third approximation, the parametric resonance appears only for $n=\frac{1}{k},\,\frac{c}{k},\,k\in I\sim\{(0)\,,(1)\,,(-1)\}$.

4 Resonant Planar Oscillations of a Satellite

Now we proceed to study the asymptotic solution near the main resonances $n \cong 1, c$ by the BKM method.

4.1 Asymptotic Solution near $\mathbf{n} \cong \mathbf{c}$

For $\in = 0$, the generating solutions are

$$q = a\cos\Psi, \text{ where } \Psi = cv + \theta \tag{4.1}$$

where amplitude $'a'$ and phase θ are determined by the following equations:

$$\frac{da}{dv} = \in A_1(a,\ \theta),\ \frac{d\Psi}{dv} = n + \in B_1(a,\ \theta) = c + \frac{d\theta}{dv},\ \text{where}$$
$$\frac{d\theta}{dv} = n - c + \in B_1(a,\ \theta) \tag{4.2}$$

where $A_1(a,\ \theta)$ and $B_1(a,\ \theta)$ are particular solutions, periodic with respect to θ. Using (4.1) and (4.11), we calculate $\frac{dq}{dv}$ and $\frac{d^2q}{dv^2}$, and then by substituting the values of q, $\frac{dq}{dv}$, and $\frac{d^2q}{dv^2}$ in (3.2) and equating the coefficients of $\in$, we get

$$\cos\Psi\left[-2anB_1 + (n-c)\frac{\partial A_1}{\partial\theta}\right] - \sin\Psi\left[2nA_1 + a(n-c)\frac{\partial B_1}{\partial\theta}\right]$$
$$= 4e_1\sin v - 2e_1 an\sin v\sin\Psi + e_1 an^2\cos v\cos\Psi + \alpha a\cos\Psi - \alpha\sin(a\cos\Psi)\,. \tag{4.3}$$

Using Fourier expansion given by

$$\begin{aligned}
\sin(a\cos(\Psi)) &= 2\sum(-1)^j J_{2j+1}(a)\cos(2j+1)\ \Psi \\
\cos(a\cos(\Psi)) &= J_0(a) + 2\sum(-1)^j J_{2j}(a)\cos(2j)\,\Psi,
\end{aligned}$$

where $J_j(a)$ stands for Bessel's function of order j, in (4.3) and then comparing the coefficients of $\cos(\Psi)$ and $\sin(\Psi)$, we get

$$\begin{aligned}
-2anB_1 + (n-c)\frac{\partial A_1}{\partial\theta} &= \alpha a - 2\alpha J_1(a) \\
2nA_1 + a(n-c)\frac{\partial B_1}{\partial\theta} &= 0.
\end{aligned} \tag{4.4}$$

Solving (4.4), we get $A_1 = \cos\left[\frac{2n}{n-c}\theta + \gamma\right]$

$$B_1 = -\frac{1}{a}\sin\left(\frac{2n}{n-c}\theta + \gamma\right) + \frac{1}{2an}\left[2\alpha J_1(a) - \alpha a\right] \tag{4.5}$$

with γ being the constant of integration. Thus, the solution in the first approximation is obtained as $q = a\cos(cv + \theta)$, where amplitude $'a'$ and phase θ are the solutions of the equations

$$\frac{da}{dv} = \in \cos\left(\frac{2n}{n-c}\theta + \gamma\right) \text{ and}$$
$$\frac{d\theta}{dv} = n - c + \in \left[-\frac{1}{a}\sin\left(\frac{2n}{n-c}\theta + \gamma\right) + \frac{1}{2an}2\alpha J_1(a) - \alpha a\right]. \tag{4.6}$$

Equations (4.6) cannot be integrated in a closed form due to dependence of the right-hand side on a and θ. However, qualitative aspects of the solution can be examined with the help of Poincare theory. The stationary state of oscillation is defined by

$$\frac{da}{dv} = 0 = \frac{d\theta}{dv}. \tag{4.7}$$

Eliminating the phase θ from (4.6), we get

$$16 \in + 16a\delta + na^3 = 0 \tag{4.8}$$

$$\delta = \frac{-na^3 - 16\in}{16a} = -\frac{na^2}{16} - \frac{\in}{a}. \tag{4.9}$$

Let us now examine the relation existing between the parameters of the system for the occurrence of the effect under consideration. Proceeding with (4.8) and differentiating with respect to a, we obtain $16\delta + 3na^2 + 16a\frac{d\delta}{da} = 0$. The necessary condition for instability (jump or fall) is $\frac{d\delta}{da} = 0$, which gives $\delta = -\frac{3na^2}{16} < 0$. Also $c = n + \delta$; therefore, $c < n$. Hence, the effect occurs only at a frequency of the external force which is less than the frequency of natural oscillation of the system. Now the maximum value of amplitude is obtained by the condition $da/d\delta = 0$. From (4.9) we get $\frac{da}{d\delta} \neq 0$. Hence, there is no restriction on the value of a. With the help of (4.8) we have studied the resonance curves for the main resonance at $n = 2$ in the Earth-Moon-Artificial Satellite (1958 B2 Vanguard 1) system. For the Artifical Satellite (1958 B2 Vanguard 1) the eccentricity is 0.190 and the semi-major axis is 8676 km. In our problem $c = 2\left[1 - \Omega\sqrt{\frac{l^3}{\mu}}\right] = 1.99 \cong 2$. We have calculated the resonance curves $a =$

$a(\delta)$ for $\in = 0.001, 0.005, 0.01, 0.05, 0.1,\ 0.15, 0.2, 0.25, 0.3, 0.35, 0.4, 0.45, 0.5, 0.55,\ \ 0.6, 0.65,\ \ 0.7, 0.75, 0.8, 0.85, 0.9$ at $n = 1.99$ in the domain $-1 < \delta < 1$. In this chapter, we have drawn the resonance curves for $\in = 0.001$, 0.3, 0.8 uses (Figure 15 (9(a)–9(c)). For each value of $\in$, there are three branches of $a = a(\delta)$, which have been numbered as 1, 2, and 3. We observe that branches 1 and 2 meet at A and after that the values of a corresponding to δ are imaginary. The imaginary part of a has been shown in Figure 14. The coordinates of the critical point A have been shown in Table 2. As we move from left to right on the branches 1 or 2 of the resonance curve, there is a jump in the amplitude from A to A'. We also observe that as $\in$ increases, this jump also increases. If we move from right to left, the branch 3 is continuous throughout and there is no jump or fall in the amplitude. We have also drawn the three-dimensional (3D) plots of the function $f(a,\ \delta) = 16 \in + 16a\,\delta + na^3 = 0$ in the domain $-1 < \delta < 1$ at $n = 1.9999$ for $e = 0.001$, 0.3, 0.8, uses (Figure 15). The resonance curve may be obtained from the intersection of the plane drawn through $f(a,\ \delta) = 0$. The analysis regarding resonance curves is also confirmed in 3D plots.

Table 2: For $n = 1.99 \cong 2$.

ϵ	$A_\delta = A_\delta$	A_α	A_α	Jump = $A_a - A_a$
0.001	0	-0.200	0.100	0.300
0.005	-0.025	-0.5254	0.2627	0.7881
0.010	-0.025	-0.5811	0.2905	0.8716
0.050	00.125	-1.1597	0.5799	1.7396
0.100	-0.200	-1.4650	0.7325	2.1975
0.150	-0.250	-1.6513	0.8256	2.4769
0.200	-0.300	-1.8119	0.9060	2.7179
0.250	-0.350	-1.99552	0.9776	2.9328
0.300	-0.400	-2.0859	1.0429	3.1288
0.350	-0.450	-2.2066	1.1033	3.3099
0.400	-0.500	-2.3195	1.1597	3.4792
0.450	-0.550	-2.4258	1.2129	3.6387
0.500	-0.575	-2.4912	1.2456	3.73368
0.550	-0.625	-2.5885	1.2942	3.8827
0.600	-0.650	-2.6482	1.3241	3.9723
0.650	-0.700	-2.7385	1.3692	4.1077
0.700	-0.725	-2.7937	1.3969	4.1906
0.750	-0.775	-2.8784	1.4392	4.3176
0.800	-0.800	-2.9300	1.4650	4.3950
0.850	-0.825	-2.9803	1.4901	4.4704
0.900	-0.875	-3.0585	1.5293	4.5878

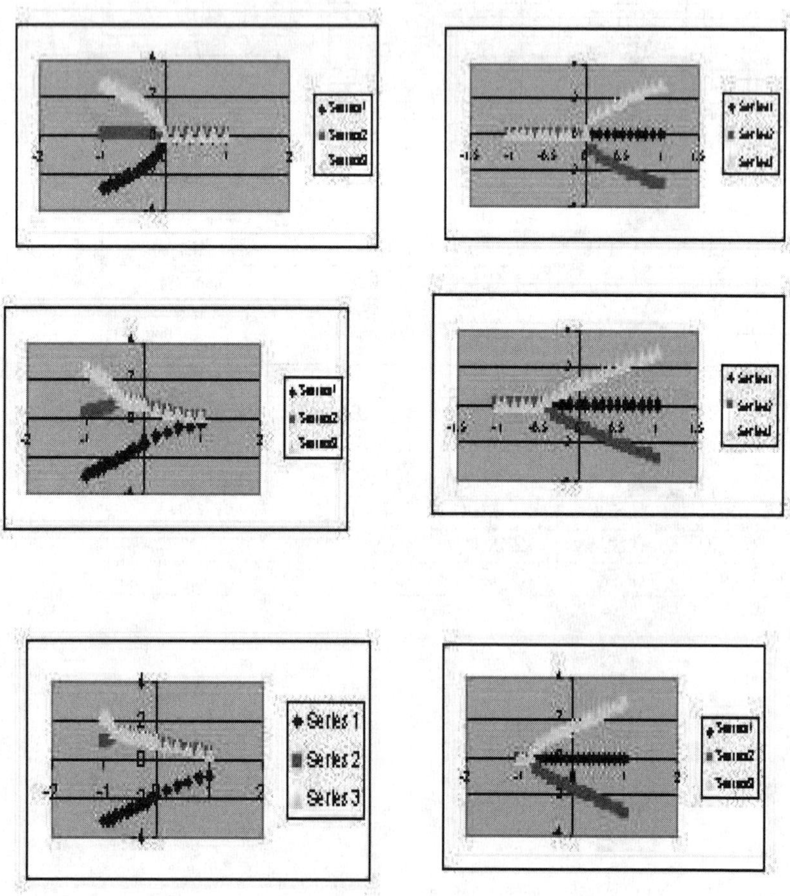

Figure 14: Resonance curves in case of main resonance at $n = c = 1.99$.

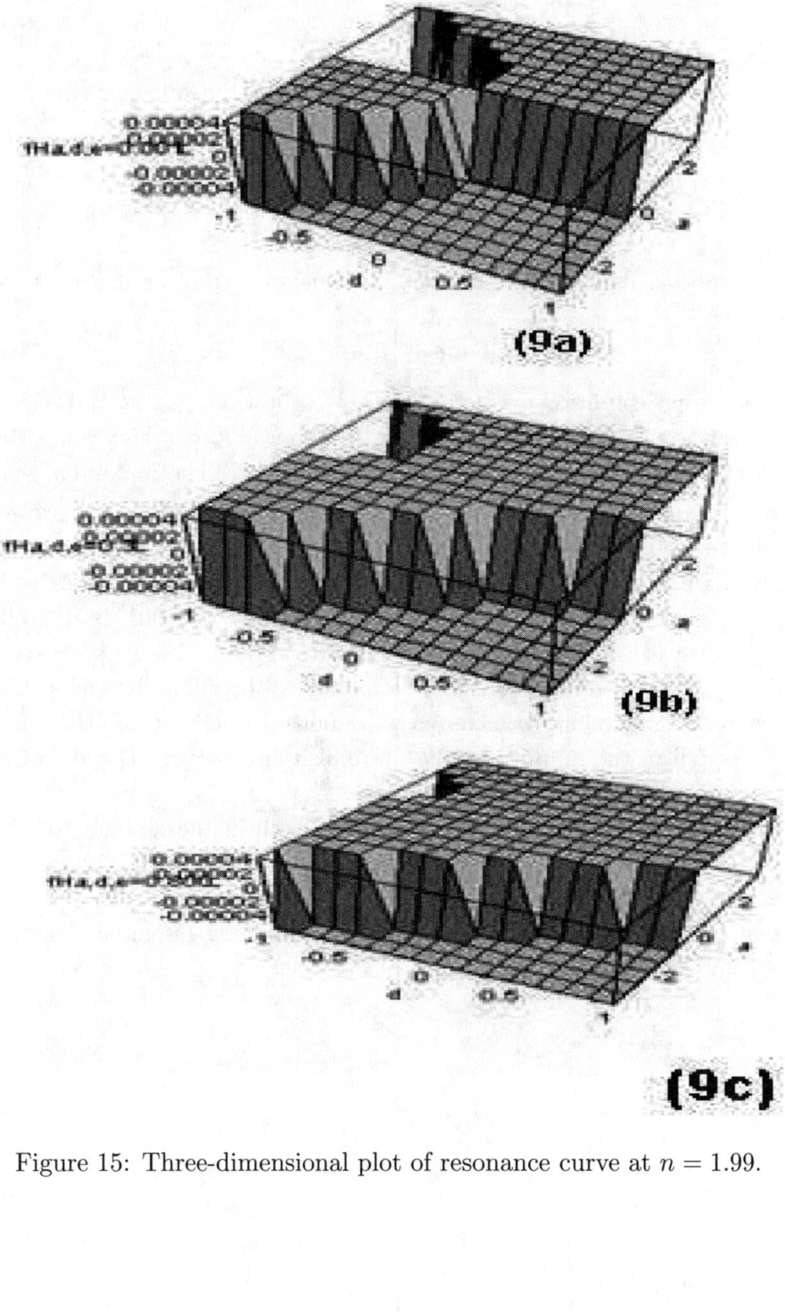

Figure 15: Three-dimensional plot of resonance curve at $n = 1.99$.

4.2 Asymptotic Solution near n ≅ 1

For $\in = 0$, the generating solutions are

$$q = a\cos\Psi, \text{ where } \Psi = v + \theta \tag{4.10}$$

where amplitude $'a'$ and phase θ are determined by the following equations:

$$\frac{da}{dv} = \in A_1(a, \theta), \frac{d\Psi}{dv} = n + \in B_1(a, \theta) = 1 + \frac{d\theta}{dv},$$

$$\frac{d\theta}{dv} = n - 1 + \in B_1(a, \theta). \tag{4.11}$$

Proceeding as in Case (4.1) the stationary state of oscillation is obtained as

$$8a\delta^2 + 16an\delta + \left(n^2a^3 - 32e\right) = 0. \tag{4.12}$$

The necessary condition for instability is $\frac{d\delta}{da} = 0$, which gives $\delta < 0$. Hence, the effect occurs only at a frequency of the external force which is less than the frequency of natural oscillation of the system. The maximum value of amplitude is obtained by $\frac{d\delta}{da} = 0$, which gives $a_{\max} = 2\sqrt{\frac{2}{3}}$, so e is restricted to 0.034 as shown in Table 3.

Now with the help of (4.2.6) we construct the resonance curves. In this chapter, we have drawn the resonance curves for $e = 0.001, 0.018, 0.034$ was (Figure 16). The coordinates of the critical point A have been shown in Table 3. We also observe that as e increases, this jump increases. If we move from right to left, branch 3 is continuous throughout and there is no jump or fall in the amplitude. We have also drawn the 3-D plots of the function.

$f(a, \delta) = 8a\delta^2 + 16an\delta + \left(n^2a^3 - 32e\right) = 0$ in the domain $-0.15 < \delta < 0.15$ at $n = 1$ for $e = 0.001, 0.018, 0.034$ uses (Figure 17 (11.a–11.c)). The resonance curve may be obtained from the intersection of the plane drawn through $f(a, \delta) = 0$. The analysis regarding resonance curves is also confirmed in 3-D plots.

Table 3: For $n = 1$.

E	$A_\delta = A_\delta$	A_α	A_α	Jump = $A_a - A_a$
0.001	-0.01	0.4759	-0.2379	0.7138
0.002	-0.015	0.5889	-0.2945	0.8834
0.004	-0.030	0.7960	-0.3980	1.1940
0.006	-0.040	0.9149	-0.4574	1.3723
0.008	-0.045	0.9821	-0.4910	1.4731
0.010	-0.055	1.0741	-0.5371	1.6112
0.012	-0.060	1.1277	-0.5638	1.6915
0.014	-0.070	1.2052	-0.6026	1.8078
0.016	-0.075	1.2507	-0.6254	1.8761
0.018	-0.080	1.2938	-0.6469	1.9407
0.020	-0.090	1.3588	-0.6794	2.0382
0.022	-0.095	1.3970	-0.6985	2.0955
0.024	-0.100	1.4338	-07169	2.1507
0.026	-0.110	1.4904	-0.7452	2.2356
0.028	-0.115	1.5239	-0.7619	2.2858
0.030	-0.120	1.5562	-0.7781	2.3343
0.032	-0.125	1.5875	-0.7937	2.3812
0.034	-0.130	1.6178	-0.8089	2.4267

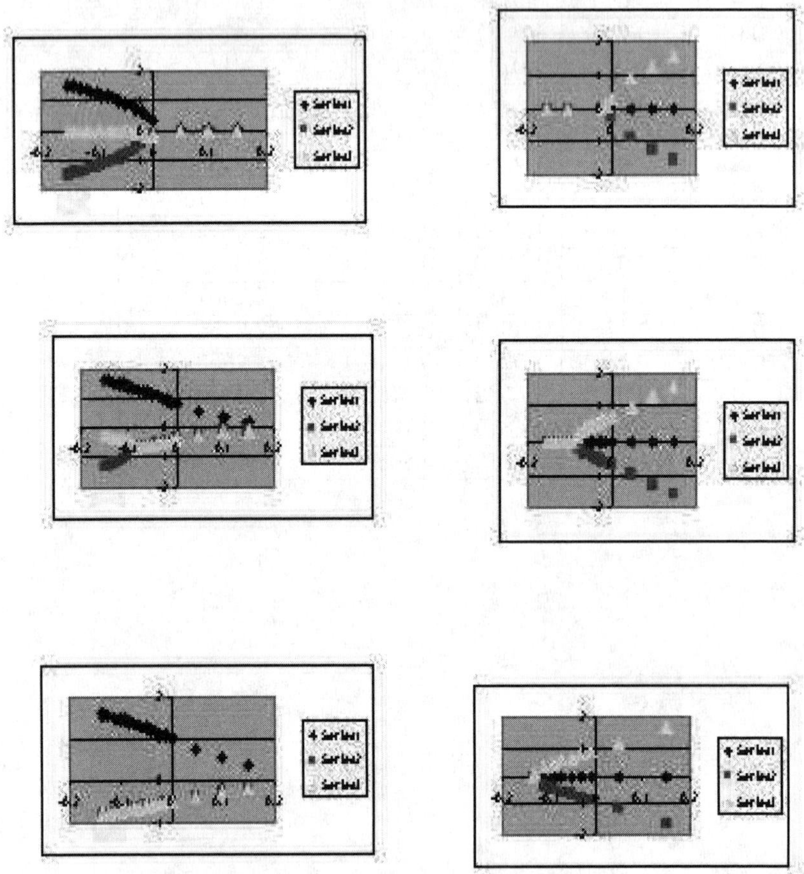

Figure 16: Resonance curves in case of main resonance at $n = 1$.

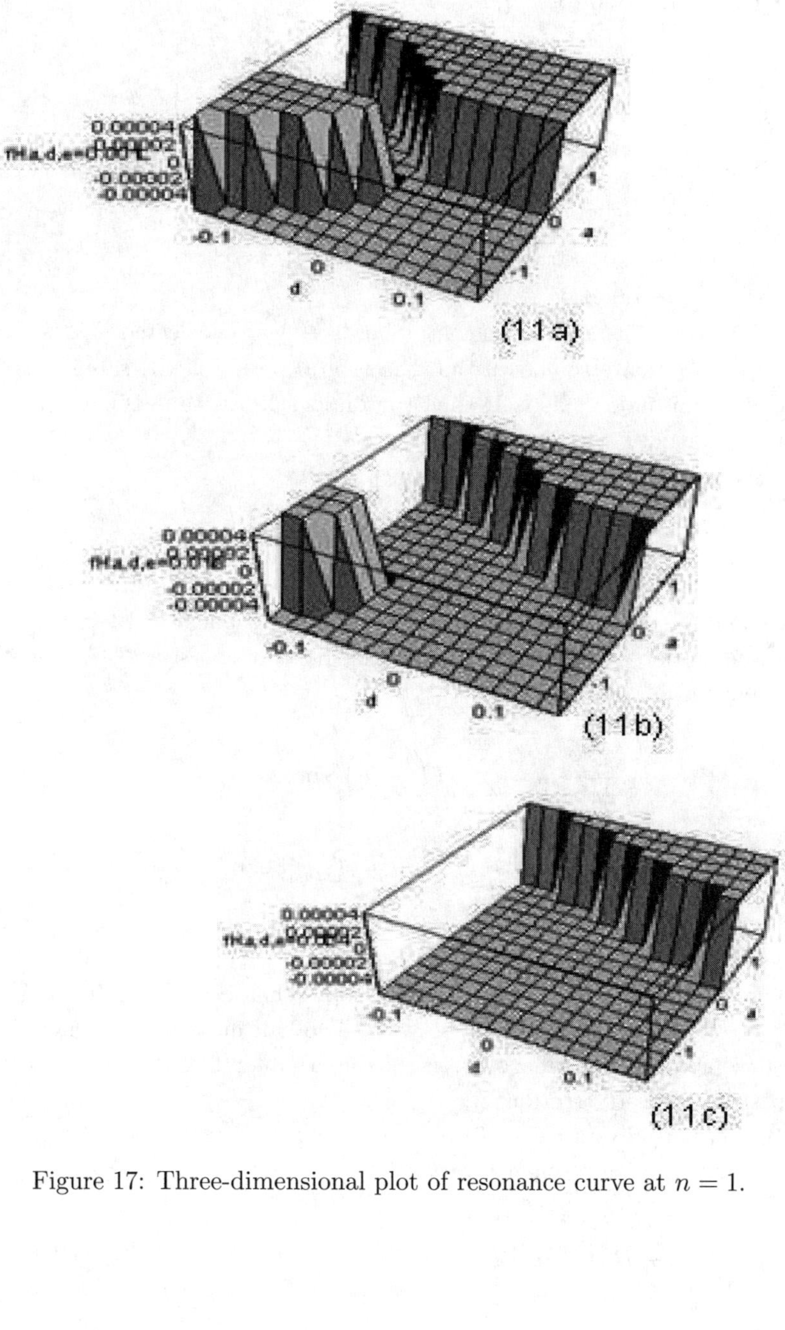

Figure 17: Three-dimensional plot of resonance curve at $n = 1$.

5 Estimation of Resonance Width

The equation of motion can be written as

$$\frac{d^2\theta}{dt^2} = -3\frac{B-A}{2C}\frac{\mu}{r^3}\left[\sin\delta + \frac{\mu'}{\mu}\frac{r^3}{R^3}\sin\delta'\right].$$

Now, $\frac{r^3}{R^3} = \frac{\Omega_2^2}{\Omega_1^2}\frac{\left(1-e^2\right)^3}{(1+e\cos V_1)^3} = \frac{\Omega_2^2 l^3(1+3e\cos V_1)}{\mu}$, neglecting terms containing $O\left(e^2\right)$ and higher and using the fact $\Omega_1^2 a^3 = GM = \mu$. The above equation can be written as $\frac{d^2\theta}{dt^2} = -n^2\frac{\mu}{2r^3}\left[\sin\delta + \frac{m}{M}\frac{\Omega_2^2 l^3}{\mu}(1-3e\cos V_1)\sin\delta'\right] = -n^2\frac{\mu}{2r^3}\left[\sin\delta - \in \sin\left(V-\delta\right)\right]$.

Since $e = e_1, \in$ and terms containing $0\left(\in^2\right)$ are neglected.

If the units are so chosen that the orbital period of the satellite is 2π and its semi-major axis is 1, then the dimensionless time is equal to the mean longitude.

Hence, the above equation becomes

$$\frac{d^2\theta}{dt^2} + \frac{n^2}{2r^3}\left[\sin\left[2\left(\theta - V_1\right)\right] + \in \sin\left[2\left(\theta - V_1\right) - 2\left(V_1 - \Omega_2 t\right)\right]\right] = 0.$$

As r and V_1 one 2π periodic in time, the second and third terms in the above equation can be written as Fourier-like Poisson series [46], and the above equation becomes

$$\frac{d^2\theta}{dt^2} + \frac{n^2}{2r^3}\left[m = \sum_{-\infty}^{\infty} H\left(\frac{m}{2}, e\right)\left[\sin\left(2\theta - mt\right) + \in \sin\right.\right.$$

$$\left.\left.\left(2\theta - mt - mt + 2\Omega_2 t\right)\right]\right] = 0.$$

$$= 0$$

The coefficients $H\left(\frac{m}{2}, e\right)$ are proportional to $e^{2\left|\frac{m}{2}-1\right|}$ and tabulated by Cayley [13], and Goldreich and Peale [16]. When e is small, $H\left(\frac{m}{2}, e\right) \cong -\frac{e}{2}, 1, \frac{7e}{2}$ for $\frac{m}{2} = \frac{1}{2}, 1, \frac{3}{2}$, respectively. The half integer $\frac{m}{2}$ is denoted by p. Resonance occurs whenever one of the arguments of the sine or cosine functions is nearly stationary, i.e., whenever $\left|\frac{d\theta}{dt} - p\right| << \frac{1}{2}$. In such a situation it is advantageous to rewrite the equation of motion in terms of the slowly varying resonance variable $\gamma_p = \theta - pt$.

$$\frac{d^2\gamma_p}{dt^2} + \frac{n^2}{2}H\left(p, e\right)\sin 2\gamma_p + \frac{n^2}{2}\sum_{l}^{0} H\left(p + \frac{l}{2}, e\right)\sin\left(2\gamma_p - lt\right) +$$

$$\in \frac{n^2}{2}\left[H\left(p,\,e\right)\sin\left(2\gamma_p-2\Omega_2 t\right)+\sum_{l}^{0} H\left(p+\frac{l}{2},\,e\right)\sin\left(2\gamma_p-2nt-2\Omega_2 t\right)\right]$$
$$=0,$$

where $'n'$ is sufficiently small so the terms in the sum oscillate rapidly compared to the much slower variation of γ_p determined by the other terms, and, consequently, will give little net contribution to the motion. As a first approximation for small $'n'$, these high frequency terms may be removed by holding γ_p fixed and averaging the above equation over an orbital period. In such a case the above equation can be approximated by

$$\frac{d^2\gamma_p}{dt^2}+\frac{n^2}{2}H\left(p,\,e\right)\sin\left(2\gamma_p\right)+\in\frac{n^2}{2}H\left(p,\,e\right)\sin\left(2\gamma_p-2\Omega_2 t\right)=0.$$

The above is the equation of a pendulum perturbed by the force $\frac{n^2}{2}.\in$ $H\left(p,\,e\right)$
$\sin\left(2\gamma_p-2\Omega_2 t\right)$. The above equation can be studied for cases $\in=0$ and $\in\neq 0$. The case $\in=0$ has been studied in [33]. When $\in\neq 0$ equation represents the equation of motion of disturbed pendulum given by
$\frac{d^2X_p}{dt^2}+f'\left(X_p\right)=M_p g'\left(X_p,\,t\right)$ where $f'\left(X_p\right)=K_{1p}^2\sin\left(X_p\right)$

$K_{1p}^2=n^2H\left(p,\,e\right)$

$m_p=-K_{1p}^2\in<<1,\;X_p=2\gamma_p,\,g'\left(X_p,t\right)=\sin\left(X_p-2\Omega_2 t\right)$

$\left(\frac{dX_p}{dt}\right)^2=C_p+2k_{1p}^2\left(\cos\left(X_p\right)\right)$. For this equation we have

$\left(\frac{dX_p}{dt}\right)^2=C_p+2k_{1p}^2\cos\left(X_p\right)$. Where $C_p's$ are constant of integration.

There are three categories of motion depending upon $C_p>=<2k_{1p}^2$.

Category (i): $C_p>2k_{1p}^2$.

In this case the unperturbed solution is

$$X_p=L_p+C_{1p}\sin\left(L_p\right)+O\left(C_{1p}^2\right),\text{where } L_p=n_p t+\varepsilon_{1p},\,C_{1p}=\frac{k_{1p}^2}{n_p^2},$$

and

$$\frac{1}{n_p}=\frac{1}{2\pi}\int_0^{2\pi}\frac{dX_p}{\left(C_1^2+2K_{1p}^2\cos\left(X_p\right)\right)^{1/2}}.$$

$C_{1p},\,\varepsilon_{1p}$ are arbitrary constants and L_p is an argument. The periodic portion of this series can be regarded as an oscillation about the mean state of motion which is revolution with a period $\frac{2\pi}{n_p}$; the half amplitude of this motion is evidently less than π.

Here, we may observe that $\frac{dX_p}{dt} \neq 0$ and the motion is said to be that of revolution. In case of the perturbed equation by the theory of variation of parameters [5], we have

$$\frac{dC_{1p}}{dt} = \frac{m_p}{V_p}\frac{\partial X_p}{\partial L_p} g'(X_p, t), \quad \frac{dL_p}{dt} = n_p - \frac{m_p}{V_p}\frac{\partial X_p}{\partial C_{1p}} g'(X_p, t),$$

where

$$V_P = \frac{\partial}{\partial C_{1p}}\left(n_p \frac{\partial X_p}{\partial L_p}\right)\frac{\partial X_p}{\partial L_p} - n_p \frac{\partial^2 X_p}{\partial L_p^2}\frac{\partial X_p}{\partial C_{1p}},$$

and we get

$$\frac{\partial X_p}{\partial L_p} = 1 + C_{1p}\cos(L_p), \quad \frac{\partial^2 X_p}{\partial L_p^2} = -C_{1p}\sin(L_p), \quad \frac{\partial X_p}{\partial C_{1p}} = \sin(L_p),$$

$$\frac{\partial^2 X_p}{\partial L_p \partial C_{1p}} = \cos(L_p).$$

$$\begin{aligned} V_p &= \frac{\partial}{\partial C_{1p}}\left[\frac{k_{1p}}{\sqrt{C_{1p}}}(1 + C_{1p}\cos(L_p))\right](1 + C_{1p}\cos(L_p)) \\ &\quad + \frac{k_{1p}}{\sqrt{C_{1p}}} C_{1p}\sin^2(L_p) \\ &\cong -\frac{n_p}{2C_{1p}}. \end{aligned}$$

$$\frac{dC_{1p}}{dt} = -2\frac{m_p k_{1p}^2}{n^3}\sin(X_p - 2\Omega_2 t).$$

Since both m_p and k_{1p}^2 are small quantities, the term $m_p k_{1p}^2$ is of the third order and is therefore rejected. Thus, $\frac{dC_{1p}}{dt} \cong 0$ or $C_{1p} =$ constant up to second order of approximation. $\frac{dL_p}{dt} = n_p + 2\frac{m_p C_{1p}}{n_p}\sin(L_p)\sin(x_p - 2\Omega_2 t)$ and $\frac{d^2 L_p}{dt^2} \cong m_p \sin\left(\frac{2\Omega_2 \varepsilon_{1p}}{n_p} + L_p\left(1 - \frac{2\Omega_2}{n_p}\right)\right)$. In the first approximation, taking $2\Omega_2 = \Omega_0$, $n_p = n_{op}$ and rejecting $2\Omega_2\varepsilon_{1p}$ being a term of second order, the above equation can be written as $\frac{d^2 L_p}{dt^2} \cong m_p \sin\left(L_p\left(1 - \frac{\Omega_0}{n_{op}}\right)\right)$. If we take $L_p\left(1 - \frac{\Omega_0}{n_{op}}\right) = X_p$, then the above equation becomes $\left(\frac{dX_p}{dt}\right)^2 = C_{2p} + 2k_{2p}^2\cos(X_p)$. We get again three types of motions. Type (i) is that in which $\frac{dX_p}{dt}$ is never zero. Type (ii) is that in which $\frac{dX_p}{dt} = 0$ at $X_p = 0$ or π. For Type (i), our solution is

$$X_p = N_p t + \varepsilon_{2p} + \frac{k_{2p}^2}{N_p^2} \sin\left(N_p t + \varepsilon_{2p}\right),$$

where

$$\frac{1}{N_p} = \frac{1}{2\pi} \int_0^{2\pi} \frac{dX_p}{\left(C_{2p} + 2K_{2p}^2 \cos\left(X_p\right)\right)^{1/2}}, \; C_{2p}$$

and ε_{2p} are arbitrary constants. In the first approximation $N_p = N_{op}$ and the solution is $X_p = X_{op} + \frac{k_{2p}^2}{N_{op}^2} \sin\left(X_{op}\right)$, where $X_{op} = N_{op} t + \varepsilon_{2p}$. This is the case of revolution.

For Type (ii), the solution is $X_p = \lambda_p \sin\left(p^1 t + \lambda_0\right)$, [where $p^1 = n\sqrt{\left(\varepsilon H\left(p,\, e\right)\left(1 - \frac{\Omega_0}{n_{op}}\right)\right)}$, with λ_p and λ_0 being arbitrary constants. This is the case of liberation.

Type (iii) occurs when $C_{2p} = 2k_{2p}^2 = 2n^2 H\left(p,\, e\right) \varepsilon \left(1 - \frac{\Omega_0}{n_{op}}\right)$. The solution is $X_p + \pi = 4\tan^{-1} \exp\left(k_{2p} t + \alpha_0\right)$, where α_0 is an arbitrary constant. When $t \to \pm\infty$, $X_p = \pm\pi$ and at both places $\frac{dX_p}{dt} = 0$ and all higher derivatives of X_p approaches to zero. Near this point X_p one of the limit tends to $\pm\pi$, t tends to become an indeterminate function of X_p. This is the case of infinite period separatrix which is asymptotic backward and forward to unstable equilibrium. Thus, the results of Type (i), Type (ii), and Type (iii) enable us to conclude that the third-body torque plays a significant role. It may change a revolution to liberation (Type ii) or to intinite period separatrix (Type iii)

Category (ii): $C_p < 2k_{1p}^2$.

In this case the uperturbed solution is $X_p = C_{1p} \sin\left(L_p\right) + O\left(C_{1p}^3\right)$, where $L_p = n_p t + \varepsilon_{1p}$, $n_p = k_{1p}\left(1 - \frac{1}{16} C_{1p}^2 + O\left(C_{1p}^4\right)\right)$. C_{1p} and ε_{1p} are arbitrary constants. In case of the perturbed equation, again using the theory of variation of parameters, we get

$$\frac{dC_{1p}}{dt} \cong \frac{m_p}{k_{1p}} \cos\left(L_p\right) \sin\left(X_p - 2\Omega_2 t\right),$$

$$\frac{dL_p}{dt} \cong k_{1p} - \frac{m_p}{k_{1p} C_{1p}} \sin\left(L_p\right) \sin\left(X_p - 2\Omega_2 t\right),$$

$$\frac{d^2 L_p}{dt^2} \cong \frac{m_p 2\Omega_2}{k_{1p} C_{1p}} \sin\left(L_p\right) \cos\left(2\Omega_2 \frac{L_p - \varepsilon_{1p}}{n_p}\right).$$

As a special case, let us assume that $\Omega_0 \frac{L_p - \varepsilon_{1p}}{n_{op}} = n_1 \pi$, $n_1 \in I$. When n_1 is even, $\frac{d^2 L_p}{dt^2} + k_{3p}^2 \sin(L_p) = 0$, where $k_{3p}^2 = -\frac{m_p \Omega_0}{k_{1p} C_{op}} > 0$, which is again the equation of disturbed pendulum. As in the previous case this equation also gives us revolution, libration, and infinite period separatrix motion. On the other hand, if n_1 is odd, then $\frac{d^2 L_p}{dt^2} \cong k_{3p}^2 \sin(L_p)$. When L_p is small, the solution of the above equation is $L_p = e^{k_{3p} t} e^{-k_{3p} t}$.

Category (iii): $C_p = 2k_{1p}^2$.

The unperturbed solution is $X_p + \pi = 4 \tan^{-1} e^{k_{1p} t + \alpha_0}$, where α_0 is an arbitrary constant. This is the case of infinite period separatrix which is asymptotic forward and backward to unstable equilibrium. We are mainly concerned in this category of motion. In this category the nature of the unperturbed solution does not change by taking into account the third-body torque. Near the infinite period separatrix broadened by the high frequency term into narrow chaotic band [15], for small n, the half-width of the chaotic separatrix is given by $\omega_p = \frac{I_p - I_p^s}{I_p^s} = 4\pi \varepsilon_1 \lambda^3 e^{-\pi\lambda/2}$, where ε_1 is the ratio of the coefficient of the nearest perturbing high frequency term to the coefficient of the perturbed term, and $\lambda = \frac{\Omega}{\omega}$ is the ratio of the frequency difference between the resonant term and the nearest non-resonant term (Ω) to the frequency of small-amplitude liberations (ω). For the synchronous spin-orbit state perturbed by the third-body torque, we have $\lambda = \frac{1}{n}$, $\omega_1 = \frac{I_1 - I_1^s}{I_1^s} = 4\pi\varepsilon \frac{1}{n^3} e^{-(\pi/2n)}$. Here, ω_1 increases both with ε and n. Chrikov's overlap criterion states that when the sum of two unperturbed half-widths equals the separation of resonance centers, large-scale chaos ensues. In the spin-orbit problem the two resonances with the largest widths are the $p = 1$ and $p = 3/2$ states. For these two states the resonance overlap criterion becomes $n^{RO}\sqrt{|H(1, e)|} + n^{RO}\sqrt{|H(3/2, e)|} = \frac{1}{2}$, $n^{RO} = \frac{1}{2+\sqrt{14e}}$. For $e = 0.1$ the mean eccentricity of Hyperion, the critical value of n above which large-scale chaotic behavior is expected is $n^{RO} = 0.31$. (T.C. Duxbury, 1983, personal communication). It is expected then that for Hyperion there is a large chaotic zone surrounding (at least) the $p = 1$ and $p = 3/2$ states, and possibly more.

6 The Spin-Orbit Phase Space

We have drawn the Poincare surface-section for our model. From Figures 18 and 19 we may observe that as the third-body torque is introduced (comparing with Wisdom case, $\varepsilon = 0$) the regular curves start disintegrating and this disintegration increases with the increase in ε. From Figures 20, and 21, we may observe that as the eccentricity increases, the regular curves disintegrate, and this disintegration increase with the increase in e. From Figures 22, and 23 we may observe that as n increases, the regular curves disintegrate and this disintegration increases with the increase in n. In Figures 24 and 25 we have plotted two different figures for different values of ε (i.e., for $\varepsilon = 0.0$ and $\varepsilon = 0.8$) at $e = 0.1$ and $n = 0.89$. The value of e and n taken are appropriate for Hyperion. Due to third-body torque, we observe that the chaotic zone is further increased.

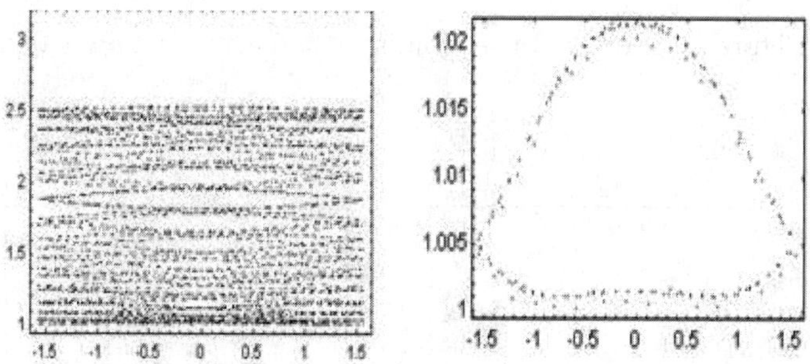

Figure 18: Surface of section for $\in = 0.3$, $n = 0.2$, $e = 0.1$.

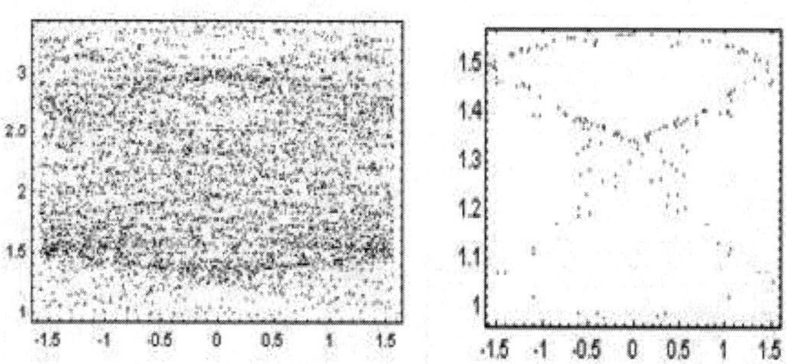

Figure 19: Surface of section for $\in = 0.9$, $n = 0.2$, $e = 0.1$.

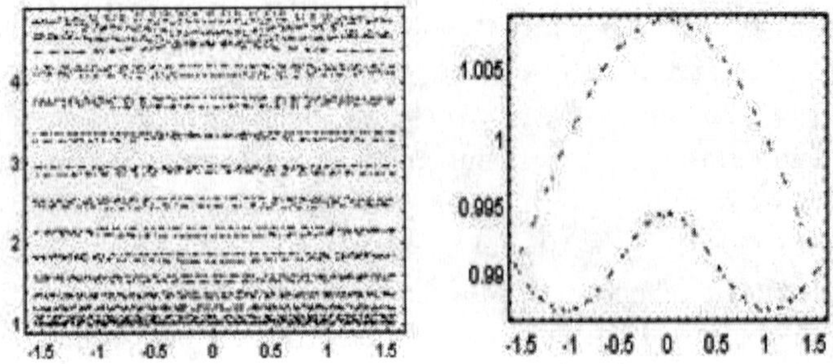

Figure 20: Surface of section for $\in = 0.5$, $n = 0.2$, $e = 0.2$.

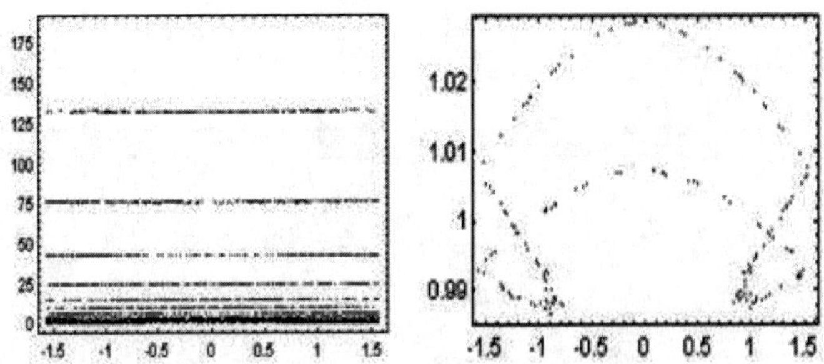

Figure 21: Surface of section for $\in = 0.5$, $n = 0.2$, $e = 0.8$.

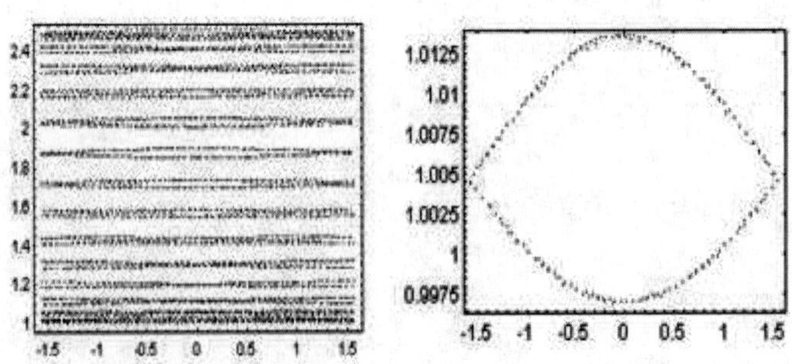

Figure 22: Surface of section for $e = 0.1$, $n = 0.1$, $\in = 0.5$.

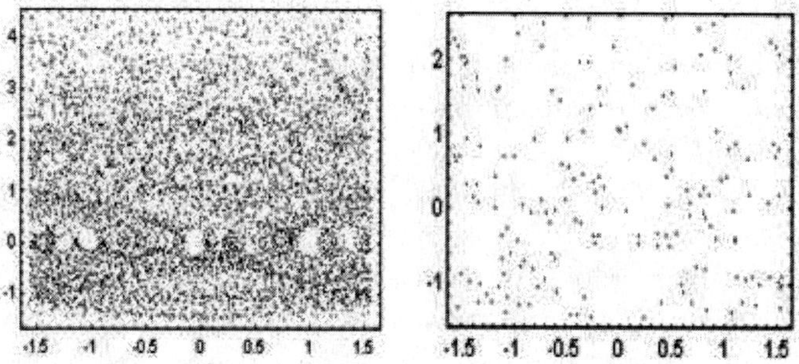

Figure 23: Surface of section for $e = 0.1$, $n = 0.6$, $\in = 0.5$.

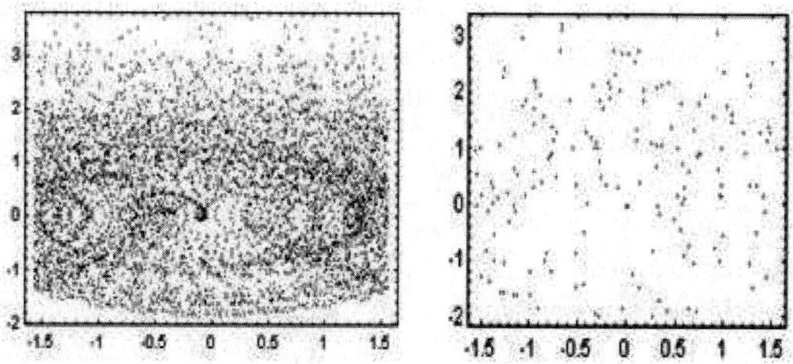

Figure 24: Surface of section for $e = 0.1$, $n = 0.89$, $\in = 0.0$.

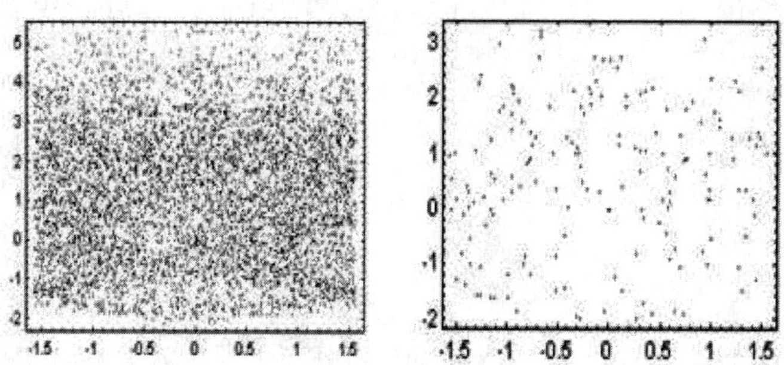

Figure 25: Surface of section for $e = 0.1$, $n = 0.89$, $\in = 0.8$.

7 Conclusion

We thus conclude that the non-linear rotational equations of motion of the planar oscillation of a satellite in an elliptic orbit under the influence of third-body torque are non-integrable. Also, we have observed graphically that in the Earth-Moon-Artifical Satellite (1958 B2 Vanguard 1) system the Melnikov's function has a simple zero, and hence, the equations of motion are non-integrable. We have seen that the amplitude of the oscillation remains constant up to the third order of approximation even with the third-body torque (non-resonance case). The resonances have been shown to exist and have been studied by the BKM method. We have further observed that the main resonances occur at $n = \pm 1, \pm 2, \pm c$ and the parametric resonances occur at $n = \frac{1}{k}, \frac{c}{k}, k \in I^{\sim} \{(0), (1), (-1)\}$. It is further observed that stability of the stationary planar oscillation of the satellite near the resonance frequency shows that discontinuity occurs in the amplitude of the oscillation at a frequency of the external periodic force which is less than the frequency of the natural oscillation. The jump in the amplitude at the critical point in the resonance curves increases with the increase in the third-body torque, and the eccentricity and this discontinuity in the amplitude of the oscillation may cause a noticeable change in the orbital parameter of the satellite. The third-body torque plays a very significant role in changing the motion of revolution into libration or infinite period separatrix. The half-width of the chaotic seperatrices estimated by the Chrikov's criterion is not affected by the third-body torque. We further conclude that in the spin-orbit phase space the regular curves start disintegrating due to third-body torque [the increase in the eccentricity and the irregular mass distribution of the satellite and this disintegration increases with the increase in ε, n, and e.] The theory developed has been applied to the values of the parameters appropriate to Hyperion. In this case, it is seen that Hyperion spin-orbit phase-space is dominated by a chaotic zone which increases further due to third-body torque.

References

[1] C. Beauge, S. Ferraz-Mello, and T. Michtchenko, Extrasolar planets in mean motion resonance: Apses alignment and asymmetric stationary solutions *Ap J.*, submitted (2002).

[2] V.V. Beletskii, *Motion of Artificial Satellite Relative to the Centre of Mass*, Nauka, Moscow, 1965, (in Russian).

[3] K.B. Bhatnagar, A. Khan, and L.M. Saha, Non-linear planar oscillation of a satellite in elliptical orbit under the influence of solar radiation pressure I, *Bull. Astr. India,* **22** (1994), 47–58.

[4] K.B. Bhatnagar and Rashmi Bhardwaj, Non-linear planar oscillation of a satellite in a circular orbit under the influence of magnetic torque I, *Indian J. Pure Appl. Math.*, **26** (1995), 1225–1240.

[5] B. Brown and A. Shook, *Planetary Theory,* Dover Publications, Inc., New York, 1964.

[6] R. Bhardwaj and K.B. Bhatnagar, Chaos in non-linear planar oscillation of a satellite in an elliptical orbit under the influence of third bosy torque, *Indian J. Pure Appl. Math.*, **28(3)** (1997), 391–442.

[7] R. Bhardwaj and R. Tuli, Chaos in attitude motion of a satellite in an elliptical orbit under the influence of third body torque I, *Indian J. Pure Appl. Math.*, **34(1)** (Feb 2003), 277–289.

[8] B.G. Bills, *Tidal Dissipation in Mercury Lunar and Planetary Sciences,* XXX111, (2002).

[9] N.N. Bogoliubov and Y.A. Mitropolsky, *Asymtotic Methods in the Theory of Nonlinear Oscillation*, Hindustan Publishing Corporation, Delhi, 1961.

[10] E.I. Chiang, D. Fischer, and E. Thommes, Excitation of orbital eccentricities of extrasolar planets by repeated resonance crossings, *Astrophysical,* **564(2)**, L105–L109.

[11] M. Cuk, J.A. Burns, V. Carruba, P.D. Nicholson, and R.A. Jacobson, New secular resonance involving the irregular satellite of Saturn: American Astronomical Society, *DDA 33rd Meeting*, Mt. Hood, OR, (April 2002).

[12] C. Phillips, High tide on Europa, *Seti Thursday, Posted* (30 ct. 2002).

[13] A. Cayley, Tables of the developments of functions in the theory of elliptic motion, *Mem. Roy. Astron. Soc.*, **29** (1859), 1991–306.

[14] B.V. Cherousko, V.V. Vecheslavov, Chaotic dynamics of comet Halley, *Astronomy and Astrophysics (ISSN 0004-6361)*, **221(1)** (1989) 146–154.

[15] S.F. Dermott, R. Malhotra, and C.D. Murray, Dynamics of the Uranian and Saturnian satellite systems - A chaotic route to melting Miranda: *Icarus*, **76** (1988), 295–334.

[16] P. Goldreich and S.J. Peale, Spin orbit coupling in the solar system 1. The rotation of Venus: *Astron. J.*, **72** (1967), 662.

[17] N. Haghighipour, Resonance's and stability of restricted three body system, *DPS 34th meeting*, 2002.

[18] A. Henon and C. Heiles, The applicability of the third integral of motion, some numerical experiments, *Astron. J.* **69** (1964), 73–79.

[19] Z. Knežević, A. Milani, and P. Farnella, The dangerous border of the 5:2 mean motion resonances, *Planetary and Space Sciences*, (1997), **45** 1581–1585.

[20] M. Konacki, A. Maciejewcki and A. Wolszczan, Resonances in PSR B1257+12 planetary system, *Ap J.*, **513** (1999), 471.

[21] N.M. Krylov and N.N. Bogoliubov, *Introduction to Nonlinear Mechanics,* Publication of the Academic Science Ukr. SSR, 1937.

[22] W. Lotko and A.V. Streltsov, Magnetospheric resonance, Auroral structure and multipoint measurements, *Adv. Space Res.*, (1997) 201067.

[23] G. Marcy, P. Butler, D. Fischer, S. Vogt, J. Lissaeur, and E. Rivera, A pair of resonant planets orbiting GJ 876, *Ap J.*, 556, (2001), 296.

[24] T. Michtchencko, C. Beauge, and F. Roig, Resonant structure of the outer solar system in the neighbourhood of the planets, *Astron J.*, **122** (2001), 3485.

[25] V.K. Melnikov, On the stability of the center for time periodic perturbations, Trusty Moskov, *Mat. Obshch,* **12** (1963), 3–52 (translation in *Trans. Moscow Math. Soc.,* **12** (1998), 1–57).

[26] A.J. Maciejewski, Instability, *Chaos and predictability in Celestial Mechanics and Stellar Dynamics*, (ed.: K.B. Bhatnagar), Nova Science, New York, 1992, pp. 23–38.

[27] M. Nauenberg, Stability and eccentricity for two planets in a 1:1 resonances and their possible occurrence in extrasolar planetary system, *Astron, J.* **124** (2002), 2332.

[28] G. Novak, The co-orbital resonance and extrasolar planets, DDA *33rd meeting, BAAS*, **34** (2002).

[29] H. Poincare, Les Methods Nauvelles De La Mechanique Celest, *Moscow (in Russian)*, (1972), **I,II**, Nauk.

[30] S. Smale, Stable manifolds of diffeomorphism and differential equations, *Annati Della Scuola Normati Superioro di Pisa Series III, XVII*, (1963), pp. 97–116.

[31] R.B. Singh, Non-linear planar oscillation of a satellite in elliptical orbit under the influence of external forces of general nature, *Space Dynamics and Celestial Mechanics* (ed. K.B. Bhatnagar), D. Reidel Publishing Company, Dordreicht, (1983), pp. 295–307.

[32] F. Varadi, Periodic orbits in the 3:2 orbital resonance and their stability, *Astron J.,* **118**, (1999), 2526.

[33] S.J. Widso Peale, and F. Mignard, The chaotic rotation of Hyperion, *Icarus,* **58** (1984), 137–52.

[34] S. Wiggins, *Introduction to Applied Non-linear Dynamical Systems and Chaos*, Springer-Verlag, Heidelberg, 1990.

[35] S. Wiggins, Global bifurcation and chaos, *Analytical Methods* Springer-Verlag, New York, Berlin, 1988.

[36] V.A. Zlatanstov, A.P. Markeev, Stability of Planar oscillation of a satellite in an elliptic orbit, *Celestial Mechanics*, **7** (1973), pp. 31.

Chapter 14

CHOAS USING MATLAB IN THE MOTION OF A SATELLITE UNDER THE INFLUENCE OF MAGNETIC TORQUE

R. Bhardwaj and P. Kaur

Guru Gobind Singh Indraprastha University

Abstract

Magnetic disturbance torque results from the interaction between the spacecraft's residual magnetic field and the geomagnetic field. The primary sources of magnetic disturbance torques are (i) spacecraft magnetic moments, (ii) eddy currents, and (iii) hysteresis. Of these, the spacecraft's magnetic moment is usually the dominant source of disturbance torques. Let us consider a rigid satellite S moving around the central body C. We further suppose that the orbital plane of satellite coincides with the equatorial plane of the central body. For this problem, we derived the non-linear planar oscillation of a satellite and its equation of motion and showed the non-integrability of equation of motion of satellite. We have discussed the non-resonance and resonance cases for the above equation of motion and also have found that the amplitude of the oscillation remains constant even up to the second order of approximation. Finally, we have studied the chaotic motion of a satellite by taking the effect of magnetic torque in the Earth-Moon system and the Saturn-Hyperion system. Using Matlab, in the Earth-Moon system and the Saturn-Hyperion system, we have shown the non-integrability of equation of motion and discussed the non-resonance and resonance cases and explained the chaotic motion.

1 Introduction

Newton [20] explained the exact proportionality between a particular force (the weight) and the general dynamical measure of inertia (the mass) in the planetary system. Laplace [17] considered the Earth-Moon system to derive a rough estimate of the observable effect of the ratio weight over the mass on the angular motion of the moon. Planar oscillation of a satellite has been studied by Zlatanstov [25], and Beletskii [3]. Childs and W. Dara [12] studied a movable mass Attitude-Stabilization System for Artificial-g Space Station. Singh [23], Maciejewski [18], Bhatnagar et al. [4–6] studie dthe nonlinear oscillation of satellite under periodic forces. Prussing and Conway [21] described the relative motion of two spacecrafts in similar near-circular orbits using Hill's equations. Hardison [15] studied it for Cable-connected Artificial-g Space Station. Kechichian [16] derived an exact formulation of the relative motion of satellites in presence of J_2 potential and atmospheric drag.Alfriend, Schaub, and Gim [1] and Gim and Alfriend [14], studied stte transition matrix which is employed to account for the effects of J_2 potential and slight eccentricities in the reference orbits. Schweighart and Sedwick [22], describes the mean motion of the satellites in defining the reference orbit. Bhatnagar and Bharadwaj [6–8], studied the effect of third body torque on planar motion of a satellite. Bhatnagar and Bharadwaj [9,10], have taken into account the effect of magnetic torque, but the satellite is assumed to move in a circular orbit. Bharadwaj and Tuli [11], studied the effect of third body torque in elliptic orbit for the nonlinear motion of a satellite. Magnetic disturbance torque results from the interation the spacecraft's residual magnetic field and the geomagnetic field. The primary sources of magnetic disturbance torques are (i) spacecraft magnetic moments, (ii) eddy currents and (iii) hysteresis. Of these, the spacecraft's magnetic moment is usually the dominant source of disturbance torques. The spacecraft is usually designed of a material selected to make disturbances from the other source negligible. Bastow [2], and Droll and Iuler [13], provide a survey of the problems associated with minimizing the magnetic disturbances in spacecrafts designs and development. The instantaneous magnetic disturbance torque, $\bar{N}_{\text{mag}}$ due to the spacecraft effective magnetic moment $\bar{m}$ (in $A : m^2$) is given by $\bar{N}_{\text{mag}} = m\bar{m} \times \bar{B}$; where B is the geocentric magnetic flux density (in Wb/m^2) and $\bar{m}$ is the sum of the individual magnetic moments caused by permanent and induced magnetism and the spacecraft generated current loops [24]. The magnetic torque is applicable below synchronous orbit less than 35,000 km. For a 40,000 pole-cm electromagnet at an altitude of 550 km the magnetic torque is approximately 0.001 Nm (Table 15.1 in [24]). This means that the magnetic torque perturbation parameter is in general very small (<< 1). We have determined hyperbolic equilibrium solution and double asymptotic solutions corresponding to unperturbed Hamiltonian H_0. The non-integrability of the system has been shown through Melnikov's [19], integral and also it has been discussed graphically in the Earth-Moon system and the Saturn-Hyperion system. We have found out the solution is non-resonance case by BKM method, which is described briefly below.

BKM Method: This method is also known as "Development of Asymptotic Solutions". Consider oscillations defined by

$$\frac{d^2x}{dt^2} + \omega^2 x = \in f\left(x, \frac{dx}{dt}\right), \tag{1.1}$$

where ε is a small positive parameter. When perturbation is absent, i.e., when $\varepsilon = 0$, the oscillation will evidently be purely harmonic,

$$x = a\cos(\Psi),$$

with a constant amplitude and a uniformly rotating phase angle given by

$$\frac{da}{dt} = 0, \frac{d\Psi}{dt} = \omega, (\Psi = \omega t + \theta)$$

(the amplitude a and the phase θ will be constants over time, depending on the initial conditions). The existence of non-linear perturbation ($\varepsilon \neq 0$) results in the appearance of overlap in the solution of (1.1), a factor that establishes dependence between the instantaneous frequency $\frac{d\Psi}{dt}$ and the amplitude, and finally gives rise to the systematic increase or decrease in the amplitude of the oscillations depending upon whether the energy is expelled or absorbed by the perturbing forces. A general solution of (1.1) is of the form

$$x = a\cos(\Psi) + \varepsilon u_1(a, \Psi) + \varepsilon^2 u_2(a, \Psi) + ..., \tag{1.2}$$

where $u_1(a, \Psi), u_2(a, \Psi), ...$ are periodic functions of the angle Ψ with a period 2π and the quantities a, Ψ are functions of time defined by the differential equations

$$\frac{da}{dt} = \varepsilon A_1(a) + \varepsilon^2 A_2(a) + ..., \frac{d\Psi}{dt} = \omega + \varepsilon B_1(a) + \varepsilon^2 B_2(a) + ... \tag{1.3}$$

We have to choose suitable expressions for the functions $u_1(a, \Psi), u_2(a, \Psi), A_1(a), A_2(a), B_1(a), B_2(a)$......, in such a way that (1.2), after replacing a and Ψ, by the functions defined in (1.3), would serve as a solution of (1.1). The above method of finding the solution of (1.1) in the form (1.2) is called the BKM method.

2 Perturbed Planar Oscillation of a Rigid Satellite

2.1 Equation of Motion

Let us consider a rigid satellite S moving in an elliptic orbit around the Earth E. We further suppose that the orbital plane coincides with the equatorial plane of the Earth (Figure 1). The satellite is assumed to be a triaxial body with principal moments of inertia $A < B < C$ at its center of mass, and C is the moment of inertia about the spin axis which is perpendicular to the orbital plane. These principal axes are taken as our coordinate axes x, y, z with the z-axis being perpendicular to the orbital plane. Let $\bar{r}$ be the radius vector of the center of mass of the satellite, θ be the angle that the long axis of the satellite makes with a fixed line EF lying in the orbital plane, and $\delta/2$ be the angle between the radius vector and the long axis. The magnetic torque is calculated in the l, b, n system as defined in [9].

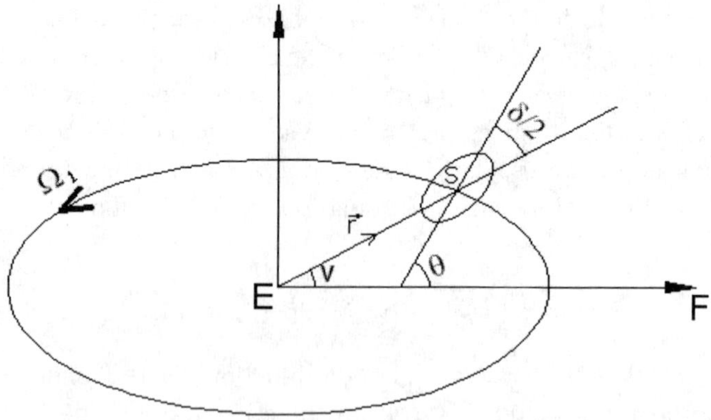

Figure 1: Satellite planar oscillation in elliptic orbit with magnetic perturbation.

Euler's dynamical equation of motion of the satellite about the z-axis is given by

$$C\frac{d\omega_3}{dt} + (B - A)\omega_1\omega_2 = G_z + N_z,$$

where

$$\begin{aligned} G_z &= z - \text{component of the gravitational torque,} \\ N_z &= z - \text{component of the magnetic torque,} \\ \omega_1, \omega_2, \omega_3 &= \text{angular velocities about the principal axes at the center of} \\ &\quad \text{mass of the satellite.} \end{aligned}$$

Here,

$$\omega_1 = \omega_2 = 0 \text{and} \omega_3 = \frac{d\theta}{dt}.$$

So, the Euler's dynamical equation of motion becomes

$$C\frac{d^2\theta}{dt^2} = G_z + N_z, \tag{2.1}$$

where [6]

$$\begin{aligned} G_z &= -\frac{3\mu}{2r^3}(B - A)\sin\delta, \\ \mu &= G(m_E + m_S) = Gm_E, \\ m_E &= \text{mass of the earth,} m_S = \text{mass of satellite.} \end{aligned}$$

We have

$$\bar{N} = \bar{m} \times \bar{B}, \tag{2.2}$$

where ([24], pp. 783–785)

$$\bar{B}(\bar{r}) = \frac{a^3 H_0}{r^3}(3(\widehat{m}.\widehat{r})\widehat{r} - \widehat{m}). \tag{2.3}$$

Since the equatorial plane of the Earth coincides with the orbital plane of the satellite, so $i = 0$. The components of $\bar{m}$ in the (l, b, n) system are

$$\begin{aligned} m_l &= m\sin(\theta'_m)\cos(\Omega - \alpha_m), \\ m_b &= -m\sin(\theta'_m)\sin(\Omega - \alpha_m), \\ m_n &= m\cos(\theta'_m). \end{aligned}$$

Therefore,

$$\widehat{m} = \begin{pmatrix} \sin(\theta'_m)\cos(\Omega - \alpha_m) \\ -\sin(\theta'_m)\sin(\Omega - \alpha_m) \\ +\cos(\theta'_m) \end{pmatrix}$$

$$\widehat{r} = \begin{pmatrix} \cos(\nu) \\ \sin(\nu) \\ 0 \end{pmatrix},$$

$$\begin{aligned}
a^3 H_0 &= \text{total dipole strength}, \\
\theta'_m &= \text{coelevation of the dipole}, \\
\phi'_m &= \text{east longitude of the dipole}, \\
\alpha_{G0} &= \text{right ascension of the Greenwich meridian at some reference time}, \\
\frac{d\alpha_G}{dt} &= \text{average rotation rate of the Earth}, \\
t &= \text{time since reference}, \\
m &= \text{magnetic strength of the spacecraft}, \\
v &= \text{true anomaly measured from the ascending node}, \\
\Omega &= \Omega_0 + \frac{d\Omega}{dt}t = \text{right ascension of ascending node ([24], p. 68)} \\
\frac{d\Omega}{dt} &= -2.06474 \times 10^{14}\frac{a^{-7/2}\cos i}{(1-e^2)^2} = 0.9856^0/day, \\
i &= \text{orbital inclination.}
\end{aligned}$$

All these quantities are constants for a particular year. Therefore,

$$\widehat{m}.\widehat{r} = \sin(\theta'_m)\cos(\Omega - \alpha_m)\cos v - \sin(\theta'_m)\sin(\Omega - \alpha_m)\sin v.$$

Substituting this value into (2.3) and simplifying, we get

$$\begin{aligned}
B_l &= \frac{a^3 H_0}{r^3}\Big[\sin(\theta'_m)\cos(\Omega - \alpha_m)(3\cos^2 v - 1) - \frac{3\sin 2v}{2} \\
&\quad (\sin(\theta'_m)\sin(\Omega - \alpha_m))\Big] \\
B_b &= \frac{a^3 H_0}{r^3}\Big[\frac{3}{2}\sin 2v\sin(\theta'_m)\cos(\Omega - \alpha_m) - (3\sin^2 v - 1) \\
&\quad (\sin(\theta'_m)\sin(\Omega - \alpha_m))\Big] \\
B_n &= -\frac{a^3 H_0}{r^3}\left[\cos(\theta'_m)\right].
\end{aligned}$$

A negative sign in the orbit normal component B_n assures the northward direction of the field lines.

From (2.2), the z-component of N is

$$N_z = m\,[m_l B_b - B_l m_b]$$
$$= \frac{3a^3 m H_0}{r^3}[\cos 2v \sin^2(\theta'_m)\cos(\Omega - \alpha_m)\sin(\Omega - \alpha_m) +$$
$$+\frac{1}{2}\sin 2v\left(\sin^2(\theta'_m)\cos^2(\Omega - \alpha_m) - \sin^2(\theta'_m)\sin^2(\Omega - \alpha_m)\right)].$$

Now (2.1) becomes

$$\frac{d^2\theta}{dt^2} = -\frac{\mu}{2r^3}n^2\sin\delta + \frac{\varepsilon\mu}{2r^3}\sin 2\left(\Omega - \alpha_m + \nu\right), \qquad (2.4)$$

where

$$\Omega = \Omega_0 + \frac{d\Omega}{dt}\frac{\nu}{\Omega_1} = \Omega_0 + \beta_2\nu\,\beta_2 = \frac{1}{\Omega_1}\frac{d\Omega}{dt},$$

$$\alpha_m = \alpha_1 + \beta_1\nu\,\alpha_1 = \alpha_{G0} + \Phi'_m,\ \beta_1 = \frac{1}{\Omega_1}\frac{d\alpha_G}{dt},$$

$$n^2 = \frac{3\,(B-A)}{C},\ \varepsilon = \frac{3a^3 m H_0}{\mu C}\sin^2\theta'_m,$$

ε is the parameter due to magnetic torque, and Ω_1 is the angular velocity of the moon. Since the orbit of the rigid satellite is elliptic, we have

$$r^2\dot{\nu} = hand\frac{l}{r} = 1 + e\cos\nu.$$

If we take true anomaly v as the independent variable and $\delta = q$ as the generalized coordinate, then the equation of motion (2.4) becomes

$$(1 + e\cos\nu)\frac{d^2q}{d\nu^2} - 2e\sin\nu\frac{dq}{d\nu} - 4e\sin\nu + n^2\sin q = \varepsilon\sin\left(a + b\nu\right), \quad (2.5)$$

where

$$a = 2(\Omega_0 - \alpha_{G0} - \Phi'_m),$$
$$b = 2(1 + \frac{1}{\Omega_1}(\frac{d\Omega}{dt} - \frac{d\alpha_G}{dt})).$$

2.2 Hamilton's Equation

Equation (2.5) is equivalent to the Hamilton's equation

$$\frac{dq}{d\nu} = \frac{\partial H}{\partial p}, \frac{dp}{d\nu} = -\frac{\partial H}{\partial q}, \tag{2.6}$$

where

$$\begin{aligned} H &= H_0 + \varepsilon H_1 + \ldots , \\ H_0 &= \frac{p^2}{2} - 2p - n^2 \cos q, \\ H_1 &= -p^2 \cos \nu - n^2 \cos q cos\nu - \varepsilon_1 \sin (a + b\nu)\, q, \end{aligned}$$

$p =$ generalized momenta.

2.3 Equilibrium and Double Asymptotic Solution

The equilibrium solution corresponding to H_0 is given by

$$\frac{dq}{d\nu} = 0 = \frac{dp}{d\nu}.$$

From (2.6), we obtained the hyperbolic equilibrium solution as $p(\nu) = 2$, $q(\nu) = 0, \Pi, \ldots$. The unperturbed double asymtotic solutions are given by

$$\begin{aligned} p^{\pm}(\nu) &= 2 \pm \frac{2n}{\cosh n\nu}, \\ \sin(q^{\pm}(\nu)) &= \pm \frac{2\sinh(n\nu)}{\cosh^2(n\nu)}, \\ \cos(q^{\pm}(\nu)) &= \frac{2}{\cosh^2(n\nu)} - 1. \end{aligned}$$

2.4 Evaluation of Melnikov's Integral

Melnikov's result has allowed us to formulate theorems about non-integrability of system with transversal homoclinic (hetroclinic) orbits. The condition for transversal crossing of the separatrices can be expressed in terms of Melnikov's integral. Namely, if Melnikov function

$$M^{\pm}(\nu_0) = \int_{-\infty}^{+\infty} (H_0, H_1)(q^{\pm}(\nu - \nu_0), p^{\pm}(\nu - \nu_0))d\nu$$

has simple zero, then the perturbed asymptotic surfaces cross transversely and the Hamiltonian system is non-integrable. In our case,

$$M^{\pm}(\nu_0, n, \varepsilon_1, a) = \pm\, 8\Pi \sin(\nu_0) \sec j \left(\frac{\Pi}{2n}\right) \pm 6\Pi \sin(\nu_0 \cos ech \left(\frac{\Pi}{2n}\right)$$
$$\pm\, 2\varepsilon_1 \Pi \sin(a + b\nu_0) \sec h \left(\frac{\Pi b}{2n}\right). \qquad (2.7)$$

It is easy to observe that for any value of mass parameter n and magnetic torque, parameters ε_1 and a ($0.105092 \leq a \leq 0.220081$) the above function has a simple zero. Thus, both pairs of asymptotic surfaces cross transversely and (2.6) is non-integrable.

2.5 Graphical Representation of Melnikov's Function

Earth-Moon System

We have studied a graphical representation of Melnikov's function in the Earth-Moon system. For moon, the data are

$$\Omega_1 = \frac{2\Pi}{30 \times 24 \times 60 \times 60} rad/s.$$

Eccentricity

$$e = 0.0549,$$
$$\varepsilon = 0.001,$$
$$\varepsilon_1 = 0.018214936, \text{ and}$$
$$b = 2.164264351851.$$

For this fixed value of b, we have studied the graphs of Melnikov's function. We are assuming $\Omega_0 = 210°$.

Figure 2 illustrates the graphs of $M^+(\nu_0, n, a)$ and $M^-(\nu_0, n, a)$ for $0 \leq \nu_0 \leq 12$ and for fixed value $a = 0.10916685405$.

n	$M^+(\nu_0, n, a)$		$M^-(\nu_0, n, a)$	
	Max. value	**Min. value**	**Max. value**	**Min. value**
0.100	1.31153E-05	-1.3256E-05	1.3256E-05	-1.31153E-05
0.105	2.77097E-05	-2.8007E-05	2.8007E-05	-2.77097E-05
0.110	5.46954E-05	-5.5283E-05	5.5283E-05	-5.46954E-05
0.115	1.01763E-04	-1.-286E-04	1.0286E-04	-1.01763E-04
0.120	1.79788E-04	-1.8172E-04	1.8172E-04	-1.79788E-04

In both cases, the Melnikov's function behaves almost like sine functions, with one corresponding to $M^+(v_0)$ and the other to $M^-(v_0)$. They have simple zeros. Also, we observe that in each graph when $3 \leq v_0 \leq 3.5$, $M^+(v_0)$ changes sign from positive to negative and when $6 \leq v_0 \leq 6.5$, $M^+(v_0)$ changes sign from negative to positive. Thus, in all these graphs abscissas almost remain the same. We also observe from the above table that as n varies from 0.1 to 0.12, $M^+(v_0)$ and $M^-(v_0)$ elongate along the ordinate, with the abscissa remaining almost the same.

Figure 3 illustrates the graph of $M^{\pm}(\nu_0, n, a)$ for a fixed value of $v_0 = 0.1$, $a = 0.10916685405$. It has been observed that as the mass distribution parameter of satellite changes from 0.1 to 0.99, the value of Melnikov's function $M^+(\nu_0, n, a)$ increases very slowly for $0.1 \leq n \leq 0.3$. After that it increases very rapidly for $0.31 \leq n \leq 0.99$ from 0.055333 to 1.79157 and $M^-(\nu_0, n, a)$ decreases very slowly for $0.1 \leq n \leq 0.3$. After that it decreases very rapidly for $0.31 \leq n \leq 0.99$ from -0.055333 to -1.79157.

Figure 4 illustrates the graph of $M^{\pm}(\nu_0, n, a)$ for a fixed value of $v_0 = 0.1$, $n = 0.1$. It has been observed that as the parameter a due to magnetic torque effect changes, i.e., for $0.105092 \leq a \leq 0.220081$, the value of Melnikov's function $M^+(\nu_0, n, a)$ and $M^-(\nu_0, n, a)$ remains almost constant to 0.00000132343 and -0.00000132343, respectively. (The different values of a for which the graph is drawn are given in Table I).

These values of a are calculated from the expression $a = 2(\Omega_0 - \alpha_{G0} - \Phi'_m)$ for the year 2000 to 2050. Here, $\alpha_{G0} = 98.9479°$ at $0^h UT$, December 2002.

Table 1: Values of a from 2000 to 2050 at O^h UT, December 31.

Year	$g_{1'}$ (nT)	$h_{1'}$ (nT)	α_{G0} (in deg.)	$\varphi_{m'}$ (in deg.)	$\alpha_1 = \alpha_{00} + \varphi_m$ (in Rad.)	a (in Rad.)
2000	-1786	5480	98.9379	108.0514	3.612645	0.105092
2001	-1776	5469/8	98/9429	107/9882	3/611629	0.107125
2002	-1766	5459.6	98.9479	107.9247	3.610608	0.109168
2003	1756	5449.4	98.9529	107,8609	3.609581	0.11122
2004	-1746	5439.2	98.9579	107.7968	3.60855	0.113283
2005	-1736	5429	98.9629	107.7324	3.607514	0.115355
2006	-1726	5418.8	98.9679	107.6678	3.606473	0.117438
2007	-1716	5408.6	98.9729	107.6028	3.605426	0.119531
2008	-1706	5398.4	98.9779	107.5376	3.604375	0.121634
2009	-1696	5388.2	989829	107.472	3.603318	0.123747
2010	-1686	5378	98.989879	107.4062	3.602256	0.12587
2011	-1676	5367.8	98.9929	107.34	3.601189	0.128004
2012	-1666	5357.6	98.9979	107.2736	3.600117	0.130149
2013	-1656	5347.4	99.0029	107.2069	3.59904	0.132304
2014	-1646	5337.2	99.0079	107.1398	3.597957	0.134469
2015	-1636	5327	99.0129	107.0725	3,596869	0.136646
2016	-1626	5316.8	99.0179	107.0048	3.595775	0.138833
2017	-1616	5306.6	99.0229	106.9369	3.594676	0.14103
2018	-1606	5296.4	99.0279	106.8686	3.593572	0.143239
2019	-1596	5286.2	99.0329	106.8	3.592462	0.145458
2020	-1586	5276	99.0379	106.7311	3.591347	0.147689
2021	-1576	5265.8	99.0429	106.6619	3.590226	0.149931
2022	-1566	5255.6	99.0479	106.5924	3.5891	0.152183

Table 1 cont.

2023	-1556	5245.4	99.0529	106.5225	3.587968	0.154447
2024	-1546	5235.2	99.0579	106.4523	3.58683	0.156723
2025	-1536	5225	99.0629	106.3818	3.585687	0.159009
2026	-1526	5214.8	99.0679	106.311	3.584538	0.161307
2027	-1516	5204.6	99.0729	106.2398	3.583383	0.163617
2028	-1506	5194.4	99.0779	106.1683	3.582223	0.165938
2029	-1496	5184.2	99.0829	106.0965	3.581056	0.16827
2030	-1486	5174	99.0879	106.0243	3.579884	0.170615
2031	-1476	5163.8	99.0929	105.9518	3.578706	0.172971
2032	-1466	5153.6	99.0979	105.879	3.577522	0.175339
2033	-1456	5143.4	99.1029	105.8058	3.576332	0.177719
2034	-1446	5133.2	99.1079	105.7323	3.575136	0.180111
2035	-1436	5123	99.1129	105.6584	3.573934	0.182515
2036	-1426	5112.8	99.1179	105.5842	3.572726	0.184931
2037	-1416	5102.6	99.1229	105.5096	3.571511	0.18736
2038	-1406	5092.4	99.1279	105.4347	3.570291	0.189801
2039	-1396	5082.2	99.1329	105.3594	3.569064	0.192254
2040	-1386	5072	99.1379	105.2838	3.567832	0.19472
2041	-1376	5061.8	99.1429	105.2078	3.566592	0.197198
2042	-1366	5051.6	99.1479	105.1314	3.565374	0.199689
2043	-1356	5041.4	99.1529	105.0547	3.564095	0.202193
2044	-1346	5031.2	99.1579	104.9776	3.562837	0.204709
2045	-1336	5021	99.1629	104.9002	3.561572	0.207238
2046	-1326	5010.8	99.1679	104.8223	3.560301	0.209781
2047	-1316	5000.6	99.1729	104.7441	3.559024	0.212336
2048	-1306	4990.4	99.1779	104.6655	3.557739	0.214904
2049	-1296	4980.2	99.1829	104.5866	3.556448	0.217486
2050	-1286	4970	99.1879	104.5072	3.555151	0.220081

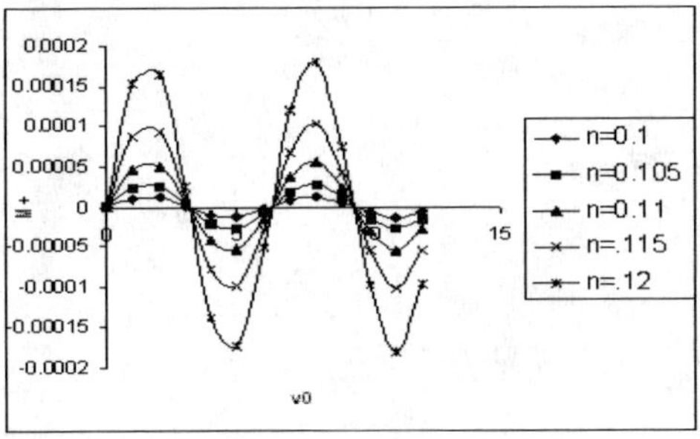

Figure 2: Melnikov's function $M^{+}(v_0, n, a)$ for $b = 2.16426351851$, $0 \leq v_0 \leq 12$, $a = 0.10916685405$, $n = 0.1,\ 0.105,\ 0.11,\ 0.115,\ 0.12$.

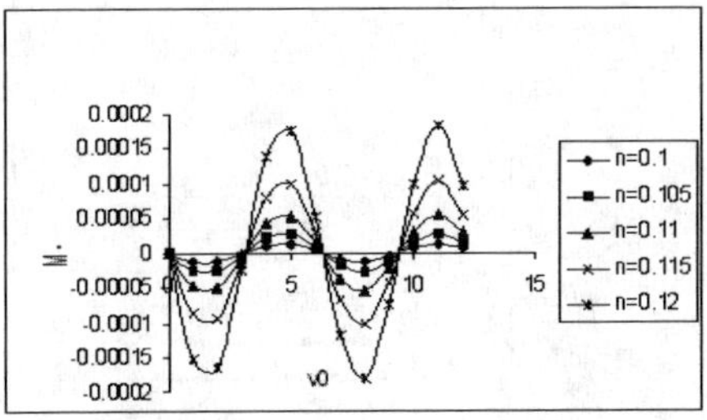

Figure 3: Melnikov's function $M^{-}(v_0, n, a)$ for $b = 2.16426351851$, $0 \leq v_0 \leq 12$, $a = 0.10916685405$, $n = 0.1,\ 0.105,\ 0.11,\ 0.115,\ 0.12$.

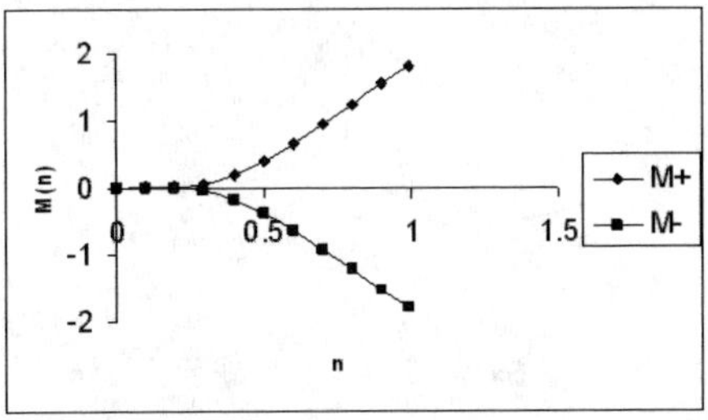

Figure 4: Melnikov's function $M^{\pm}(v_0, n, a)$ for $b = 2.16426351851$, $0.1 \leq n \leq 0.99$, $v_0 = 0.1$, $a = 0.10916685405$.

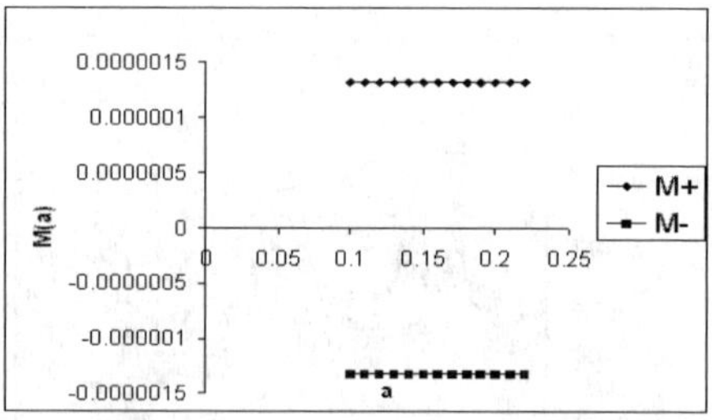

Figure 5: Melnikov's function $M^{\pm}(v_0, n, a)$ for $b = 2.16426351851$, $v_0 = 0.1$, $n = 0.1$, $0.105092 \leq a \leq 0.220081$.

3 Non-Resonant Planar Oscillations of a Satellite

Taking $q = \eta$ and $a = a_1$ in (2.5), we get

$$\frac{d^2\eta}{dv^2}+n^2\eta = -e\cos v\frac{d^2\eta}{dv^2}+2e\sin v\frac{d\eta}{dv}+4e\sin v+\varepsilon\sin(a_1+b\nu)+n^2(\eta-\sin\eta). \tag{3.1}$$

In (3.1), the non-linearity $(\eta - \sin\eta)$ is taken sufficiently weak, and therefore, it can also be taken of the order of e. So, by taking $n^2 = \alpha e$ and $\epsilon = e\epsilon_1$, (3.1) becomes

$$\frac{d^2\eta}{dv^2} + n^2\eta = ef(\nu, \eta, \frac{d\eta}{dv}, \frac{d^2\eta}{dv^2}),$$

where

$$f(\nu, \eta, \eta', \eta'') = 4\sin\nu+2(\sin\nu)\eta'-(\cos\nu)\eta''+\alpha(\eta-\sin\eta)+\varepsilon_1\sin(a_1+b\nu). \tag{3.2}$$

The dynamical system described by (3.2) moves under forced vibrations due to the presence of the periodic sine forces on the right-hand side of the equation. We are benefited by the smallness of the eccentricity e in (3.2), and hence, the solution can be obtained by exploiting the BKM method. For $e = 0$, the generating solution of the zeroth order is given by

$$\eta = a\cos\psi, \psi = n\nu + \psi^*,$$

where the amplitude a and the phase ψ^* are constants, which can be determined from the initial conditions. Proceeding exactly as in Bhatnagar and Bhardwaj, the solution of (3.2) is obtained in the form

$$\eta = a\cos\psi + eu_1(a, \nu, \psi) + e^2u_2(a, \nu, \psi) + ..., \tag{3.3}$$

where the amplitude a and the phase ψ are determined by the differential equations

$$\frac{da}{dv} = eA_1(a) + e^2A_2(a) + ... \tag{3.4}$$

$$\frac{d\psi}{dv} = n + eB_1(a) + e^2B_2(a) + ... \tag{3.5}$$

and where $A_1(a), B_1(a), u_1(a,\nu,\psi), A_2(a), B_2(a)$ etc. are found to be

$$A_1(a) = 0, \quad B_1(a) = \frac{\alpha(2J_1(a) - a)}{2na},$$

$$u_1(a,\nu,\psi) = \frac{4\sin\nu}{n^2-1} - (na + \frac{n^2a}{2})\frac{\cos(\nu+\psi)}{2n+1} + (\frac{n^2a}{2} - na)\frac{\cos(\nu-\psi)}{2n-1} +$$

$$+\frac{\varepsilon_1 \sin(a_1 + b\nu)}{n^2 - b^2} + \frac{\alpha}{2n^2}\sum_{k=1}^{\infty}\frac{(-1)^k J_{2k+1}(a)\cos(2k+1)\psi}{k(k+1)},$$

$$A_2(a) = 0, B_2(a) = \frac{\alpha^2}{2an^3}\sum_{k=1}^{\infty}\frac{J_{2k+1}(a)J'_{2k+1}(a)}{k(k+1)} - \frac{\alpha^2}{8a^2n^3}(2J_1(a) - a)^2$$

$$+\frac{3n(n^2-1)}{4(4n^2-1)}$$

where $J'_{2k+1}(a) = \frac{dJ_{2k+1}(a)}{da}$ and $J_k(a)$ is the Bessel's function of the kth order. Thus, in the first approximation, the solution is given by

$$\eta = a\cos\psi, \tag{3.6}$$

where the amplitude a and the phase ψ are given by

$$\frac{da}{dv} = 0 \Rightarrow a = \text{constant, and} \quad \frac{d\psi}{dv} = n + \frac{n}{2a}(2J_1(a) - a), \tag{3.7}$$

and in the second approximation, the solution is obtained as

$$\eta = a\cos\psi + (e\frac{4\sin v}{n^2-1} - \left(na + \frac{n^2a}{2}\right)\frac{\cos(\nu+\psi)}{2n+1} + \left(\frac{n^2a}{2} - na\right)\frac{\cos(\nu-\psi)}{2n-1}$$

$$+\frac{\varepsilon_1\sin(a_1 + b\nu)}{n^2-b^2} + \frac{\alpha}{2n^2}\sum_{k-1}^{\infty}\frac{(-1)^k J_{2k-1}(a)\cos(2k+1)\psi}{k(k+1)}), \tag{3.8}$$

where the amplitude a and the phase ψ are given by $\frac{da}{dv} = 0 \Rightarrow a =$ constant and,

$$\frac{d\psi}{dv} = n + \frac{n}{2a}(2J_1(a) - a) + \tag{3.9}$$

$$+\frac{n}{2a}\sum_{k=1}^{\infty}\frac{J_{2k+1}(a)J'_{2k+1}(a)}{k(k+1)} - \frac{n}{8a^2}(2J_1(a) - a)^2 + e^2\frac{3n(n^2-1)}{4(4n^2-1)}. \tag{3.10}$$

It may be observed that the amplitude of the oscillation remains constant even up to the second order of approximation. Moreover, it is observed that the main resonance occurs at $n \cong b$ and $n \cong 1$. Up to the second approximation, the parametric resonance appears only for $n \cong \frac{1}{2}$.

4 Resonant Planar Oscillation of a Satellite

Now, we proceed to study the asymptotic solution near the main resonance $n \cong b$and $n \cong 1$ parametric resonance $n \cong \frac{1}{2}$ by the BKM method.

Case I: Let $n = \frac{1}{k}$, k=1, -1, 2,-2.

For $e = 0$, the generating solutions are

$$\eta = a\cos\psi, \psi = \frac{1}{k}\nu + \theta, \tag{4.1}$$

where the amplitude a and phase θ are determined by the following equations:

$$\frac{da}{dv} = eA_1(a,\theta)\ \frac{d\theta}{dv} = n - \frac{1}{k} + eB_1(a,\theta)$$

$$\frac{d\psi}{dv} = \frac{1}{k} + \frac{d\theta}{dv} = n + eB_1(a,\theta), \tag{4.2}$$

where $A_1(a,\theta)$ and $B_1(a,\theta)$ are particular solutions, periodic with respect to θ. Using (4.1) and (4.2), we calculate $\frac{d\eta}{dv}$ and $\frac{d^2\eta}{dv^2}$, and then substituting the values η, $\frac{d\eta}{dv}$ and $\frac{d^2\eta}{dv^2}$ in (3.1) and equating the coefficients of e, we get

$$\left(\left(n - \frac{1}{k}\right)\frac{\partial A_1}{\partial\theta} - 2anB_1\right)\cos\psi - \left(\left(n - \frac{1}{k}\right)a\frac{\partial B_1}{\partial\theta} + 2nA_1\right)\sin\psi =$$

$$4\sin\nu - 2an\sin\nu\sin\psi + an^2\cos\nu\cos\psi + \alpha a\cos\psi - \alpha\sin(a\cos\psi) +$$
$$+\varepsilon_1\sin(a_1 + b\nu). \tag{4.3}$$

Using Fourier expansion given by

$$\sin(a\cos\psi) = 2\sum_{k-0}^{\infty}(-1)^k J_{2k+1}(a)\cos(2k+1)\psi,$$

$$\cos(a\cos\psi) = J_0(a) + 2\sum_{k-1}^{\infty}(-1)^k J_{2k}(a)\cos(2k)\psi,$$

where $J_k(a)$ stands for Bessel's function of order k, in (4.1.3), we get

$$\left(\left(n - \frac{1}{k}\right)\frac{\partial A_1}{\partial\theta} - 2anB_1\right)\cos\psi - \left(\left(n - \frac{1}{k}\right)a\frac{\partial B_1}{\partial\theta} + 2nA_1\right)\sin\psi$$
$$= 4\sin k(\psi - \theta)$$

$$-2an\sin k(\psi-\theta)\sin\psi + an^2\cos k(\psi-\theta)\cos\psi + \alpha(a - 2J_1(a))\cos\psi -$$

$$2\alpha\sum_{j-1}^{\infty}(-1)^j J_{2j+1}(a)\cos(2j+1)\psi + \varepsilon_1 \sin(a + bk\psi - bk\theta). \tag{4.4}$$

Subcase 1.1 when $k = 1$, $n \cong 1$.
Substituting these values in the equation and comparing the coefficients of like powers of $\cos\psi$ and $\sin\psi$ to zero, so that $u_1(a, \nu, \psi)$ should not contain the resonant term, we get

$$(n-1)\frac{\partial A_1}{\partial\theta} - 2anB_1 = -4\sin\theta + \alpha(a - 2J_1(a))$$

$$(n-1)\,a\frac{\partial B_1}{\partial\theta} + 2nA_1 = -4\cos\theta. \tag{4.5}$$

Solving (4.5), we get

$$A_1 = \frac{-4}{n+1}\cos\theta \; B_1 = \frac{4}{a\,(n+1)}\sin\theta - \frac{\alpha}{2an}(a - 2J_1(a)). \tag{4.6}$$

Thus, the solution in the first approximation is given by

$$\eta = a\cos(\nu + \theta),$$

where the amplitude a and phase θ are the solutions of the equations

$$\frac{da}{d\nu} = \frac{-4e}{n+1}\cos\theta \; \frac{d\theta}{d\nu} = (n-1) + \frac{4e\sin\theta}{a(n+1)} - \frac{n}{2a}(a - 2J_1(a). \tag{4.7}$$

The equations in (4.7) cannot be integrated in a closed form due to dependence of the right-hand side on a and θ. However, qualitative aspects of the solution can be examined with the help of Poincare's Theory. The stationary state of the oscillation is defined by

$$\frac{da}{d\nu} = \frac{d\theta}{d\nu} = 0. \tag{4.8}$$

However, (4.7) may be represented, correct to the second order, in the form defined by

$$2\frac{da}{d\nu} = -4e\cos\theta forn = 1$$

$$\frac{d\theta}{d\nu} = n_e - 1 + \frac{4e\sin\theta}{a(n+1)}, \tag{4.9}$$

where

$$n_e = n\left(1 - \frac{\alpha^2}{16}\right)$$

or

$$n_e \cong nsince\alpha \ll 1.$$

Taking $n \cong n_e$ and $n = 1$, the system of (4.9) may be represented correct to the second order in the form

$$2\frac{da}{d\nu} = -4e\cos\theta 2a\frac{d\theta}{d\nu} = a(n_e^2 - 1) + 4e\sin\theta. \tag{4.10}$$

Now, the stationary regimes are given by

$$4e\cos\theta = 0$$

$$a(n_e^2 - 1) + 4e\sin\theta = 0.$$

Eliminating the phase θ, we get

$$n_e^2 - 1 = \frac{-4e}{a}. \tag{4.11}$$

Putting $1 = n + \delta$, $\delta \ll 1$, we obtain

$$8\delta^2 + 16n\delta + n^2a^2 - \frac{32e}{a} = 0 \tag{4.12}$$

or

$$\delta = -n \pm \frac{1}{2\sqrt{2}}\sqrt{\frac{32e}{a} - n^2(a^2 - 8)}.$$

Let us now examine the relation existing between the parameters of the system for the occurrence of the effect under consideration.

The necessary condition for instability (jump and fall) is $\frac{d\delta}{da} = 0$. Proceeding with the equation, we get

$$\frac{d\delta}{da} = -\frac{8\delta^2 + 16n\delta + 3n^2a^2}{16a(\delta + n)} = 0\delta^2 + 16n\delta + 3n^2a^2 = 0. \tag{4.13}$$

Thus, the two values represented by the equation are both negative, so that the effect occurs only at a frequency of external periodic force which is less than the frequency of the natural oscillation of the system.

Now, maximum value of the amplitude is obtained by the condition $\frac{da}{d\delta} = 0$, which gives

$$a(\delta + n) = 0$$

since $a > 0 \Rightarrow \delta = -n$.

Substituting this value in the equation and putting $n = 1$, we get

$$a^3 - 8a - 32e = 0.$$

Now with the help of (4.11) and (4.13), we construct the resonance curve. It is easy to analyze the stable and unstable zones of amplitudes with the help of this curve.

We have calculated the above resonance curves for various values of e.

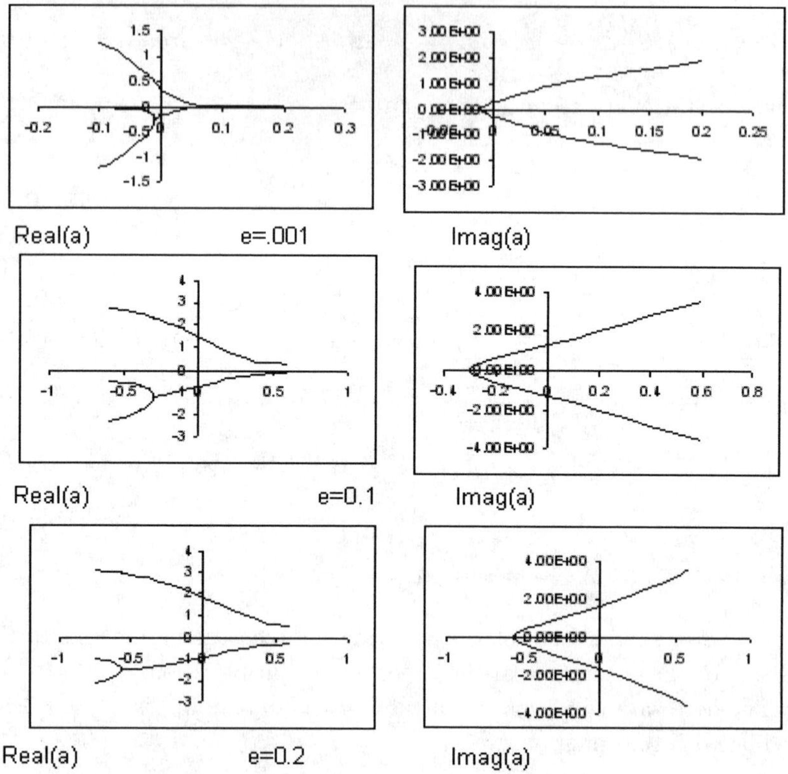

Figure 6: Resonance curve in case of main resonance at $n = 1$.

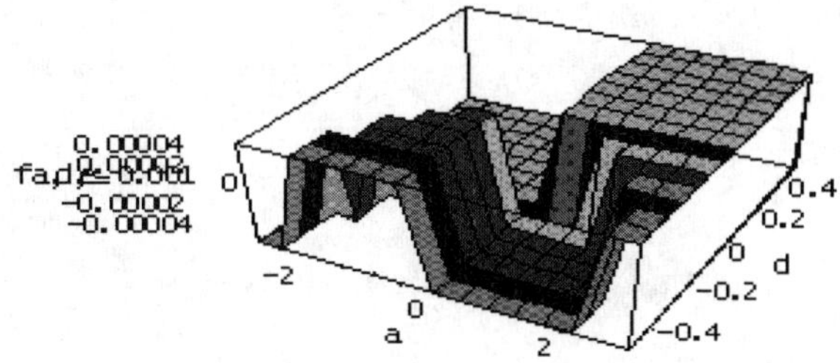

(a) e=.001

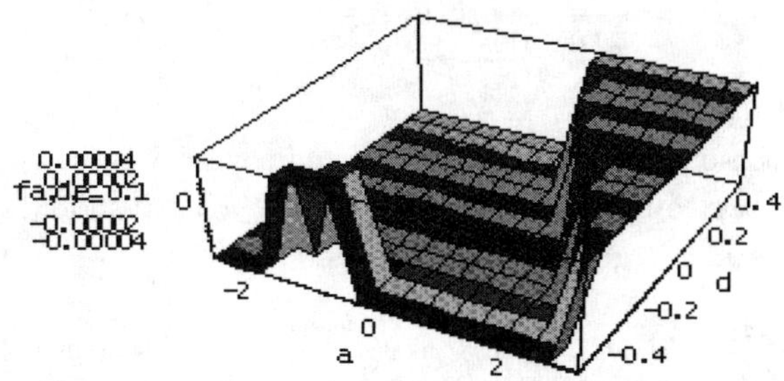

(b) e=0.1

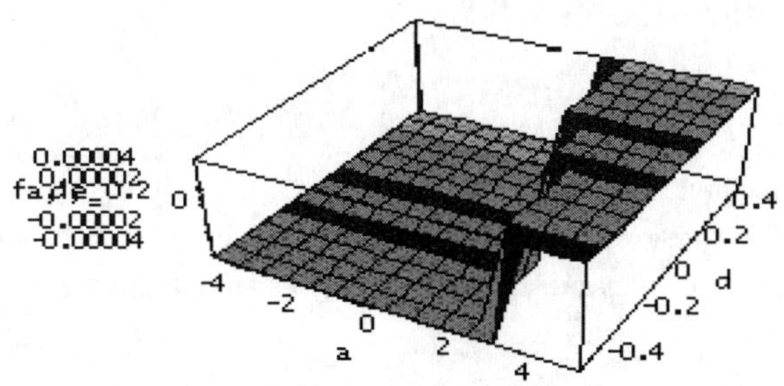

(c) e=0.2

Figure 7: Three-dimensional plot of resonance curve at $n = 1$.

Subcase 1.2 when $k = -1, n \cong -1$.
Substituting these values in the equation and comparing the coefficients of like powers of $cos\psi$ and $sin\psi$ to zero, so that $u_1(a, v, \psi)$ should not contain the resonant term, we get

$$(n+1)\frac{\partial A_1}{\partial \theta} - 2anB_1 = 4\sin\theta + \alpha(a - 2J_1(a))$$

$$(n+1)a\frac{\partial B_1}{\partial \theta} + 2nA_1 = 4\cos\theta. \tag{4.14}$$

Solving (4.14), we get

$$A_1 = \frac{4}{n-1}\cos\theta$$

$$B_1 = \frac{-4}{a(n-1)}\sin\theta - \frac{\alpha}{2an}(a - 2J_1(a)).$$

Subcase 1.3 when $k = 2$, $n \cong \frac{1}{2}$.
Substituting these values in the equation and comparing the coefficients of like powers of $cos\psi$ and $sin\psi$ to zero, so that $u_1(a, v, \psi)$ should not contain the resonant term, we get

$$(n-\frac{1}{2})\frac{\partial A_1}{\partial \theta} - 2anB_1 = -an\cos 2\theta + \frac{an^2\cos 2\theta}{2} + \alpha(a - 2J_1(a))$$

$$(n-\frac{1}{2})a\frac{\partial B_1}{\partial \theta} + 2nA_1 = an\sin 2\theta - \frac{an^2\sin 2\theta}{2} \tag{4.15}$$

Solving eqns, (4.15), we get

$$A_1 = \frac{-an(n-2)}{2}\sin 2\theta B_1 = \frac{-n(n-2)}{2}\cos 2\theta - \frac{\alpha}{2an}(a - 2J_1(a)).$$

Subcase 1.4 when $k = -2, n \cong -\frac{1}{2}$.

Substituting these values in the equation and comparing the coefficients of like powers of $cos\psi$ and $sin\psi$ to zero, so that $u_1(a, v, \psi)$ should not contain the resonant term, we get

$$(n+\frac{1}{2})\frac{\partial A_1}{\partial \theta} - 2anB_1 = an\cos 2\theta + \frac{an^2\cos 2\theta}{2} + \alpha(a - 2J_1(a))$$

$$(n+\frac{1}{2})a\frac{\partial B_1}{\partial \theta} + 2nA_1 = -an\sin 2\theta - \frac{an^2\sin 2\theta}{2}. \tag{4.16}$$

Solving (4.16), we get

$$A_1 = \frac{an(n+2)}{2}\sin 2\theta B_1 = \frac{n(n+2)}{2}\cos 2\theta - \frac{\alpha}{2an}(a - 2J_1(a)).$$

Case II: Let $n = kb$, $k = \pm 1$.
For $e = 0$, the generating solutions are

$$\eta = a\cos\psi\psi = kb\nu + \theta, \tag{4.17}$$

where the amplitude is determined by the following equations:

$$\frac{da}{d\nu} = eA_1(a,\theta)$$
$$\frac{d\theta}{d\nu} = n - kb + eB_1(a,\theta)$$
$$\frac{d\psi}{d\nu} = kb + \frac{d\theta}{d\nu} = n + eB_1(a,\theta), \tag{4.18}$$

where $A_1(a,\theta)$ and $B_1(a,\theta)$ are particular solutions, periodic with respect to θ. Using (4.2.1) and (4.2.2), we calculate $\frac{d\eta}{d\nu}$ and $\frac{d^2\eta}{d\nu^2}$, and then substituting the values of $\eta, \frac{d\eta}{d\nu}$ and $\frac{d^2\eta}{d\nu^2}$ in (3.1) and equating the coefficients of e, we get

$$\left((n-kb)\frac{\partial A_1}{\partial\theta} - 2anB_1\right)\cos\psi - \left((n-kb)\,a\frac{\partial B_1}{\partial\theta} + 2nA_1\right)\sin\psi =$$
$$4\sin\nu - 2an\sin\nu\sin\psi + an^2\cos\nu\cos\psi + \alpha a\cos\psi - \alpha\sin(a\cos\psi) +$$
$$\varepsilon_1\sin(a_1 + b\nu). \tag{4.19}$$

Using Fourier expansion given by

$$\sin(a\cos\psi) = 2\sum_{k-D}^{\infty}(-1)^k J_{2k-1}(a)\cos(2k+1)\psi$$

$$cos(a\cos\psi) = J_0(a) + 2\sum_{k-1}^{\infty}(-1)^k J_{2k}(a)\cos(2k)\psi, \tag{4.20}$$

where $J_k(a)$ stands for Bessel's function of order k, in (4.2.4), we get

$$\left((n-kb)\frac{\partial A_1}{\partial\theta} - 2anB_1\right)\cos\psi - \left((n-kb)\,a\frac{\partial B_1}{\partial\theta} + 2nA_1\right)\sin\psi =$$
$$4\sin\left(\frac{\psi-\theta}{kb}\right) - 2an\sin\left(\frac{\psi-\theta}{kb}\right)\sin\psi + an^2\cos\left(\frac{\psi-\theta}{kb}\right)\cos\psi$$
$$+\alpha(a - 2J_1(a))\cos\psi - 2\alpha\sum_{j-1}^{\infty}(-1)^j J_{2j+1}(a)\cos(2j+1)\psi$$

$$+\varepsilon_1 \sin\left(a + \frac{\psi - \theta}{k}\right). \tag{4.21}$$

Subcase 2.1 when $k = 1,\ n \cong b$.
Substituting these values in the (4.21) and comparing the coefficients of like powers of Subcase 2.1 when $k = 1$, Substituting these values in (4.21) and comparing the coefficients of like powers of $\cos\psi$ and $\sin\psi$ to zero, so that $u_1(a, \nu, \psi)$ should not contain the resonant term, we get

$$(n-b)\frac{\partial A_1}{\partial\theta} - 2anB_1 = \varepsilon_1 \sin(a-\theta) + \alpha(a - 2J_1(a))(n - \frac{b}{2})a\frac{\partial B_1}{\partial\theta} - 2nA_1 =$$

$$-\varepsilon_1 \cos(a - \theta). \tag{4.22}$$

Solving (4.22), we get

$$A_1 = \frac{-\varepsilon_1}{n+b}\cos(a-\theta), B_1 = \frac{-\varepsilon_1}{a\,(n+b)}\sin(a-\theta) - \frac{\alpha}{2an}(a - 2J_1(a)). \tag{4.23}$$

Thus, the solution in the first approximation is given by

$$\eta = a\cos(\nu + \theta), \tag{4.24}$$

where the amplitude a and phase θ are the solutions of the equations

$$\frac{da}{d\nu} = \frac{-\varepsilon_1}{n+b}\cos(a-\theta),$$

$$\frac{d\theta}{d\nu} = (n-b) - \frac{\varepsilon_1}{a\,(n+b)}\sin(a-\theta) - \frac{n}{2a}(a - 2J_1(a)). \tag{4.25}$$

Equation (4.25) cannot be integrated in a closed form due to dependence of the right-hand side on a and θ. However, qualitative aspects of the solution can be examined with the help of Poincare theory.

The stationary state of the oscillation is defined by

$$\frac{da}{d\nu} = \frac{d\theta}{d\nu} = 0. \tag{4.26}$$

However, (4.25) may be represented, correct to the second order, in the form

$$2b\frac{da}{d\nu} = -\varepsilon_1 \cos(a-\theta)\, forn = b,$$

$$\frac{d\theta}{d\nu} = n_e - b - \frac{\varepsilon_1}{a\,(n+b)}\sin(a-\theta), \tag{4.27}$$

where $n_e = n\left(1 - \frac{a^2}{16}\right) \cong n$ since $a \ll 1$.

Taking $n \cong n_e$ and $n = b$, the system of (4.27) may be represented correct to the second order in the form

$$2b\frac{da}{d\nu} = -\varepsilon\cos(a-\theta),$$

$$2ab\frac{d\theta}{d\nu} = a(n_e^2 - b^2) - \varepsilon\sin(a-\theta).$$

Now, the stationary regimes are given by

$$-\varepsilon\cos(a-\theta) = 0$$

$$a(n_e^2 - b^2) - \varepsilon\sin(a-\theta) = 0.$$

Eliminating the phase, we get

$$n_e^2 - b^2 = \frac{\varepsilon}{a}.$$

Putting $b = n + \delta$, $\delta \ll 1$, we obtain

$$8\delta^2 + 16n\delta + n^2a^2 + \frac{8\varepsilon}{a} = 0 \tag{4.28}$$

or

$$\delta = -n \pm \frac{1}{2\sqrt{2}}\sqrt{n^2(8-a^2) - \frac{8\varepsilon}{a}}.$$

Let us now examine the relation existing between the parameters of the system for the occurrence of the effect under consideration.

The necessary condition for instability (jump and fall) is $\frac{d\delta}{da} = 0$. Proceeding with the equation, we get

$$\frac{d\delta}{da} = -\frac{8\delta^2 + 16n\delta + 3n^2a^2}{16a(\delta + n)} = 0$$

$$\Rightarrow 8\delta^2 + 16n\delta + 3n^2a^2 = 0.$$

Thus, the two values represented by the equation are both negative, so that the effect occurs only at a frequency of external periodic force which is less than the frequency of the natural oscillation of the system.

Now, maximum value of the amplitude is obtained by the condition $\frac{da}{d\delta} = 0$, which gives

$$16a(\delta + n) = 0.$$

Since $a \succ 0 \Rightarrow \delta = -n$ Substituting this value in the equation and putting $n = b$, we get

$$b^2a^3 - 8ab^2 + 8\varepsilon = 0.$$

Now we construct the resonance curve. It is easy to analyze the stable and unstable zones of amplitudes with the help of this curve.

We have calculated the above resonance curves for various values of e.

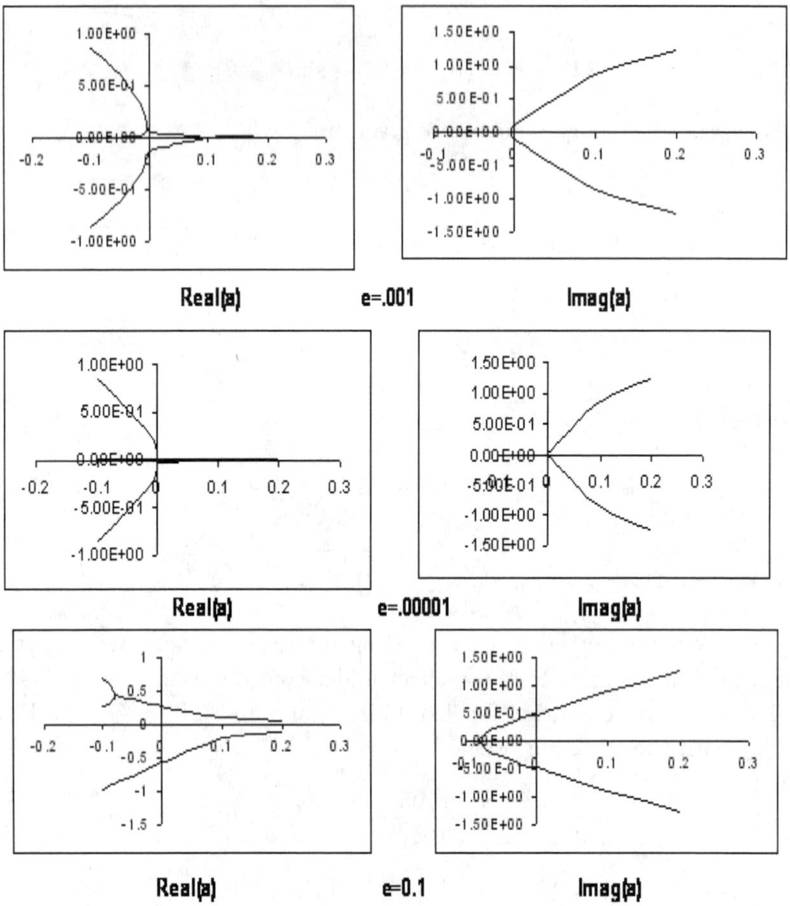

Figure 8: Resonance curve in case of main resonance at $n = b$.

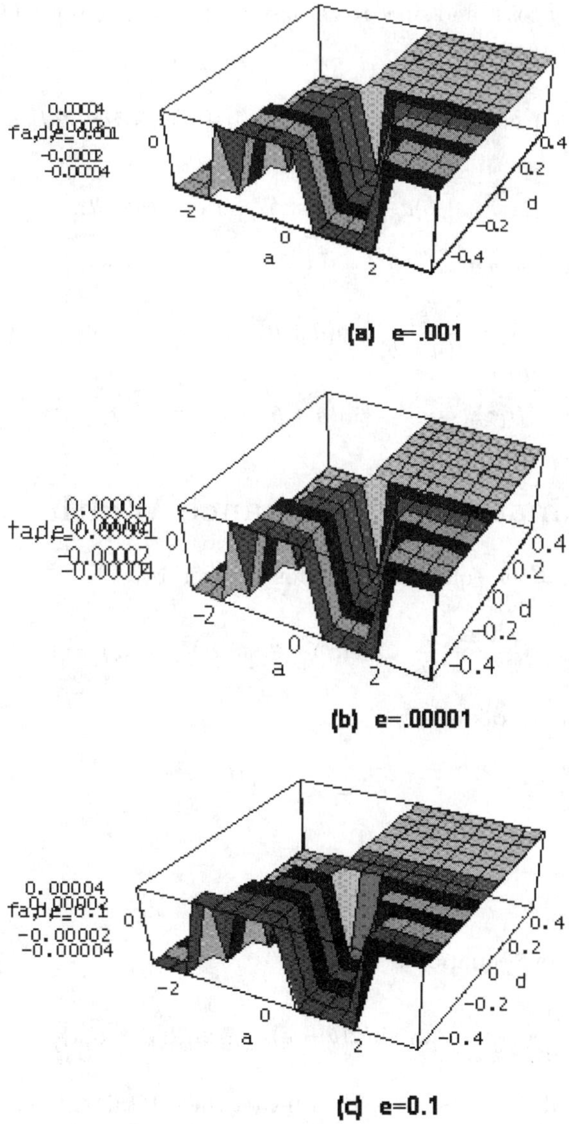

Figure 9: Three-dimensional plot of resonance curve at $n = b$.

Subcase 2.2 when $k = -1,\ n \cong -b$.
Substituting these values in the equation and comparing the coefficients of like powers of $\cos\psi$ and $\sin\psi$ to zero, so that $u_1(a, \nu, \psi)$ should not contain the resonant term, we get

$$(n+b)\frac{\partial A_1}{\partial\theta} - 2anB_1 = \varepsilon_1 \sin(a+\theta) + \alpha(a - 2J_1(a))$$

$$(n+b)a\frac{\partial B_1}{\partial\theta} + 2nA_1 = \varepsilon_1 \cos(a+\theta).$$

Solving (4.4), we get

$$A_1 = \frac{\varepsilon_1}{(n-b)}\cos(a+\theta)$$

$$B_1 = \frac{-\varepsilon_1}{(n-b)}\sin(a+\theta) - \frac{\alpha}{2an}(a - 2J_1(a)).$$

5 Estimation of Resonance Width

We know that the equation of motion is given by

$$\frac{d^2\theta}{dt^2} + \frac{\mu}{2r^3}\left[n^2 \sin\delta - \in \sin 2(V - \alpha_m)\right] = 0.$$

In this equation taking

$$n^2 = \omega_0^2 = 3\frac{B-A}{C}$$

Also,

$$\theta = v + \frac{\delta}{2} \Rightarrow \delta = 2(\theta - v).$$

Then the above equation becomes

$$\frac{d^2\theta}{dt^2} + \frac{\mu}{2r^3}\left[\omega_0^2 \sin 2(\theta - v) - \in \sin 2(V - \alpha_m)\right] = 0. \tag{5.1}$$

In (5.1), if the units are so chosen that the orbital period of the satellite is 2π and its radius be unity, then the dimensionless time is equal to the mean longitude by the true anomaly v and v or t is 2π periodic and $\mu = 1$.

Also, we have

$$\alpha_m = \alpha_{G0} + \frac{d\alpha_G}{dt}t + \phi'_m = \alpha_1 + \beta t$$

and

$$b = 1 - \beta,$$

Then Equation (5.1) becomes

$$\frac{d^2\theta}{dt^2} + \frac{\omega_0^2}{2}\sin 2(\theta - t) - \frac{\in}{2}\sin 2(bt - \alpha_1) = 0. \tag{5.2}$$

This is an equation of a pendulum perturbed by a force $\frac{\epsilon}{2}\sin 2(bt - \alpha_1)$. This equation can be studied under the cases $\epsilon = 0$ and $\epsilon \neq 0$.

When $\epsilon \neq 0$, (5.2) represents the motion of a disturbed pendulum given by

$$\frac{d^2x}{dt^2} + f'(x) = m\phi^{'}(x,t), \tag{5.3}$$

where

$$\begin{aligned} x &= 2(\theta - t), \\ f'(x) &= k_1^2 \sin(x), \\ k_1^2 &= \omega_0^2, \\ m &= -\epsilon, \\ \phi^{'}(x,t) &= \sin 2(bt - \alpha_1). \end{aligned}$$

The unperturbed part of (5.3) is

$$\frac{d^2x}{dt^2} + f'(x) = 0.$$

For this equation, we have

$$\left(\frac{dx}{dt}\right)^2 = C + 2k_1^2\cos(x),$$

where C is a constant of integration.

The motion to be real, if $C + 2k_1^2\cos(x) \geq 0$. There are three categories of motion depending upon $C > 2k_1^2$, $C < 2k_1^2$, and the intermediate case $C = 2k_1^2$.

Category I: $C > 2k_1^2$.

As $C > 2k_1^2$, so $\frac{dx}{dt}$ never vanishes in this case, it is always either +ve or –ve, and the pendulum is making a complete revolution in one sense or other.

In this case the unperturbed solution is

$$x = L + C_1 \sin(L) + O(C_1^2), \tag{5.4}$$

where

$$L = nt + \epsilon_1,$$
$$C_1^2 = \frac{K_1^2}{n^2},$$

and

$$\frac{1}{n} = \frac{1}{2\pi} \int_0^{2\pi} \frac{dx}{\left(C + 2k_1^2 \cos(x)\right)^{1/2}}$$

C_1 and ϵ_1 are arbitrary constants and L is an argument.

The periodic portion of this series can be regarded as an oscillation about the mean state of motion which is revolution with a period $\frac{2\pi}{n}$. The half amplitude of the oscillation is evidently less than π, and it decreases as n increases.

Here, we may observe that $\frac{dx}{dt} \neq 0$ and the motion is said to be of the type I, i.e., revolution.

In case of perturbed pendulum by making use of the theory of variation of parameters, we have

$$\begin{aligned} \frac{dC_1}{dt} &= \frac{m}{v}\frac{\partial x}{\partial L}\phi'(x,t), \\ \frac{dL}{dt} &= n - \frac{m}{v}\frac{\partial x}{\partial C_1}\phi'(x,t), \end{aligned} \tag{5.5}$$

where

$$v = \frac{\partial}{\partial C_1}\left(n - \frac{\partial x}{\partial L}\right)\frac{\partial x}{\partial L} - n\frac{\partial^2 x}{\partial L^2}\frac{\partial x}{\partial C_1}. \tag{5.6}$$

We already have calculated that

$$v \cong -\frac{n}{2C_1}.$$

Substituting the value of v in (5.5), we get

$$\frac{dC_1}{dt} \cong -\frac{2mk_1^2}{n^3}\sin 2(bt - \alpha_1) \cong 0.$$

Since, both m and k_1^2 are small quantities, the term mk_1^2 is of the third order and therefore rejected.

Thus,

$$\frac{dC_1}{dt} \cong 0$$

$\Rightarrow C_1 =$ constant up to the second order of approximation,
and

$$\frac{dL}{dt} \cong n + \frac{2mC_1}{n}\sin(L)\sin 2(bt - \alpha_1),$$

$$\frac{d^2L}{dt^2} \cong m\sin 2(bt - \alpha_1).$$

Since

$$L = nt + \in_1 \Rightarrow t = \frac{L - \epsilon_1}{n}.$$

Thus,

$$\frac{d^2L}{dt^2} \cong 2m\sin\left(\frac{Lb}{n}\right).$$

In the first approximation taking $n = n_0$, we get

$$\frac{d^2L}{dt^2} \cong 2m\sin\left(\frac{Lb}{n_0}\right).$$

If we take $\frac{Lb}{n_0} = x$, then the above equation becomes

$$\frac{d^2x}{dt^2} + k_2^2\sin x = 0, \tag{5.7}$$

where

$$k_2^2 = 2\frac{\in b}{n_0}.$$

Integrating (5.7), we get

$$\left(\frac{dx}{dt}\right)^2 = C_2 + 2k_2^2\cos(x),$$

where C_2 is a constant of integration.

Equation (5.7) describes the motion of a pendulum. We get again three types of motion. Type I is that in which $\frac{dx}{dt}$ is never zero. Type II is that in which $\frac{dx}{dt} = 0$, at $x = 0$ or π.
For type I, our solution is

$$x = Nt + \in_2 + \frac{k_2^2}{N^2}\sin(Nt + \epsilon_2) + \ldots,$$

where

$$\frac{1}{N} = \frac{1}{2\pi}\int_0^{2\pi}\frac{dx}{(C_2 + 2k_2^2\cos(x))^{1/2}}.$$

In the first approximation $N = N_0$, and the solution is

$$x = x_0 + \frac{k_2^2}{N^2}\sin(x_0),$$

where

$$x_0 = N_0 t + \epsilon_2.$$

This is the case of revolution.

For type II, the solution is

$$x = \lambda \sin(pt + \lambda_0),$$

where

$$p^2 = \left|\frac{ma}{v}\left(\frac{\partial n}{\partial C}\right)_0\right|,$$

so

$$p^2 \cong k_2^2$$
$$\Rightarrow p = \sqrt{\frac{2b\epsilon}{n_0}}$$

with λ and λ_0 being arbitrary constants. This is the case of libration.

Type III occurs when $C_2 = 2k_2^2 = 4\frac{b\epsilon}{n_0}$. The solution is

$$x + \pi = 4\tan^{-1}\left(\exp\left(\sqrt{\frac{2b\epsilon}{n_0}}t + \alpha_0\right)\right),$$

where α_0 is an arbitrary constant and the other having a particular value.

When $t \to \pm\infty, x \to \pm\pi$, at both places, $\frac{dx}{dt} = 0$ and all higher derivatives of x approach to zero. Near this point, while x approaches to one of the limits $\pm\pi, t$ tends to become an indeterminate function of x. This is the case of infinite period separatrix which is asymptotic forward and backward in time to the unstable equilibrium. Thus, the results of type I, type II, and type III enable us to conclude that the magnetic torque perturbation plays a significant role. It may change a revolution to libration or to infinite period separatrix.

Category II: $C < 2k_1^2$.

In this case, the unperturbed solution is

$$x = C_1 \sin(L) + O(C_1^3),$$

where

$$L = nt + \epsilon_1,$$
$$n = k_1\left[1 - \frac{1}{16}C_1^2 + O(C_1^4)\right].$$

C_1 and ϵ_1 are arbitrary constants.

In case of a perturbed equation, again using the theory of variation of parameters, we get

$$\frac{dC_1}{dt} \cong \frac{m}{k_1}\cos(L)\sin 2(bt-\alpha_1),$$
$$\frac{dL}{dt} \cong k_1 - \frac{m}{k_1C_1}\sin L\sin 2(bt-\alpha_1),$$
$$\frac{d^2L}{dt^2} \cong -\frac{2mb}{k_1C_1}\sin L\cos 2(b\frac{L-\epsilon_1}{n}-\alpha_1).$$

In the first approximation if $n = n_0, C_1 = C_{10}$, we get

$$\frac{d^2L}{dt^2} \cong -\frac{2mb}{k_1C_{10}}\sin L\cos 2(b\frac{L-\epsilon_1}{n_0}-\alpha_1). \tag{5.8}$$

As a special case, let us assume that

$$b\frac{L-\epsilon_1}{n_0}-\alpha_1 = n_1\frac{\pi}{2}, n_1 \in I,$$

when n_1 is odd, then (5.8) becomes

$$\frac{d^2L}{dt^2} + K_3^2\sin L \cong 0,$$

where

$$K_3^2 = -\frac{2mb}{k_1C_{10}} > 0 \text{ as } m < 0,$$

which is again the equation of the pendulum.

As in the previous case, this equation gives us revolution, libration, and infinite period separatrix motion.

On the other hand, if n_1 is even, then

$$\frac{d^2L}{dt^2} \cong K_3^2\sin L.$$

When L is small, the solution of the above equation is

$$L = e^{k_3t} + e^{-k_3t}.$$

Category III: $C = 2k_1^2$.

The unperturbed solution is

$$x + \pi = 4\tan^{-1}e^{k_1t} + \alpha_0,$$

where α_0 is an arbitrary constant and other having a specific value. This is the case of infinite period separatrix which is asymptotic forward and backward in time to the unstable equilibrium. We are mainly concerned with this category of motion. In this category the nature of the unperturbed solution does not change by taking into account the magnetic torque.

6 The Spin Orbit Phase Space

In our spin orbit problem which is 2π periodic in dimensionless time, we have computed the mapping on the surface of a section by looking at the trajectories stroboscopically with period 2π. The section has been drawn with $\frac{d\theta}{dt}$ versus θ at every periapse passage. In the case of quasi-periodic trajectories the points are contained in smooth curves, while for the chaotic trajectories they appear to fill up the area in the phase space in a random manner. Since the orientation is denoted by $n + \theta$, we have restricted the interval from 0 to π. The spin orbit states that are determined in the previous section are states where a resonance variable librates. For each of these states $\frac{d\theta}{dt}$ has an average value precisely equal to p, and θ rotates through all values. Consequently, quasi-periodic successive points will trace a simple curve on the section near $\frac{d\theta}{dt} = \theta$. This will cover only a fraction of the interval from 0 to π. In the case of non-resonant quasi-periodic trajectories, all resonant states rotate and successive points on the surface of section will trace a simple curve which covers all values of θ.

7 Conclusion

From these studies, we conclude that the magnetic torque parameter plays a significant role in changing the motion of revolution into the motion of libration and infinite period of revolution into the motion of libration and infinite period separatrix. We also observe that the regular motion changes into a chaotic one for some values of the magnetic torque parameter and the values of b. The changes in the values do not play a significant role, even over twenty years, i.e., from 1975 to 1995. We thus conclude that the non-linear rotational equations of motion of the planar oscillation of a satellite in an elliptic orbit under the influence of magnetic torque are non-integrable. Also, we have observed graphically this in the Earth-Moon system. The Melnikov's function has a simple zero, and hence, the equations of motions are non-integrable.

References

[1] K. Alfriend, H. Schaub, and D. Gim, Gravitational perturbations, nonlinearity and circular orbit assumption effects on formation flying control strategies, AAS 00-12, Presented at *AAS Rocky Mountain Conference,* Brecinridge, CO, Feb 2–6, 2000.

[2] J.G. Bastow (Ed.), Proceedings of the magnetic workshop, *JPL Tech. Memo.,* (1965), 32–316.

[3] V.V. Beletskii, *The Motion of Artificial Satellite around a Centre of Mass,* Nauka, Moscow, 1965.NASA TT (1966), F-429.

[4] K.B. Bhatnagar, L.M. Saha, and Ayub Khan, non-linear planar oscillation of a satellite in elliptical orbit under the influence of solar radiation pressure (1) *Bull. Astr. Soc. India,* **22** (1994), 47–58.

[5] K.B. Bhatnagar, L.M. Saha, and Ayub Khan, Non-linear planar oscillation of a satellite in elliptical orbit under the influence of solar radiation pressure (2), *Bull. Astr. Soc. India,* **22** (1994), 275–290.

[6] K.B. Bhatnagar and R. Bhardwaj, Rotational motion of a satellite in an elliptic orbit under the influence of third body torque (I), *Bull. Astr. Soc. India,* **22** (1994), 359–367.

[7] K.B. Bhatnagar and R. Bhardwaj, Nonlinear planar oscillation of a satellite in a circular orbit under the influence of magnetic torque (I), *Indian J. Pure Appl. Math.,* **26(12)** (1995), 1225–1240.

[8] K.B. Bhatnagar and R. Bhardwaj, Resonance on nonlinear planar oscillation of a satellite under the influence of magnetic torque, in *Proceedings of Workshop on Space Dynamics and Celestial Mechanics,* Edited by K.B. Bhatnagar, Dept. of Mathematics, Muzaffarpur, Bihar, (1995), pp. 57–63.

[9] R. Bhardwaj and K.B. Bhatnagar, Chaos in nonlinear planar oscillation of a satellite in an elliptic orbit under the influence of third body torque, *Indian J. Pure Appl. Math.,* **28(3)** (1997), 391–492.

[10] R. Bhardwaj and K.B. Bhatnagar, Nonlinear planar oscillation of a satellite in a circular orbit under the influence of magnetic torque (II), *Indian J. Pure Appl. Math.,* **29(2)** (1998), 139–150.

[11] R. Bhardwaj and R. Tuli, Chaos in Attitude motion of a satellite under a third body torque in an elliptic orbit(I), accepted for publication, *Indian J. Pure Appl. Math.,* (2002).

[12] D.W. Childs, A movable-mass attitude-stabilization system for artificial-g space station, *J. Spacecraft,* **8** (1971), 829–834.

[13] P.W. Droll and E.J. Iuler, Proceedings of symposium in Space Magnetic Exploration and Technology, *Engineering, Report No.9,* (1967), 189–197.

[14] D. Gim and K. Alfriend, The state transition matrix of relative motion for the perturbed non-circular reference orbit, *AAS Paper 01–222, Presented at the AAS/AIAA Space Flight Mechanics Meeting* (2001), Santa Barbara, CA, February 11-15.

[15] T.L. Hardison, A movable-mass attitude-stabilization system for cable-connected artificial-g space station, *J. Spacecraft,* **11** (1874), 165–172.

[16] J.A. Kechichian, Motion in general elliptic orbit with respect to a dragging and precessing coordinate frame, *J. Astronautical Sci.,* **46** (1998), 25–45.

[17] P.S. Laplace, Traite de Mecanique Celeste, Volume V(Bachelier, Paris. Laplace's work is contained in Book XVI, Chapter IV. Reprinted in Oeuvres de Laplace, Tome Cinquieme(Gauthier-Villars, Paris, 1882), (1825), p. 445–452.

[18] A.J. Maciejewski, Instability, chaos and predictability in celestial mechanics and stellar dynamics (Ed.: K.B. Bhatnagar), *Nova Science, New York, 1992*, (1973), 23–38.

[19] V.K. Melnikov, Trudy Moscow Mat. Obshch., 12, 3–52. *(Translation in Trans. Moscow Math.Soc., 12 (1983)*, (1963), 1–57).

[20] I. Newton, *Philosophiae Naturalis Principia Mathematica*, (London, 1687); English translation by F. Cajori, Newton's Principial (University of California Press, Berkeley, (1987).

[21] J.E. Prussing and B.A. Conway, *Orbital Mechanics*, Oxford University Press, New York, (1993).

[22] S.A. Schweighart and R.J. Sedwick, A Perturbative Analysis of geopotential disturbances for satellite formation flying, *Proceedings of the IEEE Aerospace Conference*, **2** (2001), 1001–1019.

[23] R.B. Singh, *Space Dynamics and Celestial Mechanics* (Ed.: K.B. Bhatnagar), D. Reidel Publishing Company, (1983), pp. 291–307.

[24] J.R. Wertz (Ed.), *Space Craft Attitude Determination and Control*, D. Reidel Publishing Company, Holland, (1978).

[25] V.A. Zlatanstov and A.P. Markeev, Stability of planar oscillation of a satellite in an elliptic orbit, *Celestial Mechanics*, **7** (1973), pp. 31.

Chapter 15

A NEW ANALYSIS APPROACH TO POROUS MEDIA TEXTURE – MATHEMATICAL TOOLS FOR SIGNAL ANALYSIS IN A CONTEXT OF INCREASING COMPLEXITY

F. Nekka* and J. Li

Université de Montréal

Abstract

We draw a general picture of our recent research in the context of texture analysis, in particular, porous media. We propose new strategies to deal with different structures according to their degree of complexity. First, we discuss the porosity within the context of the autocorrelation function. Then, we suggest an original way to extract complementary parameters from the autocorrelation function beyond porosity. When complexity is increased, we combine the regularization dimension, a method proposed to estimate the curve variation, with the autocorrelation function. This leads to a more robust classification of porous media. Finally, we report on our previous results concerning sets of extreme complexity and of porosity zero, namely, thin fractals. We show how the method we developed to handle this case gives access to characteristic parameters beyond the fractal dimension.

*Corresponding author

1 Introduction

In the research field of porous media, it is well known that microstructure fluctuations have an important impact on bulk physical and rheological properties. Hence, a structural characterization through an accurate quantification of microstructure is necessary to predict mechanical performance. Relevant aspects of the complex structure of these media include porosity as well as pore shape and size. Recently, synthetic polymer design has been revolutionized by the new achievements in high-resolution characterization tools, such as the broad-mass-range spectrometry. These new measurement technologies make it possible to get access to finer details of the structures, so that more sophisticated data analyses are needed. As a matter of fact, the obtained experimental signals are generally very complex, making it necessary to develop more advanced mathematical tools for their analysis. For example, high-resolution mass spectra of synthetic polymers can contain a large number of peaks, so that one cannot still consider these spectra in the traditional continuum way. This calls for more appropriate and powerful analysis methods. The autocorrelation function has been proposed to handle complex time series [1]. Its Fourier transform, known as the structure factor, is used in polymer characterization by light scattering experiments to indicate which wave vector components are present in the autocorrelation function [2]. In [3], two system classes have been distinguished according to their structure factor value. However, our recent experience in this field showed the necessity to revisit many classical mathematical tools of signal analysis and update them in today's context of increasing complexity. Our research in this field is twofold. From one side, we developed a deeper analysis of some classical mathematical tools. We also proposed complementary parameters to characterize structures beyond porosity [4]. From the other side, we generalize the classical methods to make them sensitive to an exceeding complex nature of the data. In this case, we were led to develop parameters that can quantify the second-degree complexity of the geometrical structure [5–9]. Considering the fractal dimension as a first-degree complexity, the methods we developed go beyond this index and can be used to distinguish two different structures having the same fractal dimension and estimate as well their degree of homogeneity.

This chapter is organized as follows. In Section 2, we recall the definition of the autocorrelation function and its usual use in signal analysis. We also introduce a statistical porosity formalism through the autocorrelation function. In Section 3, we propose complementary parameters to porosity by exploiting the autocorrelation function of the signals. We illustrate this on periodic pattern polymers. In Subsection 3.1, we deal with fractal-like textures. First, for porous media exhibiting power law properties, we show how it is possible to classify them by combining the autocorrelation function with the regularization dimension [15]. In Subsection 3.2, we investigate the case of zero porosity (thin fractals). For these structures, we introduce what we previously called Hausdorff measure spectrum function (HMSF), which generalizes the classical autocorrelation function and which has proven to be more sensitive and suitable to describe these kind of scattered objects [5–7] In Section 4, we conclude with a discussion of some possible paths to continue each part initiated in this paper.

2 Porosity and the Autocorrelation Function

The porosity is one dominant pore structure whose aspects are used to measure porous media. It refers to the occupancy of the set, measured on samples having regular geometrical shapes. To facilitate porosity estimation, we generally select for the sampling of familiar shapes such as a cube, cylinder, sphere, etc. Hence, the porosity indicates how dense the porous media is when considered into the space where it lives.

Each chosen sample from the porous media is dealt with in a deterministic way, having a particular porosity value. However, since the sampling is carried arbitrary from the whole porous media, only the porosity obtained in a statistical sense is meaningful and representative. Using the large number theorem, two options are possible to obtain the porosity of a structure. One option is to choose a sample large enough to make the estimation of the porosity of the whole structure or, alternatively, to take a large number of samples.

The autocorrelation function is dedicated to measure the similarity in a set's geometric structure [10]. Porosity turns out to have a very close relationship to the autocorrelation function. This connection is not meant in a deterministic sense (for one sample), but rather in a statistical meaning. Indeed, we will show how we can recover porosity through the autocorrelation function, using a theoretical statistical model of homogeneous porous media that we will define in the example below.

Example 2.1 A theoretical statistical model of homogeneous porous media.

Let us consider the example of a homogeneous porous set F in one dimension that we define as follows: for the sake of simplicity, we assume the set F to be hosted into the unit interval. Subdivide the unit interval into n equal parts $[(i-1)/n, i/n]$ for $i = 1, \cdots, n$. We associate to each part $[(i-1)/n, i/n]$ a binary random variable X_i indicating the membership of the subinterval $[(i-1)/n, i/n]$ to the set F, where $F = \bigcup_{i:X_i=1}[(i-1)/n, i/n]$ and X_i is defined as

$$\mathbb{P}(X_i = 1) = p, \quad \mathbb{P}(X_i = 0) = 1 - p.$$

We use $\mathbb{P}$ to denote the probability.

Furthermore, we assume that the variables X_i are i.i.d. for all $1 \leq i \leq n$.

Consider now the autocorrelation function $I(t)$ of the set F defined as

$$I(t) = \frac{\int \mathcal{X}_F(x)\mathcal{X}_F(x+t)dx}{\int \mathcal{X}_F(x)^2 dx}, \tag{2.1}$$

where $\mathcal{X}_F$ is the indicator function of F: $\mathcal{X}_F(x) = \sum_{i=1}^{n} X_i \mathcal{X}_{[(i-1)/n, i/n]}(x)$. The normalization is applied to assure that $I(0) = 1$, preventing thus the limit cases of infinity or of zero to occur [8].

In this example, we have

$$\int \mathcal{X}_F(x)^2 dx = \frac{\sum_{k=1}^{n} X_i^2}{n}$$

and

$$\int \mathcal{X}_F(x)\mathcal{X}_F(x+t)dx = \frac{\sum_{k=i}^{n} X_k X_{k-i+1}}{n}$$

for $t = (i-1)/n$, $i = 1, \cdots, n$.

Since we have assumed a statistical definition of F, it is suitable to redefine the autocorrelation function of F in the following statistical way:

$$I_{\mathbb{E}}(t) = \frac{\mathbb{E}\int \mathcal{X}_F(x)\mathcal{X}_F(x+t)dx}{\mathbb{E}\int \mathcal{X}_F(x)^2 dx}, \tag{2.2}$$

where $\mathbb{E}$ denotes the expectation.

Then, we have $\mathbb{E}\displaystyle\int \mathcal{X}_F(x)\mathcal{X}_F(x+t)dx = (1-\frac{i-1}{n})p^2$ and $\mathbb{E}\int \mathcal{X}_F(x)^2 dx = p$. On the other hand, variances of these two expressions, involved in $I_{\mathbb{E}}(t)$, are of order $1/n$, i.e., $O(1/n)$, which guarantees their convergence when n is large enough.

It is easy to see that

$$I_{\mathbb{E}}(t) = \begin{cases} 1, & t = 0; \\ (1 - (i-1)/n)p, & t = (i-1)/n, \quad i = 2, \cdots, n. \end{cases} \tag{2.3}$$

Let n go to infinity, and we obtain

$$I_{\mathbb{E}}(t) = \begin{cases} 1, & t = 0; \\ (1 - t)p, & 0 < t < 1. \end{cases} \tag{2.4}$$

Figure 1(a) illustrates $I_{\mathbb{E}}(t)$.

The porosity of F is $\frac{1}{n}\sum_{k=1}^{n} X_i$, and statistically, it is $\frac{1}{n}\mathbb{E}\sum_{k=1}^{n} X_i = p$.

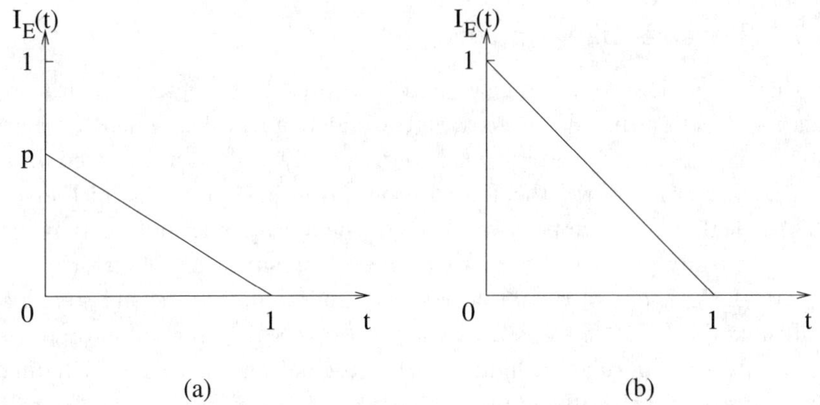

Figure 1: The autocorrelation of the considered example of porous media (left) and of the unit segment (right).

We see that p is obtained as the right limit of $I_{\mathbb{E}}(t)$ at $t = 0$. Hence, the porosity of a set F can be related to its autocorrelation function $I_{\mathbb{E}}(t)$ by considering an immediate neighborhood of $t = 0$. This proves that the autocorrelation function can also be used to study the density of the set's space occupancy. This problem will be handled from a practical point of view in the next section.

We also see that the simple case of porosity of the segment $[0, 1]$ is obtained from (2.4) as a limit of its monotonic autocorrelation function which is illustrated in Figure 1(b). We may observe that there is no jump from $t = 0$ to $t = 0+$. This means that the similarity is carried by all the points of $[0, 1]$ and that $I_{\mathbb{E}}(t)$ smoothly decreases as t increases. However, for $p < 1$, we notice that from $t = 0$ to $t = 0+$, the similarity is carried by only p percent of points of $[0, 1]$, which equals F's porosity in our example, Figure 1(a).

3 Complementary Parameters to Porosity

Properties of porous media are highly dependent on the morphology of the pore space as well as on those of its complementary part. However, porosity alone is far from enough to reflect the irregular morphology of micro-porous structures. In the following, we see how, according to the complexity degree of the problem in hand, we can develop parameters which give complementary geometrical information beyond porosity.

3.1 Thread-Like Texture

In this subsection, we use a new analysis approach to revisit the autocorrelation function in order to extract two independent components, which we relate to porosity as well as to pore frequency and extent. Indeed, in a first step, relying on the theoretical proof given in Section 2, which gives a statistical meaning to porosity through the autocorrelation function, we will practically prove how to extract images porosity using the least mean square (LMS) slope of the autocorrelation function. In a second step, we will subtract this LMS slope from the autocorrelation function in order to keep only the information hidden in the oscillating part, thus excluding the porosity. On this remaining oscillating curve, a Fourier transform will be used to get the main frequency components.

First Step

To explain our procedure, we use a set of porous images of decreasing porosities for which we apply the autocorrelation function. Figure 2 represents (from [12], Fig. 2f, p. 118) blending films of porous scaffold which having decreasing porosities in terms of the concentration of the two used polymers, PHB and PHBHHx. It reports the equivalence (matching) between the LMS slope of the autocorrelation of Figures 2 (a-f) and the porosity values calculated by the porosity formula.

Second Step

In Step 1 above, we have experimentally proved the equivalence between

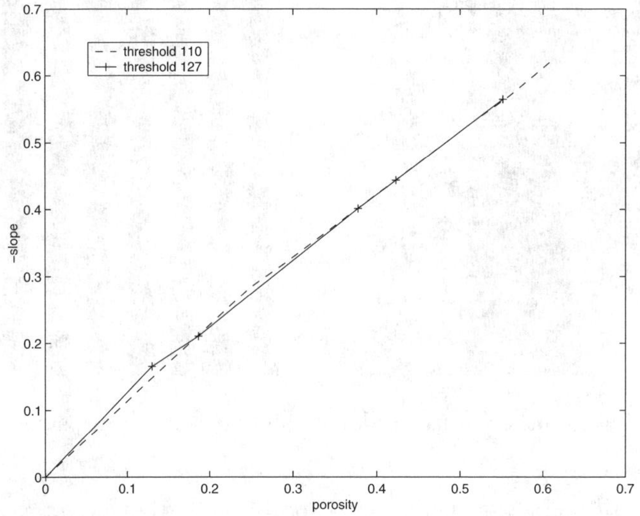

Figure 2: The matching between estimated porosity using the porosity formula and the slope of the linear part of the autocorrelation function of images in ([12], page 118).

porosity and the LMS slope of the autocorrelation function. Now, we need to analyze the remaining oscillating curves of which the porosity information has been cleared, to extract complementary parameters for periodic patterns. To do this, we take the case of two different thread-like patterns shown in Figure 3, both having the same porosity value of 0.4633. When the porosity value has been dropped, we apply the Fourier transform to the oscillating curve (Figure 4). For the considered thread-like textures, it is more convenient to estimate the Fourier transform by using a variant of the discrete Fourier transform (DFT), rather than the fast Fourier transform (FFT) algorithm implemented in the available softwares. The reason for this choice is that we are more concerned here with accuracy than with processing speed. Indeed, the range of frequency values for those quite simple textures is not large enough to require a rapid process. However, frequency values are close enough, making it necessary to use a DFT variant to obtain a good accuracy and to differentiate between textures. The results on frequency f as well as on $1/f$ give two parameters independent from porosity, representing the number of strips as well as their size, respectively.

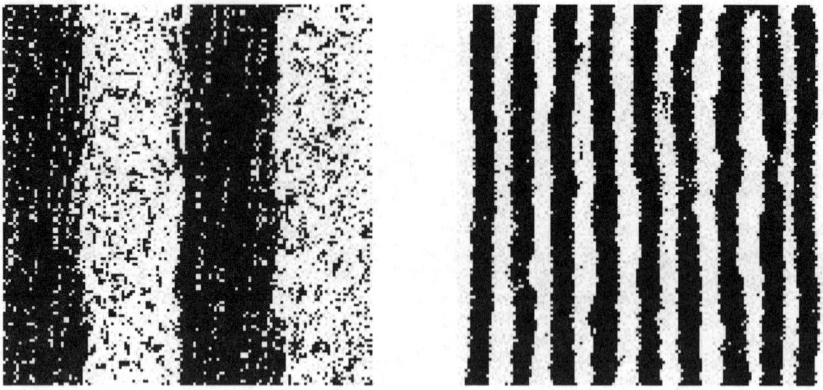

Figure 3: Two synthetic thread-like porous structures of the same porosity.

For more complex structures, an appropriate averaging procedure has to be applied to the oscillating curve to extract significant average frequencies. This aspect is under development.

We want to emphasize that without clearing the autocorrelation function from its linear part (the LMS slope), the information about the significant frequency will be tangled up in redundant information, which is due to the inclusion of the Fourier transform of this linear part.

3.2 Fractal-Like Texture

Fat Fractals: Synthetic Porous Media

When complexity of a structure increases further while it has a positive (non-zero) porosity, we propose a different approach to the autocorrelation function, combining fractal tools. A typical case for those more complex structures of positive porosity are known as fat fractals. Their main peculiarity, compared to thin fractals, consists in a finite and non-zero Lebesgue measure of their support. The empty holes of fat fractals have size-dependant power distribution similar to porous materials [13, 14], and they have been proposed as realistic models of micro-porous media [13]. Let us recall how a one-dimensional (1-D) regular fat fractal is generated. For the sake of simplicity, we usually use the unit interval as initiator. A regular fat fractal can be obtained through the following simple iterative rule: from one step to the next, we just drop the open middle intervals of length l_n from each of the remaining intervals. If the length of the

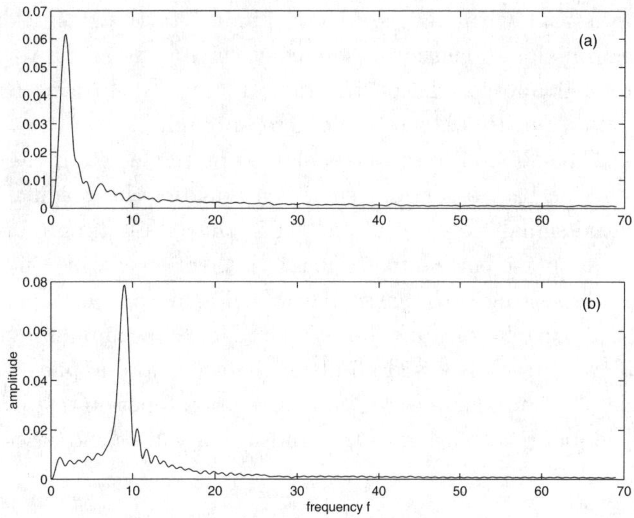

Figure 4: The Fourier transform of the the autocorrelation function of images in Figure 3, once it has been cleared from porosity.

interval at step $n-1$ is L_{n-1}, then this removed length is L_{n-1} divided by a^n. After n iterations, the obtained set is composed of 2^n intervals of lengths L_n and n subsets which are composed of $N_n(k) = 2^{k-1}$ empty holes of lengths l_k, ($k = 1, \cdots, n$), respectively. For a 1-D regular fat fractal, a is the only parameter involved in its definition. It is clear that the empty holes are symmetrically distributed in a 1-D regular fat fractal. To describe a porous media having asymmetric features, we have to use the mixed fat fractal which, intuitively, is a redistribution of the parts of the regular one, obtained by rearranging alternatively its voids and occupied intervals. We have considered eight fat fractals having the following values for the parameter a: 3, 4, 5, 6, 8, 10, 12, and 15. The autocorrelation function of these fat fractals is a more irregular curve, preventing thus a valuable application of the Fourier transform as was the case above [4, 9]. The irregularity of those curves suggests the utility of the fractal dimension usually used to quantify complexity. Thus, we use the regularization dimension (RD), introduced by Roueff and Levy-Vehel [15], since it proved to be more sensitive to variations. However, one can also raise the question of applying the RD directly to the fat fractal instead of applying it to the autocorrelation function of the set. Comparison of the obtained results shows that when RD is directly applied to the sets, a differentiation based on this dimension is less convincing compared to when it is applied to their autocorrelation function (Figure 5). In fact, RD of the autocorrelation function is a strictly increasing function of the size of the initial hole. This difference can be explained by the fact that the autocorrelation function has an "action" since it attenuates the irregularities of a signal (spatial signal representing the set in this case) and produces more uniform ones: sparse parts intersecting with themselves will still be sparse and when intersecting with denser parts, will again produce sparse parts. Also, the autocorrelation function is a similarities accumulating process: these similarities are dispersed all along the set before application of the autocorrelation function; once the autocorrelation function is applied, their amount is globally decreasing as the translation value t increases.

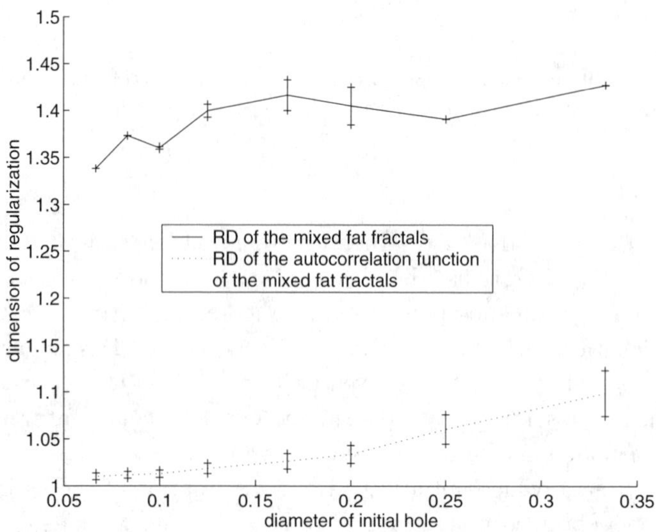

Figure 5: RD directly applied on the mixed fat fractals (top) and on their autocorrelation function (bottom).

Thin Fractals: Structures of Porosity Zero

If the set F has a zero-Lebesgue measure, as is the case for Cantor sets or, more generally, thin fractals, we have to use what we have named as the Hausdorff Measure Spectrum Function (HMSF), which involves integra-

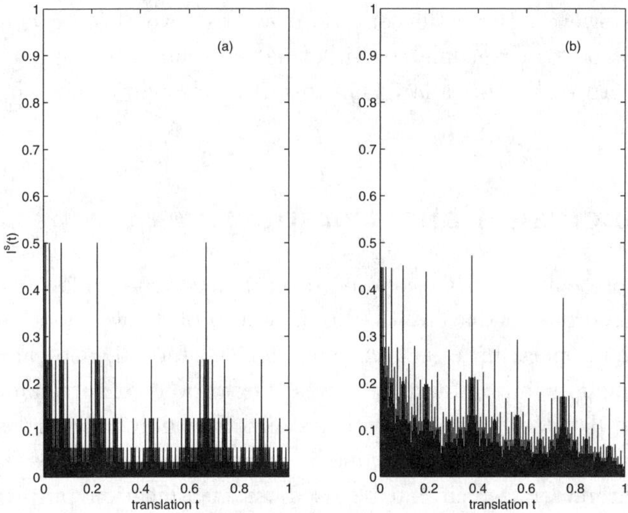

Figure 6: The HMSF of the triadic Cantor set (left) and of the considered multifractal set (right).

tion according to the Hausdorff measure instead of the Lebesgue measure involved in (2.1). In fact, we have defined the HMSF as follows:

$$I^{s_H}(t) = \int_{x \in F} \mathcal{X}_F(x)\mathcal{X}_F(x+t)d\mathcal{H}^{s_H}(x), \tag{3.5}$$

where s_H is the Hausdorff dimension of the set [11]. Figure 6 shows the HMSF of the Cantor fractal set as well as of a geometrical multifractal set. The latter example is generated on the unit interval using two similitudes, with the scaling factors equal to 1/2 and 1/4, on left and right extremities, respectively. This two-scale Cantor set gives rise to what is called a geometrical multifractal [16, 17].

In [6, 7], we have proposed two different ways to exploit the HMSF in order to distinguish between sets having the same fractal dimension. The first one is based on what we call the translation invariance based method (TIBM). The second one consists in comparing the measure values of this function for different sets at a fixed level. We call it the fixed level based method (FLBM). The latter one is necessary only when the first step in not enough. The FLBM associates an index to sets which allows for their differentiation as well as for evaluation of their degree of homogeneity. For geometrical multifractals, peaks of the HMSF have variation of different heights. However, they still converge towards several level values, from which we can extract geometrical information related to the homogeneity of the structure. Further analysis in this direction will be done in a future publication.

4 Conclusion and Perspectives

We have presented in this chapter several new ways to investigate the texture of porous media. With this new approach, we are able to deal with a broad spectrum of texture complexity. Indeed, one cannot hope for an almighty solution for all textures. For various texture complexities, we have to develop different strategies aimed at different purposes. The new descriptions of texture obtained through these strategies define the various parameters which can be used for classification purposes or as reconstruction criteria. In fact, in this chapter, we have combined some largely used mathematical tools, namely, the autocorrelation function and the Fourier transform, with more recent complexity analysis tools, i.e., the HMSF that we previously introduced along with the regularization dimension. This combination has been put in practice using computerized numerical means. This chapter proposes some possible ways to probe the fine details in texture. However, for each of the proposed strategies, further work has to be done theoretically and numerically. Parameter definitions can indeed be refined, and more advanced algorithms have to be developed.

Acknowledgments. This work has been supported by NSERC grant (RGPIN-227118) and NSERC UFA, held by F. Nekka as well as by Faculté de Pharmacie, Université de Montréal (fonds de démarrage) and by CRM (Centre de Recherches Mathématiques) group funds.

References

[1] W. E. Wallace and C. M. Guttman, Data analysis methods for synthetic polymer mass spetrometry: Autocorrelation, *J. Res. Natl. Inst. Stand. Technol.*, **107** (2002), 1–17.

[2] I. Teraoka, *Polymer Solutions: An Introduction to Physical Properties*, Wiley-Interscience, New-York, (2002).

[3] C. Allain and M. Cloitre, Spatial spectrum of a general family of self-similar arrays, *Phys. Rev. A*, **36**(12) (1987), 5751–5757.

[4] J. Li, C. Dubois, and F. Nekka, The colligation of the autocorrelation function and the regularization dimension – A new characterization of porous media, *Fractals 2004, Complexity and Fractals in Nature*, Vancouver (2004), 22-23.

[5] F. Nekka and J. Li, Intersection of triadic Cantor sets with their translates - I. Fundamental properties, *Chaos, Solitons and Fractals*, **13**(9) (2002), 1807–1817.

[6] J. Li and F. Nekka, Intersection of triadic Cantor sets with their translates -II. Hausdorff measure spectrum function and its introduction for the classification of Cantor sets, *Chaos, Solitons and Fractals*, **19**(1) (2004), 35–46.

[7] J. Li and F. Nekka, Characterization of fractal structures through a Hausdorff measure based method, *Fractals 2004. Thinking in Patterns, Fractals and Related Phenomena in Nature*, World Scientific, Vancouver (2004), 213-220.

[8] F. Nekka and J. Li, Structure complexity and scaling: Development and adaptation of advanced tools for their characterization and synthesis, *SIAM Proceedings*, Montreal, June (2003).

[9] J. Li and F. Nekka, The Hausdorff measure functions: A new way to characterize fractal sets, *Pattern Recognition Lett.*, **24**(15) (2003), 2723–2730.

[10] D.C. Champeney, *Fourier Transforms and Their Physical Applications.* Academic Press, New York, (1973).

[11] J. Li, A. Arneodo, and F. Nekka, A practical method to experimentally evaluate the Hausdorff dimension: an alternative phase-transition-based methodology, *CHAOS, An Interdisciplinary J. Nonlinear Sci.*, **214**(4) December (2004), 1004-1017.

[12] K. Zhao, Y. Deng, and G. Q. Chen, Effects of surface morphology on the biocompatibility of polyhydroxyalkanoates, *Biochemical Engg. J.*, **16**(2003), 115-123.

[13] S.A. Bulgakov, Wave interaction with a random fat fractal: Dimension of reflection coefficient, *Phys. Rev. A*, **46**(12) (1992), 8024.

[14] D.K. Umberger and J.D. Farmer, Fat fractals on the energy surface, *Physical Rev. Lett.*, **55**(7) (1985), 661–664.

[15] F. Roueff and J. Levy-Vehel, A regularization approach to fractional dimension estimation, in *Fractals and Beyond. Complexities in the Sciences*, Edited by M.M. Novak, World Scientific, Singapore, (1998).

[16] J. Feder, *Fractals*, Plenum Press, New York, (1988).

[17] T. Vicsek, *Fractal Growth Phenomena*, Worl Scientific, Singapore, (1989).

Part IV

Trends in Variational Methods

Chapter 16

A CONVEX OBJECTIVE FUNCTIONAL FOR ELLIPTIC INVERSE PROBLEMS

M. S. Gockenbach and A. A. Khan

Michigan Technological University

Abstract

The output least-squares approach for elliptic inverse problems, using a coefficient-dependent energy norm functional, results in a smooth, convex minimization problem. This convexity leads to strong convergence results for fixed point algorithms such as the auxiliary problem principle and the extragradient method. Numerical optimization methods such as Newton's method work well in conjunction with this functional, as numerical results show. Moreover, these results hold in an abstract framework that describes any elliptic inverse problem with interior data. Special cases include the standard divergence-form (scalar) elliptic equation and the equations of isotropic linear elasticity.

1 Introduction

The problem of identifying parameters in elliptic boundary value problems has applications in several fields, such as image processing, groundwater management, identifying cracks, modeling of car shields, and many others (see [3]). Therefore, it is not surprising that much work has been devoted to the theoretical and numerical investigation of these problems.

Although most of the early work imposed some smoothness assumptions on the coefficients to be recovered, some recent applications, such as in image processing, have necessitated methods allowing the recovery of discontinuous parameters. Motivated by the desire to identify discontinuous coefficients in elliptic or parabolic problems, several papers have considered the parameter space as the space of functions of bounded variations (BVs). This has also been done with some unconventional functionals for identifying the discontinuous coefficients. In this setting, since the parameter space is a nonreflexive Banach space, the analysis becomes more involved. The reader is referred to [8] and [24] for some instances whereas BV seminorm regularization has been successfully employed and discontinuous coefficients have been identified. This field is dynamic and several interesting approaches are in the making.

Recently, in [11], we presented an abstract framework for elliptic inverse problems. We have applied our theoretical results to identify discontinuous Lamé parameters in linear elasticity. In this work we discuss a few possibilities of dealing with elliptic inverse problems which are based on our abstract formulation. Before embarking more on the approach itself, we outline the general nature of the problems which can be dealt within the framework given here.

Most of the work on parameter identification has been in the context of the scalar elliptic problem

$$-\nabla \cdot (q \nabla u) = f \text{ in } \Omega, \tag{1.1a}$$

$$u = 0 \text{ on } \partial\Omega. \tag{1.1b}$$

For example, papers such as [1], [2], [10], [15], [20], [21], [23], [28], and [29] have examined this problem or variations of it. The above problem is not only mathematically interesting, but it also appears in numerous industrial applications such as reservoir and underground water resources [20]. The books by Banks and Kunisch [3] and by Engl, Hanke, and Neubauer [9] are excellent references for a detailed exposition of the subject.

In the context of (1.1), there have been mainly two approaches for attacking the corresponding inverse problem. The first approach reformulates the inverse problem as an optimization problem and then employs some suitable method for the solution. The second approach treats (1.1) as a hyperbolic partial differential equation in q. Furthermore, the approach of reformulating (1.1) as an optimization problem is divided into two possibilities, namely, either formulating the problem as an unconstrained optimization problem (by the so-called output least-square method) or treating it as a constrained optimization problem, in which the PDE itself is the constraint. In [11], we analyzed common features of elliptic inverse problems by establishing an abstract framework based on the output least-square approach. One advantage of our approach was the proposal of an objective functional which used a coefficient-dependent energy norm. This objective functional turned out to be smooth and convex, allowing us to use several powerful existing results for the theoretical and numerical treatment of the inverse problems.

The main objective of this chapter is to discuss some possibilities of exploiting the convexity of the objective functional proposed in [11] at an abstract level and to highlight the use of some new algorithms for the numerical solvability of elliptic inverse problems.

This chapter is divided into six sections. Section 2 begins by describing the abstract framework for the elliptic inverse problem. Besides discussing some particular cases which can be recovered from the abstract setting, we also discuss an abstract counterpart of the BV-regularization. This sections also includes some background for finite element approximation which is used later in full generality in the next two sections. A coupling of BV-regularization with an output least-square approach is given in Section 3. In Section 4 we extend the abstract framework for solving elliptic inverse problem by employing the augmented Lagrangian method. Section 5 presents some fixed point based iterative schemes. The methods given in this section depend on the convexity of the objective functional. This section includes, in particular, applications of the auxiliary problem principle and the extragradient method. Section 6 contains some numerical results.

2 Elliptic Inverse Problems

We begin by defining an abstract framework for the elliptic inverse problem. We suppose that V is any Hilbert space, B is a Banach space, and A is a subset of B with nonempty interior. We assume that $T : B \times V \times V \to \mathbb{R}$ is a trilinear form, with $T(a, u, v)$ symmetric in u, v, and that there exist positive constants α and β such that

$$T(a, u, v) \leq \beta \|a\|_B \|u\|_V \|v\|_V \quad \forall\, u, v \in V, \quad a \in B$$

and

$$T(a, u, u) \geq \alpha \|u\|_V^2 \quad \forall\, u \in V, \quad a \in A.$$

Finally, we assume that m is a fixed bounded linear functional on V. Then, by the Riesz representation theorem, the variational equation

$$T(a, u, v) = m(v) \quad \forall\, v \in V \tag{2.1}$$

has a unique solution u for each $a \in A$. Therefore, the mapping $F : A \to V$ defined by the condition that $u = F(a)$ is the solution to (2.1) is well-defined. Henceforth, the map F will be referred to as the solution operator.

In this abstract setting, the inverse problem associated with the direct problem (2.1) is the following. Given some measurement of u, say, z, estimate the coefficient a which together with u makes (2.1) true.

The output least-squares approach to the inverse problem seeks to minimize the functional

$$a \mapsto \frac{1}{2}\|F(a) - z\|^2 \tag{2.2}$$

defined by an appropriate norm. Here z is the data (the measurement of u). For example, Falk [10] analyzed the case of the L^2 norm applied to the scalar problem (1.1). The H^1 norm is also a possibility.

In [11], we proposed to minimize the functional

$$a \mapsto \frac{1}{2}T(a, F(a) - z, F(a) - z). \tag{2.3}$$

In this chapter we outline various possibilities of developing numerical schemes based on the above functional. We remark that (2.3) was motivated by the independent proposals of Chen and Zou [8] and Knowles [20] using the a-dependent energy norms in place of the L^2 or H^1 norm. The work of these authors was in the context of the scalar problem (cf. (1.1)).

2.1 Some Special Cases

Let us now discuss some cases of our abstract formulation. We have already mentioned that most of the research on parameter identification problems has been in the context of the elliptic boundary value problem (BVP)

$$-\nabla \cdot (q\nabla u) = f \text{ in } \Omega, \tag{2.4a}$$
$$u = 0 \text{ on } \partial\Omega. \tag{2.4b}$$

The variational form of (1.1) is

$$\int_\Omega q\nabla u \cdot \nabla v = \int_\Omega f v \;\forall\, v \in H_0^1(\Omega). \tag{2.5}$$

The existence and uniqueness of solutions to (2.5) follows, via the usual analysis, from the Riesz representation theorem, given the usual assumptions on q. There exist positive constants k_0, k_1 such that $k_0 \leq q \leq k_1$ on Ω.

We write $V = H_0^1(\Omega)$, $B = L^\infty(\Omega)$, and

$$A = \{q \in L^\infty(\Omega) : k_0 \leq q \leq k_1\},$$

where $k_1 > k_0 > 0$ are given. We notice that the trilinear form $T : B \times V \times V \to \mathbb{R}$ given by

$$T(a, u, v) = \int_\Omega q \nabla u \cdot \nabla v$$

then satisfies

$$T(q, u, v) \leq \beta \|q\|_B \|u\|_V \|v\|_V \ \forall\ u, v \in V, q \in B$$

and

$$T(q, u, u) \geq \alpha \|u\|_V^2 \ \forall\ u \in V, q \in A,$$

where $\beta = 1$ and $\alpha > 0$ comes from Poincaré's inequality. In this setting the functional m is given by

$$m(v) = \int_\Omega fv.$$

There has also been some work on extending various techniques from the papers cited above to inverse problems involving other elliptic BVPs, especially the system of linear, isotropic elasticity:

$$-\nabla \cdot \sigma = f \text{ in } \Omega, \tag{2.6a}$$

$$\sigma = 2\mu\varepsilon_u + \lambda \operatorname{tr}(\varepsilon_u) I, \tag{2.6b}$$

$$\varepsilon_u = \frac{1}{2}\left(\nabla u + \nabla u^T\right), \tag{2.6c}$$

$$u = 0 \text{ on } \Gamma_1, \tag{2.6d}$$

$$\sigma n = h \text{ on } \Gamma_2. \tag{2.6e}$$

Here $\partial\Omega = \Gamma_1 \cup \Gamma_2$ is a partition of the boundary of Ω, and n is the outward-pointing unit normal to $\partial\Omega$. The problem in this case is to estimate μ and λ, the Lamé moduli. In recent years, some interesting applications of the problem of recovering the Lamé moduli in biomedical imaging have sparked new interest in this problem. The following papers address this inverse problem: [4], [5], [14], [17], [25], [26], and [27].

The variational form of (2.6) is

$$\int_\Omega \{2\mu\varepsilon_u \cdot \varepsilon_v + \lambda \mathrm{tr}\,(\varepsilon_u)\,\mathrm{tr}\,(\varepsilon_v)\} = \int_\Omega f \cdot v + \int_{\Gamma_2} h \cdot v \;\forall\, v \in V, \tag{2.7}$$

where $V = \left\{v \in \left(H^1(\Omega)\right)^2 : v = 0 \text{ on } \Gamma_1\right\}$.

The variational equation in (2.7) can also be expressed in terms of a trilinear form by defining

$$T(\ell, u, v) = \int_\Omega \{2\mu\varepsilon_u \cdot \varepsilon_v + \lambda \mathrm{tr}\,(\varepsilon_u)\,\mathrm{tr}\,(\varepsilon_v)\},$$

where $\ell = (\mu, \lambda)$, and

$$m(v) = \int_\Omega f \cdot v + \int_{\Gamma_2} h \cdot v.$$

We write $B = (L^\infty(\Omega))^2$ and

$$A = \{\ell = (\mu, \lambda) \in B : a_0 \le \mu, \mu + \lambda \le a_1\},$$

where $a_1 > a_0 > 0$ are given. Then, for any $\ell \in A$,

$$T(\ell, u, v) = m(v) \;\forall\, v \in V$$

has a unique solution u. The trilinear form T satisfies

$$T(\ell, u, v) \le \beta \|\ell\|_B \|u\|_V \|v\|_V \;\forall\, u, v \in V, \ell \in B$$

and

$$T(\ell, u, u) \ge \alpha \|u\|_V^2 \;\forall\, u \in V, \ell \in A.$$

The constant α comes from Korn's inequality, and $\beta = 10$ will work.

Related problems that may be of interest include the matrix conductivity problem, in which the scalar function q in (1.1) is replaced by a matrix, and elasticity systems such as (2.6) but with a more general stress-strain law relating σ and ε (to replace (2.6b)).

2.2 Differentiability of the Solution Operator

In the study of elliptic inverse problems the questions regarding the continuity and differentiability of the solution operator are of paramount important. For example, in order to compute the derivative of the objective functional (2.3) (or (2.2)), among other things, one has to compute the derivative of the solution operator. Concerning the continuity of F, the following result holds.

Theorem 2.1 *The operator F is continuous, and the following bounds are all valid:*

$$\begin{aligned} \|F(a)-F(b)\|_V &\le \frac{\beta}{\alpha}\|F(a)\|_V\|b-a\|_B, \\ \|F(a)-F(b)\|_V &\le \frac{\beta}{\alpha}\|F(b)\|_V\|b-a\|_B, \\ \|F(a)-F(b)\|_V &\le \frac{\beta}{\alpha^2}\|m\|_{V^*}\|b-a\|_B. \end{aligned}$$

Proof. The proof is essentially based on the V-ellipticity and continuity of $T(a,\cdot,\cdot)$ and can be found in [11]. ■

If a belongs to the interior of A, then, for all δa sufficiently small, $a+\delta a\in A$ and $\delta w = F(a+\delta a)-F(a)$ is well-defined. By definition, if $u=F(a)$, then

$$\begin{aligned} T(a,u,v) &= m(v)\ \forall\ v\in V, \\ T(a+\delta a, u+\delta w, v) &= m(v)\ \forall\ v\in V. \end{aligned}$$

Subtracting the first equation from the second and simplifying yields

$$T(a+\delta a,\delta w,v) = -T(\delta a,u,v)\ \forall\ v\in V.$$

This suggests the form of $DF(a)$.

Theorem 2.2 [11] *For each a in the interior of A, F is differentiable at a, and $\delta u = DF(a)\delta a$ is the unique solution to the variational equation*

$$T(a,\delta u,v) = -T(\delta a,u,v)\ \forall\ v\in V, \tag{2.8}$$

where $u=F(a)$. Moreover,

$$\|DF(a)\| \le \frac{\beta}{\alpha}\|u\|_V,$$

and hence, for all a in the interior of A,

$$\|DF(a)\| \le \frac{\beta}{\alpha^2}\|m\|_{V^*}.$$

It is now easy to show that the operator F is, in fact, infinitely differentiable. We begin with the second derivative. We will write

$$\delta^2 w = DF(a+\delta a_1)\delta a_2 - DF(a)\delta a_2,$$

so that

$$DF(a+\delta a_1)\delta a_2 = DF(a)\delta a_2 + \delta^2 w.$$

By definition, then,

$$\begin{aligned} & T(a+\delta a_1, DF(a+\delta a_1)\delta a_2, v) = -T(\delta a_2, F(a+\delta a_1), v)\ \forall\, v \in V \\ \Rightarrow\ & T(a+\delta a_1, DF(a)\delta a_2 + \delta^2 w, v) = -T(\delta a_2, F(a+\delta a_1), v)\ \forall\, v \in V \\ \Rightarrow\ & T(a+\delta a_1, \delta^2 w, v) = -T(\delta a_2, F(a+\delta a_1), v) \\ & \qquad\qquad - T(a+\delta a_1, DF(a)\delta a_2, v) \quad \forall v \in V. \end{aligned}$$

Some manipulation yields

$$\begin{aligned} T(a+\delta a_1, \delta^2 w, v) &= -T(\delta a_2, DF(a)\delta a_1, v) - T(\delta a_1, DF(a)\delta a_2, v) - \\ & \quad T(\delta a_2, F(a+\delta a_1) - F(a) - DF(a)\delta a_1, v)\ \forall\, v \in V. \end{aligned}$$

This suggests the following result.

Theorem 2.3 [11] *For each a in the interior of A, F is twice-differentiable at a, and*

$$\delta^2 u = D^2F(a)(\delta a_1, \delta a_2)$$

is the unique solution to the variational equation

$$T(a, \delta^2 u, v) = -T(\delta a_2, DF(a)\delta a_1, v) - T(\delta a_1, DF(a)\delta a_2, v)\ \forall\, v \in V.$$

Moreover,

$$\|D^2F(a)\| \le \frac{2\beta^2}{\alpha^2}\|F(a)\|_V \le \frac{2\beta^2}{\alpha^3}\|m\|_{V^*}.$$

The following theorem can be proved by induction.

Theorem 2.4 [11] *For each a in the interior of A, F has derivatives of all order at a, and*

$$\delta^k u = D^kF(a)(\delta a_1, \delta a_2, \ldots, \delta a_k)$$

is the unique solution to the variational equation

$$T(a, \delta^k u, v) = -\sum_{i=1}^{k} T\left(\delta a_i, D^{k-1}F(a)\left(\delta a_1, \ldots, \delta a_{i-1}, \delta a_{i+1}, \ldots, \delta a_k\right), v\right)$$

for all $v \in V$. Moreover,

$$\|D^kF(a)\| \le \frac{k\beta^k}{\alpha^k}\|F(a)\|_V \le \frac{k\beta^k}{\alpha^{k+1}}\|m\|_{V^*}.$$

The results of this section are, in large part, well-known. Kluge [18] proved, in a similar abstract setting, that F is differentiable and that the derivative is Lipschitz continuous. His bounds on $\|F(a)\|_V$, $\|DF(a)\|$ and the Lipschitz constant correspond to ours. Other papers have used the fact that F is twice continuously differentiable; for an example, see Colonius and Kunisch [7]. We have not seen in the literature a proof of the infinite differentiability of F or a formula for $D^k F(a)$, $k > 2$.

2.3 A Smooth, Convex Objective Functional for Inversion

As already mentioned, we propose to use the objective functional $J : A \to \mathbb{R}$ defined by

$$J(a) = \frac{1}{2} T(a, F(a) - z, F(a) - z). \tag{2.9}$$

For numerical stability, regularization is required. We address regularization below, but first we analyze the properties of J.

The functional J is clearly infinitely differentiable. The first derivative is given by the chain rule:

$$DJ(a)\delta a = \frac{1}{2} T(\delta a, F(a) - z, F(a) - z) + T(a, DF(a)\delta a, F(a) - z).$$

By (2.8),

$$T(a, DF(a)\delta a, F(a) - z) = -T(\delta a, F(a), F(a) - z),$$

so

$$\begin{aligned} DJ(a)\delta a &= \frac{1}{2} T(\delta a, F(a) - z, F(a) - z) - T(\delta a, F(a), F(a) - z) \\ &= -\frac{1}{2} T(\delta a, F(a) + z, F(a) - z). \end{aligned}$$

It now follows that

$$\begin{aligned} D^2 J(a)(\delta a, \delta a) &= -\frac{1}{2} T(\delta a, DF(a)\delta a, F(a) - z) \\ &\quad -\frac{1}{2} T(\delta a, F(a) + z, DF(a)\delta a) \\ &= -T(\delta a, F(a), DF(a)\delta a) \\ &= T(a, DF(a)\delta a, DF(a)\delta a). \end{aligned}$$

In the last step, we applied (2.8) again. We notice, in particular, that the following inequality holds for all a in the interior of A:

$$D^2 J(a)(\delta a, \delta a) \geq \alpha \|DF(a)\delta a\|_V^2.$$

Thus J is a smooth and convex functional.

2.4 BV-Regularization

In elliptic inverse problems, it is of particular interest to estimate coefficients that are discontinuous or at least rapidly-varying. For this reason, we want any theory we develop to apply the case of BV-regularization, which has recently been considered in [29] and [8]. We now briefly discuss the space $BV(\Omega)$ and its norm.

By definition, the *total variation* of $f \in L^1(\Omega)$ is

$$\mathrm{TV}(f) = \sup\left\{\int_\Omega f(\nabla\cdot g)\ :\ g\in\left(C_0^1(\Omega)\right)^2,\ |g(x)|\le 1\ \forall\, x\in\Omega\right\}.$$

When f belongs to $W^{1,1}(\Omega)$, then it is easy to show (by integration by parts) that

$$\mathrm{TV}(f) = \int_\Omega |\nabla f|.$$

If $f \in L^1(\Omega)$ satisfies $\mathrm{TV}(f) < \infty$, then f is said to have *bounded variation*, and $BV(\Omega)$ is defined by

$$BV(\Omega) = \left\{f \in L^1(\Omega)\ :\ \mathrm{TV}(f) < \infty\right\}.$$

The norm on $BV(\Omega)$ is

$$\|f\|_{BV(\Omega)} = \|f\|_{L^1(\Omega)} + \mathrm{TV}(f).$$

The functional $\mathrm{TV}(\cdot)$ is a seminorm on $BV(\Omega)$ and is often called the BV-seminorm.

The (square of the) $H^1(\Omega)$-seminorm,

$$|f|^2_{H^1(\Omega)} = \int_\Omega \nabla f\cdot\nabla f = \int_\Omega |\nabla f|^2,$$

is often used as a regularization functional. By contrast with $\mathrm{TV}(\cdot)$, $|\cdot|_{H^1(\Omega)}$ imposes a large penalty on large gradients and is therefore appropriate for recovering mildy-varying coefficients. The BV-seminorm, on the other hand, has been used successfully to recover sharply-varying images in image processing and, as mentioned above, has recently been applied to elliptic inverse problems.

We use the following properties of $BV(\Omega)$ and $L^\infty(\Omega)$.

1. $L^\infty(\Omega)$ is continuously embedded in $L^1(\Omega)$ (so $\|\cdot\|_{L^\infty(\Omega)}$ is stronger than $\|\cdot\|_{L^1(\Omega)}$).

2. $BV(\Omega)$ is compactly embedded in $L^1(\Omega)$.

To continue the development of the abstract theory, we therefore assume that:

1. The Banach space B (corresponding to $L^\infty(\Omega)$ in the scalar case) is continuously embedded in a larger Banach space L.

2. There is another Banach space X (corresponding to $BV(\Omega)$ in the scalar case) that is compactly embedded in L. The norm on X is defined by
$$\|f\|_X = \|f\|_L + |f|_X,$$
where $|\cdot|_X$ is a seminorm.

3. The subset A (the domain of the solution operator F) is a closed and bounded subset of $B \cap X$, and A is also closed in L.

We now assume that we have regularization functional $R : X \to \mathbb{R}$ satisfying the following properties:

$$R \text{ is convex and lower semicontinuous.} \tag{2.10}$$
$$\text{There exists } \gamma > 0 \text{ such that } R(f) \geq \gamma |f|_X \text{ for all } f \in X. \tag{2.11}$$

We note that all of the following results hold if (2.11) is changed to $R(f) \geq \gamma |f|_{\hat{B}}^2$ for all $f \in \hat{B}$. This would be natural if, for example, the (square of) H^1-seminorm is used in place of the BV-seminorm:
$$R(q) = \int_\Omega |\nabla q|_2^2.$$

2.5 Some Assumptions on the Trilinear Form $T(\cdot,\cdot,\cdot)$

Suppose u, v are fixed elements of $H_0^1(\Omega)$, and $\{q_k\}$ is a bounded sequence in $L^\infty(\Omega)$. If $q_k \to 0$ in the $L^1(\Omega)$ norm, then, by the Lebesque dominated convergence theorem,
$$\int_\Omega q_k \nabla u \cdot \nabla v \to 0.$$
In terms of the trilinear form and the abstract framework established above, this corresponds to the following result:

If $\{a_k\}$ is a bounded sequence in B, $a_k \to 0$ in L, and u and v are fixed elements of V, then $T(a_k, u, v) \to 0$. (2.12)

We now adopt (2.12) as an assumption on the T, B, and L.

For any $q \in L^\infty(\Omega)$ and any $u, v \in H_0^1(\Omega)$, we can bound

$$\left| \int_\Omega q \nabla u \cdot \nabla v \right|$$

as follows:

$$\begin{aligned} \left| \int_\Omega q \nabla u \cdot \nabla v \right| &\leq \int_\Omega |q \nabla u \cdot \nabla v| \\ &= \int_\Omega \left(|q|^{1/2} \nabla u \right) \cdot \left(|q|^{1/2} \nabla v \right) \\ &\leq \left[\int_\Omega |q| \nabla u \cdot \nabla u \right]^{1/2} \left[\int_\Omega |q| \nabla v \cdot \nabla v \right]^{1/2}. \end{aligned}$$

This result is a generalization of the Cauchy-Schwarz inequality. Henceforth, in our abstract framework, we will assume that L is a space of real-valued functions and that

$$|T(a, u, v)| \leq [T(|a|, u, u)]^{1/2} [T(|a|, v, v)]^{1/2} \, \forall \, u, v \in V, a \in B.$$

We finally assume that the following condition holds for both $\| \cdot \|_B$ and $\| \cdot \|_L$:

$$\| |a| \| \leq \|a\| \, \forall \, a.$$

We have motivated these assumptions by the scalar elliptic problem. It is straightforward to verify that these assumptions are satisfied when T is defined by the system for isotropic elasticity, as in (2.7).

2.6 Finite-Dimensional Approximation

To solve either the forward or inverse problem, discretization is required. In the examples we have in mind, finite element discretization will be used. For the sake of the abstract development, we assume that $\{V_k\}$ is a sequence of finite-dimensional subspaces of V and, for each k, $P_k : V \to V_k$ is a projection operator. We assume that V_k and P_k have the property that

$$\forall \, v \in V, \; \|v - P_k v\|_V \to 0 \text{ as } k \to \infty.$$

We similarly assume that $\{B_k\}$ is a sequence of finite-dimensional subspaces of B and define

$$A_k = B_k \cap A.$$

Moreover, we need to assume that

$$\bigcap_{k=1}^{\infty} A_k \neq \emptyset.$$

The necessary approximation properties of the spaces A_k are described by the following assumption. For any $a \in A$, there exists a sequence $\{a_k\}$ such that

$$\begin{aligned} &a_k \in A_k \forall\ k, \\ &a_k \to a \text{ in } L, \\ &R(a_k) \to R(a). \end{aligned}$$

We define the discretized solution operator $F_k : A \to V_k$ by the condition that $u = F_k(a)$ is the unique solution of the the variational problem

$$T(a, u, v) = m(v)\ \forall\ \ v \in V_k.$$

All of the theory developed in the previous sections applies by merely restricting T to $B \times V_k \times V_k$, so F_k is smooth, and the formulas for and the bounds on the derivatives of F_k are as given above.

3 Output Least-Squares Approach

We propose the following method for estimating a from a measurement z of $u = F(a)$, Choose $a \in A$ to minimize

$$J_\rho(a) = J(a) + \rho R(a), \tag{3.1}$$

where $\rho > 0$ is a given regularization parameter, R is the regularization operator discussed in Section 2.4, and J is as in (2.9), that is,

$$J(a) = \frac{1}{2} T(a, F(a) - z, F(a) - z).$$

We have the following theorem.

Theorem 3.1 *The minimization problem (3.1) has a solution.*

Proof. The proof is based on a standard argument of choosing a minimizing sequence and extracting a convergent subsequence. The abstract conditions imposed on the trilinear form are decisive. All the details are available in [11]. ∎

As mentioned earlier, to solve (3.1) we first need to perform a finite-dimensional discretization. The inverse problem is discretized by defining $J^{(k)} : A_k \to \mathbb{R}$ by

$$J^{(k)}(a) = \frac{1}{2}T(a, F_k(a) - P_k z, F_k(a) - P_k z).$$

(The data z is replaced by $P_k z$ for computational convenience. Since $P_k z \to z$ (strongly) in V, all of the following results hold whether z or $P_k z$ is used in the definition of $J^{(k)}$.) We then define

$$J_\rho^{(k)}(a) = J^{(k)}(a) + \rho R(a)$$

and solve

$$\min_{a \in A_k} J_\rho^{(k)}(a). \tag{3.2}$$

As for the forward problem, the theory developed in the earlier sections applies to (3.2), and, in particular, Theorem 3.1 shows that (3.2) has a solution.

We can now prove the following convergence result.

Theorem 3.2 [11] *Suppose that, for each k, $a_k^* \in A_k$ is a solution of (3.2). Then every subsequence of $\{a_k^*\}$ has a subsequence that converges to a solution of (3.1).*

Some computational results depicting the efficiency of the present approach are given in Section 6.

4 Augmented Lagrangian Approach

By the Riesz representation theorem, there is an isomorphism $E : V \to V^*$ defined by

$$(Eu)(v) = \langle u, v \rangle_V \quad \text{for every } v \in V.$$

For each $(a, u) \in A \times V$, $T(a, u, \cdot) - m(\cdot)$ defines an element of V^*. We define $e(a, u)$ to be the pre-image under E of this element:

$$\langle e(a, u), v \rangle_V = T(a, u, v) - m(v) \quad \text{for every } v \in V.$$

We consider the following constrained optimization problems:

$$\min_{(a,v) \in A \times V} \quad \tilde{J}_\rho(a, v) = \frac{1}{2}T(a, v - z, v - z) + \rho R(a) \tag{4.1a}$$

$$\text{subject to} \quad e(a, v) = 0, \tag{4.1b}$$

where R is the regularizing operator introduced in Section 2.4 and $\rho > 0$ is the regularization parameter.

We begin with stating a solvability theorem for the above minimization problem.

Theorem 4.1 *The constrained minimization problem (4.1) has a solution.*

Proof. The proof can be found in [11]. ■

We intend to solve (4.1) by the augmented Lagrangian method. For $r \geq 0$, the augmented Lagrangian functional $L_r : A \times V \times V \to \mathbb{R}$ is defined as follows:

$$\mathcal{L}_r(a, v, \lambda) := \tilde{J}_\rho(a, v) + \langle e(a, v), \lambda \rangle + \frac{r}{2} \|e(a, v)\|_V^2.$$

The following result establishes the equivalence between the saddle point problem and the constrained minimization problem (4.1).

Theorem 4.2 *A pair $(a^*, v^*) \in A \times V$ is a solution to the constrained minimization problem (4.1) if and only if there exists $\lambda^* \in V$ such that $(a^*, v^*, \lambda^*) \in A \times V \times V$ is a saddle point of the augmented Lagrangian $\mathcal{L}_r$, that is, for all $(a, v, \zeta) \in A \times V \times V$, we have*

$$\mathcal{L}_r(a^*, v^*, \zeta) \leq \mathcal{L}_r(a^*, v^*, \lambda^*) \leq \mathcal{L}_r(a, v, \lambda^*).$$

Sketch of the proof. Let us define a set

$$S = \{(\tilde{J}_\rho(a, v) - \tilde{J}_\rho(a^*, v^*) + s, e(a, v)) \in \mathbb{R} \times V; \ (a, v) \in A \times V, \ s \geq 0\}.$$

In view of the optimality of (a^*, v^*), we have

$$S \cap (-\mathbb{P} \times \{0_V\}) = \emptyset,$$

where $\mathbb{P} := \{t \in \mathbb{R} : \ t > 0\}$ and 0_V is the zero element of V. The proof now is based on the arguments showing that the set S is solid and convex and hence that the well-known separation theorem can be applied. All the details are given in [12]. ■.

We consider the finite-dimensional analog e_k of the operator e defined as follows:

$$\langle e_k(a_k, v_k), w \rangle = T(a_k, v_k, w) - m(w) \quad \forall \, w \in V_k.$$

The inverse problem is discretized by defining $J^{(k)} : A_k \times V_k \to \mathbb{R}$ by

$$\tilde{J}^{(k)}(a,u) = \frac{1}{2}T(a, v-z, v-z).$$

Then, we define

$$\tilde{J}_\rho^{(k)}(a,u) = J^{(k)}(a,u) + \rho R^{(k)}(a)$$

and solve

$$\min_{(a,v)\in A_k \times V_k} \quad \tilde{J}_\rho^{(k)}(a,u) \tag{4.2a}$$

$$\text{subject to} \quad e_k(a,v) = 0. \tag{4.2b}$$

Here, $R^{(k)}$ is a finite-dimensional version of the regularizing operator, satisfying the following property:

$$R_k : B_k \to \mathbb{R} \text{ is convex and lower semicontinuous for each } k$$
$$a_k \in B_k \forall\, k \;\Rightarrow\; |R_k(a_k) - R(a_k)| \to 0.$$

The discretized analog of the augmented Lagrangian $L_r^{(k)} : A_k \times V_k \times V_k \to \mathbb{R}$ is defined as follows:

$$L_r^{(k)}(a_k, v_k, \zeta_k) = \tilde{J}_\rho^{(k)}(a_k, v_k) + \langle e_k(a_k, v_v), \zeta_k\rangle + \frac{r}{2}\|e_k(a_k, v_k)\|_V^2.$$

We have the following version of Theorem 4.2.

Theorem 4.3 *A pair $(a_k^*, v_k^*) \in A_k \times V_k$ is a solution to the discretized constrained minimization problem (4.2) if and only if there exists $\lambda_k^* \in V$ such that $(a_k^*, v_k^*, \lambda_k^*) \in A \times V \times V$ is a saddle point of the discretized augmented Lagrangian $L_r^{(k)}$, that is, for all $(a_k, v_k, \zeta_k) \in A_k \times V_k \times V_k$, we have*

$$L_r^{(k)}(a_k^*, v_k^*, \zeta_k) \le L_r^{(k)}(a_k^*, v_k^*, \lambda_k^*) \le L_r^{(k)}(a_k, v_k, \lambda_k^*).$$

Proof. The proof is the same as for Theorem 4.2. ■

The following is the main result of this section.

Theorem 4.4 *For each subsequence of the saddle points $\{(a_k^*, v_k^*, \lambda_k^*)\}_{k\in\mathbb{N}}$ of $L_r^{(k)}$ there exists a subsequence, still denoted by $\{(a_k^*, v_k^*, \lambda_k^*)\}_{k\in\mathbb{N}}$, with $a_k^* \to a^*$ in L, $v_k^* \to v^*$ weakly in V and $\lambda_k^* \to \lambda^*$ weakly in V. Moreover, (a^*, v^*, λ^*) is a saddle point of $\mathcal{L}_r$.*

Proof. See [12]. ■

5 Fixed Point Approach

In this section we shall discuss some methods which rely on the variational inequality formulation as an (necessary or necessary and sufficient) optimality condition for optimizing the objective functional (see [13]). We mention that all the algorithms given below essentially depend on the monotonicity property of the the gradient of the objective functional. This, among other things, also allows us to perform a clearer comparison of

$$a \mapsto \frac{1}{2}\|F(a) - z\|^2 \tag{5.1}$$

and

$$J(a) = \frac{1}{2}T(a, F(a) - z, F(a) - z). \tag{5.2}$$

In this section, unless the contrary is mentioned explicitly, X and Y are two real Hilbert spaces, A is a nonempty closed and convex subset of X, $T : X \times Y \times Y \to \mathbb{R}$ is a trilinear form, and $m : Y \to \mathbb{R}$ is a bounded linear functional. We assume that there exist positive constants α, β such that for all u, $v \in Y$ and for all $a \in X$, the following estimates hold:

$$\begin{aligned} T(a,u,v) &\leq \beta\|a\|_X\|u\|_Y\|v\|_Y \ \ \forall\, a \in X \\ T(a,u,u) &\geq \alpha\|u\|_Y^2 \ \ \forall\, a \in A. \end{aligned}$$

By the Riesz representation theorem, the variational equation

$$T(a,u,v) = m(v)\ \forall\, v \in Y \tag{5.3}$$

has a unique solution u for each $a \in A$. We therefore define $F : A \to Y$ by the condition that $u = F(a)$ is the solution to (5.3).

We begin with recalling a useful property of the solution operator F*.

Lemma 5.1 [18] *For $a, b \in A$, we have $F(b) - F(a) = F'(a)(b - a) + Q(a, b - a)$ with*

$$\|Q(a, b-a)\| \ \leq \ \frac{\beta^2}{\alpha^3}\|m\|\,\|a-b\|^2.$$

---

*Notice that for a fixed $(a,u) \in X \times Y$, by considering the map $v \to T(a,u,v)$ and keeping in mind the linearity and continuity of the trilinear form, we get the existence of a mapping $\mathcal{T} : X \times Y :\to \mathbb{R}$ such that

$$\langle \mathcal{T}(a,u), v\rangle_Y = T(a,u,v) \quad \forall\, v \in Y.$$

This equation allows us to interpret the results of Kluge [18], which are in the context of the parametric equation $\mathcal{T}(a,u) = m$, in the present setting.

We will henceforth assume that a (possibly noisy) measurement z of u^* is available, where u^* and a^* together satisfy (5.3). The purpose of this section is to analyze a method for estimating a^* from z. We define the output least-squares functional $J_0 : A \to \mathbb{R}$ by

$$J_0(a) = H(F(a)) \tag{5.4}$$

with

$$H(\cdot) = \frac{1}{2}\| \cdot - z\|_Y^2, \tag{5.5}$$

where $u = F(a)$ is the unique solution of (5.3) corresponding to a. In principle, it is reasonable to estimate a^* by minimizing J_0 over $a \in A$, that is, by solving the following problem. Find $a \in A$ such that

$$J_0(a) \leq J_0(b) \quad \forall\, b \in A. \tag{5.6}$$

However, since the inverse problem under consideration is ill-posed, it is necessary to regularize J_0. In the present approach the following observation also justifies the need for regularization.

A necessary optimality condition for $a^* \in A$ to be a solution of (5.6) is the following variational inequality:

$$D_+ J_0(a^*)(b - a^*) \geq 0 \quad \forall\, b \in A, \tag{5.7}$$

where

$$D_+ J_0(a)(b - a) = \lim_{t \downarrow 0} \frac{J_0(a + t(b-a)) - J_0(a)}{t} \qquad a,\, b \in A,$$

provided that the derivative exists (see [18]).

As mentioned earlier, we intend to employ some iterative methods based on the variational inequality formulation. However, it is well-known that, in the context of (5.7), the majority of such algorithms will demand some kind of strong monotonicity of $D_+ J_0(\cdot)$ for strong convergence. To check the availability of this strong hypothesis, we begin with exploiting the form of H. It is known that for $J_0(\cdot)$ given in (5.4), we have

$$D_+ J_0(a)(b - a) = \langle F'^*(a)(F(a) - z), b - a\rangle.$$

We begin by noticing that H' is monotone, that is,

$$\langle H'(u) - H'(v), u - v\rangle \geq 0 \ \ \forall\, u,\, v \in Y.$$

By choosing $u = F(a_k)$ and $v = F(a^*)$ in the above inequality, we obtain

$$\langle H'(F(a_k)) - H'(F(a^*)), F(a_k) - F(a^*)\rangle \geq 0.$$

This further implies that (cf. Lemma 5.1)

$$0 \leq \langle H'(F(a_k)), F(a_k) - F(a^*)\rangle - \langle H'(F(a^*)), F'(a^*)(a_k - a^*) + Q(a^*, a_k - a^*)\rangle,$$

and, consequently,

$$\begin{aligned}\langle H'(F(a^*)), F'(a^*)(a_k - a^*)\rangle &\leq \langle H'(F(a_k)), F(a_k) - F(a^*)\rangle \\ &\quad -\langle H'(F(a^*)), Q(a^*, a_k - a^*)\rangle \\ &= -\langle H'(F(a_k)), F'(a_k)(a^* - a_k)\rangle - \\ &\quad \langle H'(F(a_k)), Q(a_k, a^* - a_k)\rangle \\ &\quad -\langle H'(F(a^*)), Q(a^*, a_k - a^*)\rangle.\end{aligned}$$

In other words,

$$\begin{aligned}&\langle F'(a_k)^* H'(F(a_k)) - F'(a^*)^* H'(F(a^*)), a_k - a^*\rangle \geq \\ &\qquad \langle H'(F(a_k)), Q(a_k, a^* - a_k)\rangle + \langle H'(F(a^*)), Q(a^*, a_k - a^*)\rangle.\end{aligned}$$

Now by employing Lemma 5.1 once again, we have

$$\begin{aligned}\langle H'(F(a_k)), Q(a_k, a^* - a_k)\rangle &\leq \|H'(F(a_k))\|\,\|Q(a_k, a^* - a_k)\| \\ &= \|F(a_k) - z\|\,\|Q(a_k, a^* - a_k)\| \\ &\leq \frac{\beta^2}{\alpha^4}\|m\|(\|m\| + \alpha\|z\|)\|a_k - a^*\|^2.\end{aligned}$$

Similarly, we have

$$\langle H'(F(a^*)), Q(a^*, a_k - a^*)\rangle \leq \frac{\beta^2}{\alpha^4}\|m\|(\|m\| + \alpha\|z\|)\|a_k - a^*\|^2.$$

Combining the above three inequalities, we obtain

$$\langle F'(a_k)^* H'(F(a_k)) - F'(a^*)^* H'(F(a^*)), a_k - a^*\rangle \geq -\kappa\,\|a_k - a^*\|^2.$$

where

$$\kappa = \frac{2\beta^2}{\alpha^4}\|m\|(\|m\| + \alpha\|z\|).$$

Therefore, the required strong monotonicity argument is not fulfilled. Kluge [19] proposed to incorporate a strongly convex and differentiable regularizing operator $\mathcal{R}$ so that $\tau J' + \rho\mathcal{R}'$, with $\mathcal{R}'(\cdot)$ as the derivative of $\mathcal{R}$, becomes strongly monotone. Here, τ and ρ are strictly positive constants.

More precisely, we define $J : A \to \mathbb{R}$ by

$$J(a) = \tau J_0(a) + \rho \mathcal{R}(a),$$

where $\mathcal{R}$ is a strongly convex Gateaux differentiable functional, that is, $\mathcal{R}'$ is strongly monotone with modulus of monotonicity as κ_0.

Notice that the condition

$$\rho\kappa_0 - \tau\kappa > 0 \tag{5.8}$$

now ensures that $\tau F'(\cdot)^* H'(\cdot) + \rho\mathcal{R}'$ is strongly monotone, that is, for all a_1, $a_2 \in A$, we have

$$\langle \tau F'(a_1)^* H'(a_1) + \rho\mathcal{R}'(a_1) - \tau F'(a_2)^* H'(a_2) - \rho\mathcal{R}'(a_2), a_1 - a_2\rangle$$
$$\geq (\rho\kappa_0 - \tau\kappa)\|a_1 - a_2\|^2.$$

Although all the arguments here are valid for an arbitrary strongly convex functional, we shall stick to the choice

$$\mathcal{R}(a) = \frac{1}{2}\|a - \bar{a}\|^2,$$

for some $\bar{a} \in X$. That is, $\kappa_0 = 1$ in (5.8).

Therefore, instead of (5.6), we consider the following regularized problem. Find $a \in A$ such that

$$J(a) \leq J(b) \quad \forall\, b \in A. \tag{5.9}$$

It follows by standard arguments that the above problem is equivalent to the following variational inequality. Find $a \in A$ such that

$$\langle \tau\, F'(a)^*(F(a) - z) + \rho\mathcal{R}'(a), b - a\rangle \geq 0 \;\; \forall \;\; b \in A. \tag{5.10}$$

In fact, by using the employed notion of the one-sided directional derivative, it can be shown that (see [19, Lemma 4.1]) under condition (5.8) the functional $J(a)$ is strongly convex and hence the above variational inequality is necessary as well as sufficient optimality condition for the minimization problem (5.9).

In the following we propose two algorithms for variational inequalities emerging from (5.1) and (5.2), respectively. A comparison between the hypothesis needed for the convergence of these algorithms reveals that (5.2) is theoretically superior to (5.1).

5.1 Auxiliary Problem Principle

We employ the so-called auxiliary problem principle (see [6]) to solve (5.10) and hence to solve the abstract elliptic inverse problem.

Assume that $\mathcal{A} : X \to \mathbb{R}$ is convex and Gateaux differentiable with $\mathcal{A}'$ as its Gateaux derivative, and assume that $\delta > 0$. For an approximate solution to (5.10) we proceed as follows. We begin with an initial guess a_0 and solve

$$\min_{a \in A} \{\mathcal{A}a + \langle \delta(\tau F'(a_0)^*(F(a_0) - z) + \rho \mathcal{R}'(a_0)) - \mathcal{A}'(a_0), a\rangle\}.$$

Assume that the functional $\mathcal{A}$ is chosen so that the above minimization problem is uniquely solvable. We denote the unique solution by a_1 and continue further by updating a_0.

More precisely, we consider the following algorithm.

Algorithm 1

(i) At $k = 0$, start with a_0.

(ii) At step $k = n$, solve the following problem:

$$\min_{a \in A} \{\mathcal{A}a + \langle \delta(\tau F'(a_n)^*(F(a_n) - z)) + \rho \mathcal{R}'(a_n)) - \mathcal{A}'(a_n), a\rangle\}. \tag{5.11}$$

Assume that a_{n+1} is the solution.

(iii) Stop if $\|a_{n+1} - a_n\|$ is below some threshold. Otherwise, go back to the previous step.

We remark that for an important choice $\mathcal{A}(\cdot) = \frac{1}{2}\|\cdot\|^2$, (5.11) leads to the following well-known projection algorithm:

$$a_{n+1} = P_A[a_n - \delta(\tau F'(a_n)^*(F(a_n) - z) + \rho \mathcal{R}'(a_n))], \tag{5.12}$$

where P_A is the projection operator to the convex set A.

The following result deals with convergence of the above algorithm.

Theorem 5.1 [13] *Assume that $\mathcal{A} : A \to \mathbb{R}$ is proper convex and Gateaux differentiable and its Gateaux derivative $\mathcal{A}'$ is strongly monotone with modulus γ. Assume that a^* is the unique solution to (5.10). Assume that there exists $0 < s < 1$ and $r > 0$ such that for $c_1 = \frac{r+\alpha\|z\|}{r}$, $c_2 = \frac{\tau\beta^2 r^2}{\alpha^4}$ and $\|m\| \le r$ the following estimates hold:*

$$\kappa = c_1 c_2 \le \rho(1 - s) \tag{5.13a}$$

$$\delta \le \frac{s\rho\gamma}{9\rho^2 + 2c_2^2}. \tag{5.13b}$$

Then the iterative scheme developed by Algorithm 1 converges strongly to a^*.

The conditions in (5.13), although ensuring the strong convergence and uniqueness, are quite stringent. These conditions suggest that to get the strong convergence it may be necessary to over regularize, that is, to take ρ large in comparison to τ.

We have noticed that the major drawback of the functional studied in the previous section is its nonconvexity. We now consider the objective functional is $\mathcal{J} : A \to \mathbb{R}$ defined by

$$\mathcal{J}(a) = \frac{1}{2}T(a, F(a) - z, F(a) - z),$$

where z is as before.

We have shown that the functional $\mathcal{J}$ is convex and infinitely differentiable, and its first derivative is given by

$$D\mathcal{J}(a)\delta a = -\frac{1}{2}T(\delta a, F(a) + z, F(a) - z),$$

We define a regularized objective functional $\mathcal{J}_\rho : X \to \mathbb{R}$ by

$$\mathcal{J}_\rho(a) = \mathcal{J}(a) + \rho\mathcal{R}(a),$$

where $\mathcal{R}$ is a Gateaux differentiable and strongly convex functional, with $\mathcal{R}'$ as its Gateaux derivative.

We consider the following regularized problem. Find $a \in A$ such that

$$\mathcal{J}_\rho(a) \leq \mathcal{J}_\rho(b) \quad \forall\ b \in A.$$

It follows by standard arguments that a necessary optimality condition for the above problem is the following variational inequality. Find $a \in A$ such that

$$\langle \mathcal{J}_\rho'(a), b - a\rangle \geq 0 \ \ \forall \ \ b \in A. \tag{5.14}$$

With an auxiliary operator $\mathcal{A}$, as in the previous section, and a positive scalar δ, we consider the following algorithm.

Algorithm 2

(i) At $k = 0$, start with a_0.

(ii) At step $k = n$, solve the following problem: find $a \in A$ such that

$$\min_{a \in A} \mathcal{A}a + \langle \delta(\mathcal{J}_\rho'(a_k)) - \mathcal{A}'(a_k), a\rangle.$$

Assume that a_{n+1} is the solution.

(iii) Stop if $\|a_{n+1} - a_n\|$ is below some threshold. Otherwise, go back to the previous step.

The following theorem is devoted to the convergence of the above algorithm.

Theorem 5.2 *Assume that $\mathcal{A} : A \to \mathbb{R}$ is proper convex and Gateaux differentiable and its Gateaux derivative $\mathcal{A}'$ is strongly monotone with modulus γ. Assume that L is the modulus of Lispchitz continuity for $\mathcal{J}_\rho$, and assume that a^* is the unique solution to (5.14). If*

$$0 < \delta < \frac{2\rho\gamma}{L^2},$$

then the sequence $\{a_k\}$ developed by Algorithm 2 converges strongly to a^*.

The convexity of J_ρ allows much weaker conditions on ρ that guarantee convergence. Over Regularization is not required.

5.2 Extragradient Method

It has been mentioned above that for the strong convergence of algorithms such as Algorithm 1, Algorithm 2, or (5.12), the operator defining the corresponding variational inequality needs to be equipped with some kind of strong monotonicity. In the present framework of solving inverse problems we have attained the strong monotonicity by means of the regularizing operator.

In [22], Korpelevic proposed a method and showed its strong convergence without the strong monotonicity condition. The basic idea in [22] is to project twice on the underlying set of constraints. Some properties of the projection operator have played a very decisive role in obtaining the strong convergence and in dealing with the lack of strong monotonicity. Her results were substantially improved in [16]. In recent years these methods have gained some popularity, and they are, in general, referred to as extragradient methods. In the following we give an extragradient algorithm for the two objective functionals we have been discussing.

Standard Output Least-Square Functional

We again consider the functional $J_0 : A \to \mathbb{R}$ by

$$J_0(a) = H(F(a)),$$

where $H(\cdot)$ is defined in (5.5) and $u = F(a)$ is the unique solution of (5.3) corresponding to a.

The following minimzation problem is to identify the coefficient. Find $a \in A$ such that

$$\hat{J}_\rho(a) \leq \hat{J}_\rho(b) \quad \forall\, b \in A,$$

where

$$\hat{J}_\rho(a) = J_0(a) + \rho\mathcal{R}(a).$$

Here, $\mathcal{R}$ is a suitable regularization operator with $\mathcal{R}'$ as its Gateaux derivative.

The following algorithm can be used to solve the above problem. For the convergence analysis, see Remark 5.1

Algorithm 3

(i) At $k = 0$, start with a_0.
(ii) At step $k = n$, compute a_{n+1} by the following steps:

$$\begin{aligned} \bar{a}_n &= P_A[a_n - \delta(\tau F'(a_n)^*(F(a_n) - z) + \rho\mathcal{R}'(a_n))] \\ a_{n+1} &= P_A[\bar{a}_n - \delta(\tau F'(\bar{a}_n)^*(F(\bar{a}_n) - z) + \rho\mathcal{R}'(\bar{a}_n))], \end{aligned}$$

where P_A is the projection onto A and δ, ρ, τ are all positive constants.
(iii) Stop if $\|a_{n+1} - a_n\|$ is below some threshold. Otherwise, go back to the previous step.

Notice that the above algorithm, in contrast to one projection per iteration in the classical projection algorithm [cf. (5.12)], requires two projections per iteration. Since computing the projection is very tedious (unless the underlying set has a very simply geometry) the method of Korpelevich is, in general, more expensive than the classical projection algorithm. However, since in our example the set A, when discretised, results in very simple constraints, the computation of the projection is very inexpensive.

Coefficient Dependent Output Least-Square Functional

We define the objective functional $\mathcal{J} : A \to \mathbb{R}$ by

$$\mathcal{J}(a) = \frac{1}{2}T(a, F(a) - z, F(a) - z),$$

where z is as before. We consider the following. Find $a \in A$ such that

$$\mathcal{J}(a) \leq \mathcal{J}(b) \quad \forall\, b \in A. \tag{5.15}$$

Notice that we have not regularized the above problem.

We consider the following algorithm.

Algorithm 4

(i) At $k = 0$, start with a_0.
(ii) At step $k = n$, compute a_{n+1} by the following steps:

$$\begin{aligned} \bar{a}_n &= P_A[a_n - \delta(\mathcal{J}'(a_n)))] \\ a_{n+1} &= P_A[\bar{a}_n - \delta(\mathcal{J}'(\bar{a}_n))], \end{aligned}$$

where P_A is the projection onto A.
(iii) Stop if $\|a_{n+1} - a_n\|$ is below some threshold. Otherwise, go back to the previous step.

The following theorem gives the convergence for the above algorithm.

Theorem 5.3 *Assume that (5.15) is solvable, and assume that L is the modulus of Lispchitz continuity for $\mathcal{J}_\rho$. If*

$$0 < \delta < \frac{1}{L},$$

then the sequence $\{a_k\}$ developed by Algorithm 4 converges strongly to a solution of (5.15).

The proof of the above theorem relies heavily on the monotonicity of $\mathcal{J}'(\cdot)$, the gradient of $\mathcal{J}(\cdot)$ (see [22]).

Remark 5.1 A result similar to the above can be proved showing the convergence of Algorithm 3. However, for this, one has to impose an additional hypothesis ensuring that the gradient of the regularized functional is monotone. Notice that we have already seen that such conditions are quite stringent [cf. (5.8)].

6 Numerical Results

We now compare the standard output least-square functional with the coefficient-dependent energy norm functional in estimating a discontinuous coefficient in the system of isotropic elasticity.

Our example is purely synthetic. We consider an isotropic elastic membrane occupying the unit square $\Omega = (0,1)\times(0,1)$. The exact Lamé moduli are $\lambda = 1$ and $\mu = 1 + \chi_S$, where

$$S = \{(x, y) \in \Omega \, : \, y \leq x^2\}$$

and χ_S is the characteristic function of S. In other words, μ is the discontinuous function whose value is 2 on S and 1 on the rest of Ω. We perform one "experiment" of stretching the membrane by a boundary traction h and measuring the resulting displacement u. The boundary traction chosen is

$$h = \frac{1}{10} \begin{bmatrix} 1 & 1 \\ 1 & 1 \end{bmatrix} n,$$

where n is the outward point unit normal to $\partial\Omega$. This traction is applied to the bottom, left, and right edges of the membrane, while the top edge ($y = 1$) is fixed by a Dirichlet condition. In all finite element computations for this example, we use a sequence of uniform triangulations on Ω; a typical mesh (with 32 triangles) is shown in Figure 1. The curve of discontinuity of μ is also shown in Figure 1.

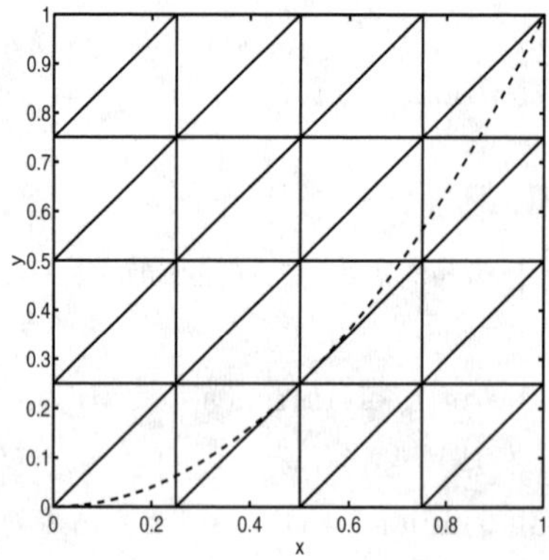

Figure 1: A typical mesh for the numerical experiment. The dashed curve is the set of discontinuities of μ.

Since this is a synthetic experiment, the data vector is computed, not measured. We used piecewise cubic finite elements on a mesh with 512 triangles to obtain an accurate solution, and we used this for the data z. As pointed out in Section 2, we use $P_k z \in V_k$ for the finite-dimensional approximations. In this case, we took P_k to be the standard nodal interpolation operator. Since, during the optimzation, we use piecewise linear finite elements to represent the solution to the BVP, the data is not exactly explicable by any model (that is, there is some modeling error present in the data, although there is no measurement error). Furthermore, the discontinuities of μ are not aligned with the triangle edges, so estimation of μ is a challenging problem. We used piecewise constant finite element functions to represent μ and λ, and we used the smoothed version [cf. [11]] of the BV-seminorm. The parameter ε_k was taken to be 10^{-2} is our experiments, and the regularization parameter was taken to be $\rho = 10^{-5}$.

We solved the unconstrained minimization problem (5.6) using Newton's method with a simple bactracking line search. (The bound constraints implicit in the definition of the feasible set A were ignored during this computation; both the exact and the computed solutions lie in the interior of A in this case.) In all cases, we started with the initial guess $\mu = 1.5$, $\lambda = 1.5$. The computed solutions on three increasingly refined meshes are shown in Figures 2, 3, and 4.

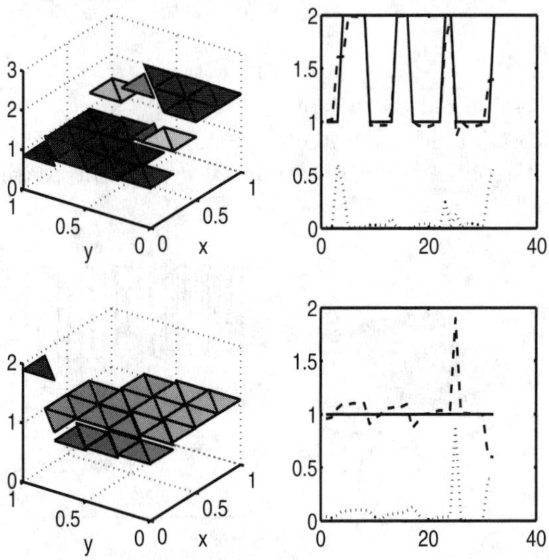

Figure 2: The estimated coefficients μ (top left) and λ (bottom left) on a mesh with 32 triangles. The plots on the right show the element-wise exact coefficient (solid line), the estimated coefficient (dashed line), and the error (dotted line).

These results seem quite satisfactory. Of course, the effect of noise in the data has yet to be explored, but we hope to do that in future work. We remark that the discontinuous coefficient was identified quite accurately, except along the curve of discontinuity, where some errors are inevitable due to the fact that this curve is not aligned with the mesh. Figure 5 shows the triangles on which the error in μ is at least 0.1 (5–10%). It can

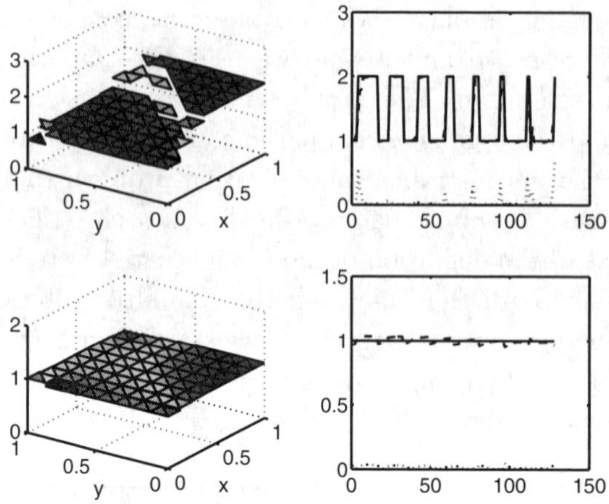

Figure 3: The estimated coefficients μ (top left) and λ (bottom left) on a mesh with 128 triangles. The plots on the right show the element-wise exact coefficient (solid line), the estimated coefficient (dashed line), and the error (dotted line).

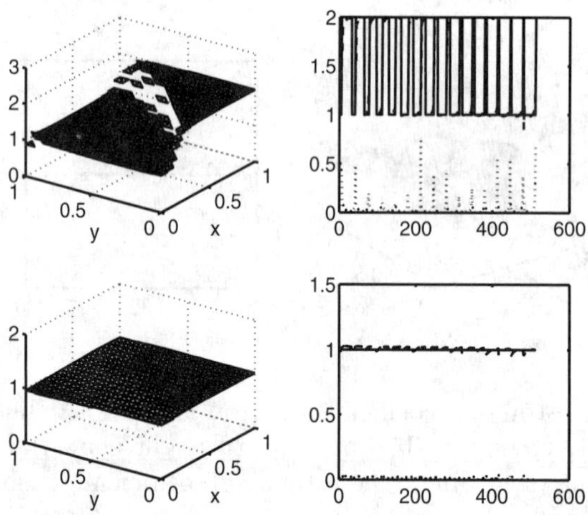

Figure 4: The estimated coefficients μ (top left) and λ (bottom left) on a mesh with 512 triangles. The plots on the right show the element-wise exact coefficient (solid line), the estimated coefficient (dashed line), and the error (dotted line).

be seen that these errors follow the curve $y = x^2$ quite closely.

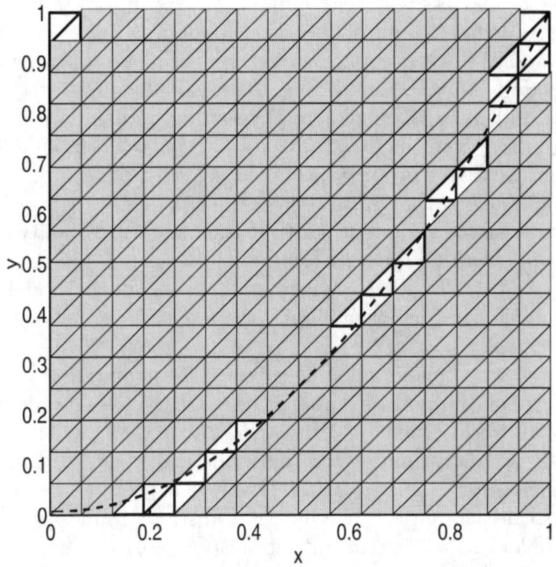

Figure 5: This figure corresponds to the computed μ displayed in Figure 4. The unshaded elements are the triangles on which the error is at least 0.1. The dashed curve is the set of discontinuities of the exact μ.

We applied the same Newton algorithm to minimize the OLS functional based on the L^2 norm (2.2). We observed that the iteration converged quite slowly, taking many more iterations than were necessary to minimize (2.3). Upon investigation, we discovered that the Hessian of (2.2) appears to be numerical singular near the solution, while the Hessian of (2.3) is safely positive definite and well-conditioned in the same region. Further investigation will be required to draw any firm conclusions, but it appears that the convexity of (2.3) may result in efficient performance of numerical algorithms that would be difficult to obtain with other methods.

References

[1] R. Acar, Identification of the coefficient in elliptic equations, *SIAM J. Control Optim.*, **31**(5) (1993), 1221–1244.

[2] G. Alessandrini, An identification problem for an elliptic equation in two variables, *Ann. Mat. Pura Appl.*, **145** (1986), 265–296.

[3] H.T. Banks and K. Kunisch, *Estimation Techniques for Distributed Parameter Systems*, Birkhauser, Boston, 1989.

[4] J. Chen and M.S. Gockenbach, A variational method for recovering planar Lame moduli, *Math. Mech. Solids*, **7** (2002), 191–202.

[5] S.J. Cox and M.S. Gockenbach, Recovering planar Lame moduli from a single-traction experiment, *Math. Mech. Solids*, **2** (1997), 297–306.

[6] G. Cohen, Auxiliary problem principle extended to variational inequalities, *J. Optim. Theory Appl.*, **59** (1988), 325–333.

[7] F. Colonius and K. Kunisch, Output least squares stability in elliptic systems, *Appl. Math.Optim.*, **29** (1989), 33–63.

[8] Z. Chen and J. Zou, An augmented Lagrangian method for identifying discontinuous parameters in elliptic systems, *SIAM J. Control Optim.*, **37** (1999), 892–910.

[9] H.W. Engl, M. Hanke, and A. Neubauer, *Regularization of Inverse Problems*, Mathematics and its Applications, 375. Kluwer Academic Publishers Group, Dordrecht, 1996.

[10] R.S. Falk, Error estimates for the numerical identification of a variable coefficient, *Math. Comp.*, **40** (1983), 537–546.

[11] M.S. Gockenbach and A.A. Khan, An abstract framework for elliptic inverse problems. Part 1: An output least-squares approach. To appear in: Mathematics and Mechanics of Solids.

[12] M.S. Gockenbach and A.A. Khan, An abstract framework for elliptic inverse problems. Part 2: An augmented Lagrangian approach, Under review.

[13] M.S. Gockenbach and A.A. Khan, Identification of Lamé parameters in linear elasticity: A fixed point approach. To appear in: *J. Ind. Mgmt. Optim.*

[14] L. Ji and J. McLaughlin, Recovery of Lamé parameter μ in biological tissues, *Inverse Problems*, **20** (2004), 1–24.

[15] K. Ito and K. Kunisch, The augmented Lagrangian method for parameter estimation in elliptic systems, *SIAM J. Control Optim.*, **28** (1990), 113–136.

[16] E.N. Khobotov, A modification of the extragradient method for solving variational inequalities and some optimization problems, (Russian) *Zh. Vychisl. Mat. i Mat. Fiz.*, **27** (1987), 1462–1473.

[17] H. Kim, and J.K. Seo, Identification problems in linear elasticity, *J. Math. Anal. Appl.*, **215** (1997), 514–531.

[18] R. Kluge, *Nichtlineare Variationsungleichungen und Extremalaufgaben,* Mathematische Monographien, 12. VEB Deutscher Verlag der Wissenschaften, Berlin, 1979.

[19] R. Kluge, Zur "Koeffizienten" bestimmung in linearen Operator- und Evolutionsgleichungen, *Math. Nachr.*, **112** (1983), 153–175.

[20] I. Knowles, Parameter identification for elliptic problems, *J. Comp. Appl. Math.*, **131** (2001), 175–194.

[21] R.V. Kohn and B.D. Lowe, A variational method for parameter identification, *RAIRO Model. Math. Anal. Numer.*, **22** (1988), 119–158.

[22] G.M. Korpelevic, An extragradient method for finding saddle points and for other problems, (Russian) *Ekonom. i Mat. Metody*, **12** (1976), 747–756.

[23] C. Kravaris and J.H. Seinfeld, Identification of parameters in distributed parameter systems by regularization, *SIAM J. Control Optim.*, **23** (1985), 217–241.

[24] R. Luce and S. Perez, Parameter identification for an elliptic partial differential equation with distributed noisy data, *Inverse Problems*, **15** (1999), 291–307.

[25] J. McLaughlin and J.R. Yoon, Unique identifiability of elastic parameters from time-dependent interior displacement measurement, *Inverse Problems*, **20** (2004), 25–45.

[26] A.A. Oberai, N.H. Gokhale, and G.R. Feijóo, Solution of inverse problems in elasticity imaging using the adjoint method, *Inverse Problems*, **19** (2003), 297–313.

[27] K.R. Raghavan and A.E. Yagle, Forward and inverse problems in elasticity imaging of soft tissues, *IEEE Trans. Nuclear Sci.*, **41** (1994), 1639–1648.

[28] G.R. Richter, An inverse problem for the steady state diffusion equation, *SIAM J. Appl. Math.*, **41** (1981), 210–221.

[29] J. Zou, Numerical methods for elliptic inverse problems, *Int. J. Comput. Math.*, **70** (1998), 211–232.

Chapter 17

THE SOLUTIONS OF BBGKY HIERARCHY OF QUANTUM KINETIC EQUATIONS FOR DENSE SYSTEMS

M. Yu. Rasulova

The Institute of Nuclear Physics

A. H. Siddiqi

King Fahd University of Petroleum & Minerals

U. Avazov and M. Rahmatullaev

The Institute of Nuclear Physics

Abstract

The soliton solution and solution via Bethe anzats of BBGKY hierarchy of quantum kinetic equations are defined through the solutions of the nonlinear Schrödinger equation.

1 Introduction

Suppose we are given a system of particles. Suppose that the particles interact through a two-body potential ϕ. In the framework of quantum statistical physics, we consider for the given system the problem of solving the hierarchy of BBGKY kinetic equations [3]:

$$\frac{\partial}{\partial t} f_n(t) = [H_n, f_n(t)] + \frac{1}{v} \int \sum_{1 \le i \le n} [\phi(x_i - x), f_{n+1}(t)]\, dx, \qquad (1.1)$$

where f_n is the density matrix, x is the particle coordinate, [,] is the Poisson bracket, $2m = 1$, and $h = 1$. $0 \le t$ is the time, $n \in N$, N is the number of particles, V is the volume of the system, $N \to \infty, V \to \infty$, $v = \frac{V}{N} =$ constant is volume per particle, and H is the Hamiltonian:

$$H_n = \sum_{1 \le i \le n} T_i + \sum_{1 \le i < j \le n} \phi(x_i - x_j), \quad T_i = \frac{\partial^2}{2\partial {x_i}^2}.$$

Introducing the notation
$(\mathcal{H}f)_n = [H_n, f_n]$;

$(\mathcal{D}_x f)_n (x_1, \cdots, x_n; x'_1, \cdots, x_n) = f_{n+1} (x_1, \cdots, x_n, x; x'_1, \cdots, x'_n, x')$;

$(\mathcal{A}_x f)_n = \frac{1}{v} \sum_{1 \le i \le n} [\phi(x_i - x), f_n]$;

$f(t) = \{f_1(t, x_1), \cdots\cdots, f_n(t, x_1, \cdots, x_n; x'_1, \cdots, x'_n), ...\}, \qquad n = 1, 2, \cdots,$
we can cast (1.1) in the form

$$\frac{\partial}{\partial t} f(t) = \mathcal{H} f(t) + \int \mathcal{A}_x \mathcal{D}_x f(t) dx. \qquad (1.2)$$

2 Derivation of Hierarchy of Kinetic Equations for Correlation Functions

Proposition 2.1 *The hierarchy of kinetic equations for the correlation functions has the form*

$$\frac{\partial}{\partial t}\varphi(t) = \mathcal{H}\varphi(t) + \frac{1}{2}\mathcal{W}(\varphi(t), \varphi(t)) + \int \mathcal{A}_x \mathcal{D}_x \varphi(t) dx + \\ + \int \mathcal{A}_x \varphi(t) \star \mathcal{D}_x \varphi(t) dx, \qquad (2.1)$$

where [6, 9, 10]

$$f(t)=\Gamma\varphi(t)=I+\varphi(t)+\frac{\varphi(t)\star\varphi(t)}{2!}+\cdots\frac{(\star\varphi(t))^n}{n!}+\cdots, \qquad (2.2)$$

$$\varphi(t)=\{\varphi_1(t,x_1;x_1'),\cdots,\varphi(t,x_1,\cdots,x_n;x_1',\cdots,x_n'),\cdots\};$$

$$(\varphi\star\varphi)(X;X')=\sum_{YCX;Y'CX'}\varphi(Y;Y')\varphi(X\setminus Y;X'\setminus Y');$$

$$I\star\varphi=\varphi;\quad (\star\varphi)^n=\underbrace{\varphi\star\varphi\star\cdots\star\varphi}\ n\ times;$$

$$X=(x_1,\cdots,x_n)=(x_{(n)});X'=(x_1',\cdots,x_n')=(x_{(n)}')$$

$$Y=(x_{(n')}),\quad Y'=(x'_{(n')})\quad n^{'}\in n;\quad n'=1,2,\cdots;$$

$$(\mathcal{U}\varphi_n)=\left[\sum_{1\le i<j\le n}\phi(x_i-x_j),\varphi_n\right],$$

$$\mathcal{W}(\varphi,\varphi)=\sum_{YCX;Y'CX'}\mathcal{U}(Y;Y',X\setminus Y;X'\setminus Y')\,\varphi(Y;Y')\varphi(X\setminus Y;X'\setminus Y').$$

Proof. To obtain (2.1), we substitute (2.2) into (1.2):

$$\frac{\partial}{\partial t}\Gamma\varphi(t)=\mathcal{H}\Gamma\varphi(t)+\int\mathcal{A}_x\mathcal{D}_x\Gamma\varphi(t)dx. \qquad (2.3)$$

We have

$$\mathcal{D}_x\Gamma\varphi(t)=\mathcal{D}_x\varphi(t)\star\Gamma\varphi(t), \qquad (2.4)$$

$$\mathcal{A}_x\Gamma\varphi(t)=\mathcal{A}_x\varphi(t)\star\Gamma\varphi(t), \qquad (2.5)$$

$$\mathcal{A}_x\mathcal{D}_x\Gamma\varphi(t)=\mathcal{A}_x\mathcal{D}_x\varphi(t)\star\Gamma\varphi(t)+\mathcal{A}_x\varphi(t)\star\mathcal{D}_x\varphi(t)\star\Gamma\varphi(t), \qquad (2.6)$$

$$T\Gamma\varphi(t)=T\varphi(t)\star\Gamma\varphi(t), \qquad (2.7)$$

$$\mathcal{U}\Gamma\varphi(t)=\mathcal{U}\varphi(t)\star\Gamma\varphi(t)+\frac{1}{2}\mathcal{W}(\varphi(t),\varphi(t)\star\Gamma\varphi(t)), \qquad (2.8)$$

$$\frac{\partial}{\partial t}\Gamma\varphi(t)=\frac{\partial}{\partial t}\varphi(t)\star\Gamma\varphi(t). \qquad (2.9)$$

Substituting (2.4)–(2.9) into (2.3), and multiplying both sides by $\Gamma\left(-\varphi(t)\right)$, we obtain (2.1). This proves the proposition. ■

To investigate our system on the basis of argument similar to those in [3], we can choose as expansion parameter v, setting

$$\phi(x_i - x_j) = v\theta(x_i - x_j) \tag{2.10}$$

and making the substitution [3, 4, 6]

$$\varphi_n(t) = v^{n-1}\psi_n(t). \tag{2.11}$$

On the basis of (2.10) and (2.11), (2.1) for n particles takes the form

$$(\frac{\partial}{\partial t}\psi)_n\,(t,X;X') = \left[\sum_{1\le i\le n} T_i, \psi\right]_n (t,X;X') + v\left(\mathcal{U}\psi\right)_n(t,X;X')$$

$$+\frac{1}{2}\left(\mathcal{W}\psi,\psi\right)_n(t,X;X') + vSp_x\left(\mathcal{A}_x\mathcal{D}_x\psi\right)_n(t,X;X') \tag{2.12}$$

$$+Sp_x\left(\mathcal{A}_x\psi \star \mathcal{D}_x\psi\right)_n(t,X;X').$$

To solve (2.12), we apply the perturbation theory. We shall seek a solution in the form of the series

$$\psi_n\left(t,X;X'\right) = \sum_{\mu} v^{\mu}\psi_n^{\mu}\left(t,X;X'\right),\ n = 1,2,3,\ldots,\ \mu = 0,1,2,\ldots \tag{2.13}$$

Substituting the series (2.13) into (2.12) and equating the coefficients of equal powers of v, we obtain

$$\left(\frac{\partial}{\partial t} + \mathcal{L}_1\right)\psi_1^o(t) = 0, \tag{2.14}$$

$$\left(\frac{\partial}{\partial t} + \mathcal{L}_1 + \mathcal{L}_2\right)\psi_2^o(t) = S_2^o, \tag{2.15}$$

....

$$\left(\frac{\partial}{\partial t} + \sum_{i=1}\mathcal{L}_i\right)\psi_n^{\mu}(t) = S_n^{\mu}, \tag{2.16}$$

where we have introduced the notation

$$\begin{aligned}
\mathcal{L}_1\psi_1^0(t,x_1,x_1') &= \frac{\Delta_x' - \Delta_x}{2m}\psi_1^0(t,x_1;x_1') - Sp_x(\Theta(x_1-x) \\
&\quad -\Theta(x_1'-x))\psi_1^0(t,x_1;x_1')\psi_1^0(t,x,x) \\
\mathcal{L}_i\psi_n^\mu(t,X;X') &= \frac{\Delta_{x_i'} - \Delta_{x_i}}{2m}\psi_n^\mu(t,X;X') \\
&\quad -Sp_x(\widetilde{\mathcal{A}}_x\psi^0)(t,x_i;x_i')(\mathcal{D}_x\psi^\mu)(t,X\setminus x_i;X'\setminus x_i') \\
S_n^\mu &= (U\psi^{\mu-1}(t))_n(X,X') + \frac{1}{2}\sum_{\delta_1+\delta_2=\mu}(V(\psi^{\delta_1}(t),\psi^{\delta_2}(t))(X,X') \\
&\quad +vSp_x(\widetilde{\mathcal{A}}_x\mathcal{D}_x\psi^{\mu-1})_n(t,X,X')) \\
&\quad +Sp_x\sum_{\substack{Y\subset X\\ Y'\subset X'}}(\widetilde{\mathcal{A}}_x\psi^{\delta_1})(t,Y,Y')(\mathcal{D}_x\psi^{\delta_2})(t,X\setminus Y;X'\setminus Y'),
\end{aligned}$$

where

$$\widetilde{\mathcal{A}}_x\psi_n(t,X;X') = \sum_{1\le i\le n}[\phi(x_i-x),\psi_n(t,X;X')].$$

Equation (2.14) is the time-dependent von Neumann equation for Hartree-Fock systems [1, 2]. Thus, the solution of (2.12) reduces to the solution of the homogeneous (2.14) and inhomogenous (2.15), (2.16) von Neumann's, equations for $\psi_1^o(t)$ and $\psi_n^\mu(t)$, accordingly. As shown in [8], the series $\psi_n(t,X;X') = \sum_\mu v^\mu\psi_n^\mu(t,X;X')$, where ψ_1^o is defined in accordance with the solution of von Neumann's equation and the ψ_n^μ, which is determined on the basis of the formula

$$\begin{aligned}
\psi_n^\mu(t,X;X') = \int dY\int dY'\int_{-\infty}^{t} dt' S_n^\mu(t,Y;Y') \\
\bigcap_{1\le i\le n} G(t-t,x_i,y_i;x_i',y_i'),
\end{aligned} \tag{2.17}$$

is a solution of (2.12). Here, $G(t,X,Y;X'Y')$ is the solution of Coushi's problem [4]:

$$\begin{aligned}
i\frac{\partial G(t-t',x_1,y_1;x_1',y_1')}{\partial t} &= -\frac{1}{2}(\Delta_{x_1}G(t-t',x_1,y_1;x_1',y_1') \\
&\quad -\Delta_{x_1'}G(t-t',x_1,y_1;x_1',y_1')) \\
&\quad +Sp_x\theta(x_1-x)\psi_1^0(t,x_1;x_1')G(t-t',x,y_1;x,y_1') \\
&\quad -Sp_x\theta(x_1'-x)\psi_1^0G(t-t',x,y_1;x_1',y_1')
\end{aligned}$$

with the initial condition

$$G(0;x_1,y_1;x_1',y_1') = \delta(x_1-y_1)\delta(x_1'-y_1').$$

Consider the von Neumann's equation:

$$i\frac{\partial\psi_1^0(t,x_1;x_1')}{\partial t} = -\frac{1}{2}(\Delta_{x_1}\psi_1^0(t,x_1;x_1') - \Delta_{x_1'}\psi_1^0(t,x_1,x_1'))+ \tag{2.18}$$
$$+Sp(\theta(x_1-x)-\theta(x_1'-x))\psi_1^0(t,x_1;x_1')\psi_1^0(t,x;x).$$

Define the density matrix as [3]

$$\psi_1^0(t,x_1;x_1') = \chi(t,x_1)\chi^*(t,x_1'). \tag{2.19}$$

Substituting (2.19) into (2.18) and taking $\theta(x_i - x_j)$ in the form of the delta function $\delta(x_i - x_j)$, we get solution of the equations [11]:

$$\begin{aligned} i\frac{\partial}{\partial t}\chi(t,x_i) &= -\frac{\partial^2}{\partial x_i^2}\chi(t,x_i) + 2c\chi(t,x_i) \mid \chi(t,x_i) \mid^2, \\ \chi(t,x_i) \mid_{t=0} &= \chi(x_i). \end{aligned} \tag{2.20}$$

$$\begin{aligned} i\frac{\partial}{\partial t}\chi^*(t,x_i') &= -\frac{\partial^2}{\partial x_i^2}\chi^*(t,x_i') + 2c\chi^*(t,x_i') \mid \chi^*(t,x_i') \mid^2, \\ \chi^*(t,x_i) \mid_{t=0} &= \chi^*(x_i). \end{aligned} \tag{2.21}$$

Equations (2.20) and (2.21) are nonlinear Schrödinger's equations. If we know the solution to (2.20) and (2.21), we shall be able to solve the von Neumann's equation. As known from [5, 7], at $c > 0$ the solution of (2.20) has the form

$$\chi_i(t,x_i) = \sqrt{\frac{2}{c}}\frac{(\lambda+i\nu)^2 + exp[2\nu(x_i - x_0 - 2\lambda t)]}{1+exp[2\nu(x_i - x_0 - 2\lambda t)]}, \tag{2.22}$$

where ν is the velocity, and parameter λ characterizes the amplitude. The velocity ν is expressed via parameter λ as $\nu = \sqrt{1-\lambda^2}$. It should be noted that two relations are valid:

$$\frac{c}{2}|\chi(t,x)|^2 = 1 - \frac{\nu^2}{ch^2\nu(x-x_0-2\lambda t)}$$

and

$$\int |\varphi(t,x)|^2 dx = n,$$

where n is the number of particles in a system. Thus, by the soliton solution (2.22) of nonlinear Schrödinger equations (2.20), a solution of von Neumann's equation ψ_1^0 can be defined as

$$\psi_1^0(t,x_1;x_1') = \chi(t,x_1)\chi^*(t,x_1').$$

Further, $\varphi_n(t,X;X')$ can be defined by (2.15), and, by using (2.6), the density matrix $f_n(t,X;X')$ can be obtained, which is a soliton solution of BBGKY's chain of quantum kinetic equations.

3 Conclusion

The soliton solution of BBGKY's chain of quantum kinetic equations, describing the system of particles, which interact by potential in form of the delta function, is defined.

Acknowledgment. We should like to acknowledge the INTAS program for grant 2001-15.

References

[1] A. Arnold, The relaxation-time von Neuman-Poisson equation, *ZAMM*, **76**(S2) (1996), 293–296.

[2] A. Arnold, R. Bosi, S. Jeschke, and E. Zorn, On global classical of the time-dependent von Neumann equation for Hartree-Fock systems, *Preprint of Munster University*, 06/03-N, 2003, Munster, Germany.

[3] N. N. Bogoluibov, *Selected Papers*, **2**, Kiev, Naukova Dumka, 1970.

[4] S. Ichimary, *Phys. Rev.*, **1** (1968), 1974.

[5] D. Ya. Petrina and V. Z. Enolskaya, *Preprint ITP*, p. 76–17, Kiev, 1976.

[6] M. Yu. Rasulova, Theoreticheskaya i mathematicheskaya, *Physika, Moskow*, **42**(1) (1980), 124–132.

[7] M. Yu. Rasulova, The soliton solution of BBGKY's chain of quantum kinetic equations for system of particles, interacting by delta potential, *Physica A.V.*, **315**(1-2) (2002), 72–78.

[8] M. Yu. Rasulova and A. H. Siddiqi, *Preprint ICTP*, IC/98/118, Trieste, 1998.

[9] M. Yu. Rasulova and A. K. Vidibida, Kinetic equations for distribition functions and density matrices, *Preprint*, ITP–27P, Kiev, 1976.

[10] D. Ruelle, *Statistical Mechanics*, Rigorous Results, W.A., Benjamin, Inc., New York, 1969.

[11] B. E. Zaharov and A. B. Shabat, *JTEP*, **61** (1971), 118.

Chapter 18

CONVERGENCE AND THE OPTIMAL CHOICE OF THE RELAXATION PARAMETER FOR A CLASS OF ITERATIVE METHODS

M. A. El-Gebeily

King Fahd University of Petroleum & Minerals

M. B. M. Elgindi

University of Wisconsin–Eau Claire

Abstract

A necessary condition for the convergence of the iterative scheme $u_{i+1} = (I - \gamma T)\, u_i + F$ is given. The existence of a value γ that minimizes the spectral radius of the iteration matrix $(I - \gamma T)$ is proved. The explicit expression of the optimizing γ in terms of the eigenvalues of T is also given.

1 Introduction

Several iterative schemes, especially those associated with domain decomposition methods, produce the iterations

$$u_{i+1} = u_i - \gamma T u_i + F, \tag{1.1}$$

where F is a fixed vector independent of u, T is an $n \times n$ matrix, and γ is a relaxation parameter (see [1], [2], [3], [4], [5], [6]).

For example, the Sequential Schwarz Dirichlet-Neumann Scheme for the iterative domain decomposition method, which is employed in coupling the finite element and boundary element method, can be described as follows [2].

1. Given the initial guess $u^I_{B,0} = \overline{u}$,
2. For $n = 1, 2, \cdots$,
   - Solve
$$\begin{bmatrix} H_{11} & H_{12} \\ H_{21} & H_{22} \end{bmatrix} \begin{bmatrix} u^B_B \\ u^I_{B,n} \end{bmatrix} = \begin{bmatrix} G_{11} & G_{12} \\ G_{21} & G_{22} \end{bmatrix} \begin{bmatrix} q^B_B \\ q^I_{B,n} \end{bmatrix}$$
   - Get $q^I_{B,n}$.
   - Solve
$$\begin{bmatrix} F_{11} & F_{12} \\ F_{21} & F_{22} \end{bmatrix} \begin{bmatrix} u^F_F \\ u^I_{F,n} \end{bmatrix} = \begin{bmatrix} f^F_F \\ -Mq^I_{B,n} \end{bmatrix}$$
   - Get $u^I_{F,n}$.
   - Iterate
$$u^I_{B,n+1} = (1-\gamma)\, u^I_{B,n} + \gamma u^I_{F,n}$$
   where γ is a relaxation parameter,
   - Until
$$\frac{\left\| u^I_{B,n+1} - u^I_{B,n} \right\|}{\left\| u^I_{B,n} \right\|} < TOL.$$

The subscript F refers to values from the finite element region and B refers to values from the boundary element region, while the superscript I refers to values from the interface region. It is easy to see that the iteration equation can be written in the form

$$u^I_{B,n+1} = ((1-\gamma)\, I + \gamma T)\, u^I_{B,n} + \gamma Q$$

for appropriate definitions of T and Q.

Convergence of the scheme (1.1) is equivalent to condition

$$\rho_\sigma (I - \gamma T) < 1. \tag{1.2}$$

Here, $\rho_\sigma (A)$ is the spectral radius of A. Often, the inequality (1.2) is satisfied for a range of values of the parameter γ, and we are naturally interested in choosing a value which minimizes the spectral radius of the matrix $I - \gamma T$ in order to speed up convergence. In this chapter we give a necessary condition for the iterative process (1.1) to converge, prove the existence of a minimizing γ, and give its explicit expression in terms of the eigenvalues of the matrix T.

We use the notation

$$\begin{aligned} \mathbf{n} &= \{1, 2, \cdots, n\}, \ \mathbf{m} = \{1, 2, \cdots, m\}, \ldots \text{ etc.}, \\ \mathbf{P} &= \{1, 2, \cdots\}. \end{aligned}$$

We also use $\mathbf{n}_1, \mathbf{n}_2, \mathbf{P}_1, \mathbf{P}_2$, ... etc. to indicate subsets of $\mathbf{n}, \mathbf{P}$, respectively.

This chapter contains two sections besides this Introduction. Section 2 is the main section in which we address the necessary conditions for the convergence of the iteration (1.1) and the existence of an optimal choice of the relaxation parameter γ, as well as its expression in terms of the eigenvalues of T. This is done in both the real case, i.e., when all the eigenvalues of T are real, and in the complex case, i.e., when T has complex eigenvalues. Two illustrative examples are given in Section 3.

2 Convergence and the Choice of the Relaxation Parameter

The eigenvalues of the iteration matrix $I - \gamma T$ are

$$\mu_j = 1 - \gamma \lambda_j, \ j \in \mathbf{n},$$

where λ_j is the jth eigenvalue of T. The condition for convergence (1.2) may be rewritten as

$$\max_{j \in \mathbf{n}} \{|\mu_j|\} < 1. \tag{2.1}$$

Lemma 2.1 *A necessary condition for the convergence of the scheme (1.1) is*

$$\operatorname{Re}(\lambda_j) > 0 \,\forall\, j \in \mathbf{n},$$

or

$$\operatorname{Re}(\lambda_j) < 0 \,\forall\, j \in \mathbf{n}.$$

Proof. If the conclusion of the lemma does not hold, then without loss of generality, we may assume that $\operatorname{Re}(\lambda_1) \geq 0$ and $\operatorname{Re}(\lambda_2) \leq 0$, which in turn implies that for $i = 1, 2$:

$$|\mu_i| = ((1 - \gamma \operatorname{Re}(\lambda_i))^2 + (\operatorname{Im}(\lambda_i))^2 \geq ((1 - \gamma \operatorname{Re}(\lambda_i))^2.$$

Thus, if $\gamma > 0$, $|\mu_2|^2 \geq 1$ and if $\gamma < 0$, $|\mu_1|^2 \geq 1$, then in either case this will violate the assumption of convergence of scheme (1.1). ■

Because of Lemma 2.1 we are going to assume that

$$\operatorname{Re}(\lambda_j) > 0 \, \forall \, j \in \mathbf{n}.$$

In this case, we need to consider only positive values of γ.

Define a vector $x(\gamma) \in \mathbb{C}^n$ by

$$x(\gamma) = \begin{bmatrix} 1 - \gamma\lambda_1 \\ 1 - \gamma\lambda_2 \\ \vdots \\ 1 - \gamma\lambda_n \end{bmatrix}, \tag{2.2}$$

then the convergence requirement (1.2) is equivalent to the condition

$$\|x(\gamma)\|_\infty < 1. \tag{2.3}$$

Also, since the entries of $x(\gamma)$ appear in conjugate pairs, we need to consider only the entries with nonnegative imaginary parts. Therefore, from now on, with a little abuse of the notation, we assume that $\operatorname{Im}(\lambda_i) \geq 0$ for $i = 1, 2, \cdots, n$.

For $z \in \mathbb{C}^n$, the inequalities

$$\frac{1}{n^{\frac{1}{p}}} \|z\|_p \leq \|z\|_\infty \leq \|z\|_p \tag{2.4}$$

give

$$\overline{\lim_{p\to\infty}} \|z\|_p = \overline{\lim_{p\to\infty}} \frac{1}{n^{\frac{1}{p}}} \|z\|_p \leq \|z\|_\infty \leq \underline{\lim}_{p\to\infty} \|z\|_p .$$

Consequently, $\lim_{p\to\infty} \|z\|_p$ exists and

$$\lim_{p\to\infty} \|z\|_p = \|z\|_\infty .$$

The next lemma establishes the main tool we are going to use to approach the problem of minimizing $\|x(\gamma)\|_\infty$, namely, by taking a limit of minimizers of $\|x(\gamma)\|_p$.

Lemma 2.2 *Suppose $\{\gamma_p\}_{p\in\mathbf{P}}$ is a sequence such that $\|x(\gamma_p)\|_p \leq \|x(\gamma)\|_p$ for all γ in an interval I. Suppose further that a subsequence $\gamma_p \to \gamma_0 \in I, p \in \mathbf{P}_1 \subset \mathbf{P}$. Then $\|x(\gamma_0)\|_\infty \leq \|x(\gamma)\|_\infty$ for all $\gamma \in I$.*

Proof. Fix $\gamma \in I$. We have by the continuous dependence of $x(\gamma)$ on γ the continuity of norms and the inequalities (2.4)

$$\begin{aligned} \|x(\gamma_0)\|_\infty &= \lim_{p \in \mathbf{P}_1} \|x(\gamma_p)\|_\infty \\ &\leq \overline{\lim_{p \in \mathbf{P}_1}} \|x(\gamma_p)\|_p \\ &\leq \lim_{p \in \mathbf{P}_1} \|x(\gamma)\|_p = \|x(\gamma)\|_\infty . \end{aligned}$$

■

Our study of the existence of an optimizing γ will be divided into two cases: the real case and the complex case. It should be noted that, although the real case can be deduced from the complex case as we shall see, the result in the real case is more explicit and simpler to warrant a separate treatment.

2.1 The Real Case

In this subsection we assume that all the eigenvalues of the matrix T are real (and positive). The main result of this subsection is stated in Theorem 2.4 below, but before we can state the theorem we need the following lemma.

Lemma 2.3 *Suppose* $\{c_{kj}\}_{j \in \mathbf{P}}$, $\{a_{kj}\}_{j \in \mathbf{P}}$, $k \in \mathbf{m}$ *are sequences with the following properties:*

1. $a_{kj} \leq 1$ *and* $0 \leq c_{kj} \leq C$, $k \in \mathbf{m}$,
2. $\underline{\lim}_{j \in \mathbf{P}} a_{k,j} \geq 0$, $k \in \mathbf{m}$,
3. $a_{k_0,j} = 1$, $j \in \mathbf{P}$ *and* $\underline{\lim}_{j \in \mathbf{P}} c_{k_0 j} = c_{k_0} > 0$,

then

$$\lim_{j \to \infty} \left(\sum_{k=1}^{m} c_{kj} a_{k,j}^j \right)^{\frac{1}{j}} = 1.$$

Proof. For sufficiently large j, we have $a_{k,j} \geq -\frac{c_{k_0}}{2(m-1)}$. The result then follows from the estimate

$$\left(\frac{c_{k_0}}{2} \right)^{\frac{1}{j}} \leq \left(\sum_{k=1}^{m} c_{kj} a_{k,j}^j \right)^{\frac{1}{j}} \leq (Cm)^{\frac{1}{j}} .$$

■

Theorem 2.4 *Let $x(\gamma)$ be given by (2.2), where we further assume that $0 < \lambda_1 \leq \lambda_2 \leq \cdots \leq \lambda_n$. Then $\|x(\gamma)\|_\infty$ is minimized by the choice $\gamma_0 = \frac{2}{\lambda_1+\lambda_n}$. Moreover, $\|x(\gamma_0)\|_\infty < 1$.*

Proof. Let

$$F(\gamma) = \sum_{k=1}^{n} (1-\gamma\lambda_k)^{2p}.$$

Differentiating with respect to γ, and equating to zero, we get

$$\sum_{k=1}^{n} \lambda_k \left(1-\gamma_p\lambda_k\right)^{2p-1} = 0, \tag{2.5}$$

where we used the notation γ_p to emphasize the dependence of γ on p. Since the right-hand side of this equation is a polynomial of odd degree in γ_p, it must have at least one real solution. Also computing $F''(\gamma)$ shows that $F'(\gamma)$ is always increasing, which means that (2.5) can have no more than one real solution. Therefore, (2.5) has exactly one real solution. It will be shown in the next subsection that $\gamma_p \in \left[\frac{\lambda_1}{\lambda_n^2}, \frac{\lambda_n}{\lambda_1^2}\right]$ [see (2.6) below]. It follows that the sequence γ_p is positive and has a convergent subsequence $\gamma_p \to \gamma_0$, $p \in \mathbf{P}_1 \subset \mathbf{P}$. Note that not all the factors $1-\gamma_0\lambda_k$ are positive; otherwise (2.5) will not be possible for sufficiently large p. So, suppose $1-\gamma_0\lambda_k > 0$ for $k = 1, 2, \cdots, j_0$, and $1-\gamma_0\lambda_k \leq 0$ for $k = j_0+1, \cdots, n$. Equation (2.5) may be rewritten as

$$\sum_{k=1}^{j_0} \lambda_k \left(1-\gamma_p\lambda_k\right)^{2p-1} = -\sum_{k=j_0+1}^{n} \lambda_k \left(1-\gamma_p\lambda_k\right)^{2p-1}, \; p \in \mathbf{P}_1.$$

The above equation yields

$$-\frac{1-\gamma_p\lambda_1}{1-\gamma_p\lambda_n} = \frac{\left(\sum_{k=j_0+1}^{n} \lambda_k \left(\frac{1-\gamma_p\lambda_k}{1-\gamma_p\lambda_n}\right)^{2p-1}\right)^{\frac{1}{2p-1}}}{\left(\sum_{k=1}^{j_0} \lambda_k \left(\frac{1-\gamma_p\lambda_k}{1-\gamma_p\lambda_1}\right)^{2p-1}\right)^{\frac{1}{2p-1}}}, \; p \in \mathbf{P}_1.$$

For sufficiently large p the terms in both the numerator and the denominator satisfy the hypothesis of Lemma 2.3. Therefore, taking the limit over $\mathbf{P}_1$, we get

$$-\frac{1-\gamma_0\lambda_1}{1-\gamma_0\lambda_n} = 1$$

so that

$$\gamma_0 = \frac{2}{\lambda_1+\lambda_n}.$$

By Lemma 2.2 $\|x(\gamma_0)\|_\infty$ minimizes $\|x(\gamma)\|_\infty$. Moreover,

$$\begin{aligned} |1-\gamma_0\lambda_j| &= \left|1-\frac{2\lambda_j}{\lambda_1+\lambda_n}\right| = \frac{|(\lambda_1-\lambda_j)+(\lambda_n-\lambda_j)|}{\lambda_1+\lambda_n} \\ &\le \frac{\lambda_j-\lambda_1+\lambda_n-\lambda_j}{\lambda_1+\lambda_n} = \frac{\lambda_n-\lambda_1}{\lambda_1+\lambda_n} < 1,\ j\in\mathbf{n}. \end{aligned}$$

Hence $\|x(\gamma_0)\|_\infty < 1$. ∎

2.2 The Complex Case

We consider here the case where the matrix T has complex eigenvalues. This subsection contains two theorems. The first one establishes the existence of an optimizing value γ_0, and the second gives the explicit expression for γ_0.

Theorem 2.5 *Let $x(\gamma)$ be given by (2.2) and I be the interval*

$$I = \left[\frac{\min\{Re(\lambda_i)\}_{i\in\mathbf{n}}}{\max\left\{|\lambda_i|^2\right\}_{i\in\mathbf{n}}}, \frac{\max\{Re(\lambda_i)\}_{i\in\mathbf{n}}}{\min\left\{|\lambda_i|^2\right\}_{i\in\mathbf{n}}}\right]. \tag{2.6}$$

There exists a $\gamma_0\in I$ such that $\|x(\gamma_0)\|_\infty$ minimizes $\|x(\gamma)\|_\infty$. Moreover, $\|x(\gamma_0)\|_\infty < 1$.

Proof. Consider the function

$$F(\gamma) = \sum_{i\in\mathbf{n}} |1-\gamma\lambda_i|^{2p}.$$

This function is differentiable, with derivative

$$F'(\gamma) = \sum_{i\in\mathbf{n}} 2p\,|1-\gamma\lambda_i|^{2p-2}\left(\gamma|\lambda_i|^2 - \mathrm{Re}(\lambda_i)\right),$$

and its minimum minimizes the norm $\|x(\gamma)\|_{2p}$. It can be argued exactly as in the case of the real eigenvalues that $F'(\gamma)=0$ has exactly one real solution γ_p which satisfies

$$\gamma_p = \frac{\sum_{i\in\mathbf{n}} \left|1-\gamma_p\lambda_i\right|^{2p-2}\mathrm{Re}(\lambda_i)}{\sum_{i\in\mathbf{n}} \left|1-\gamma_p\lambda_i\right|^{2p-2}|\lambda_i|^2}. \tag{2.7}$$

So that

$$0 < \frac{\min\{\mathrm{Re}(\lambda_i)\}_{i\in\mathbf{n}}}{\max\left\{|\lambda_i|^2\right\}_{i\in\mathbf{n}}} \le \gamma_p \le \frac{\max\{\mathrm{Re}(\lambda_i)\}_{i\in\mathbf{n}}}{\min\left\{|\lambda_i|^2\right\}_{i\in\mathbf{n}}} < \infty.$$

Let γ_0 be a limit point of the sequence $\{\gamma_p\}_{p\in\mathbf{P}}$. By Lemma 2.2, γ_0 minimizes $\|x(\gamma)\|_\infty$. We proceed to show that $\|x(\gamma_0)\|_\infty < 1$. Suppose a subsequence $\gamma_k \to \gamma_0$, $k \in \mathbf{P}_1 \subset \mathbf{P}$. By passing to a subsequence if necessary, we may assume that there is a $j_0 \in \mathbf{n}$ such that $|1-\gamma_k\lambda_{j_0}| \geq |1-\gamma_k\lambda_j|$ $\forall\, j \in \mathbf{n}$ and all $k \in \mathbf{P}_2 \subset \mathbf{P}_1$. Partition the set $\mathbf{n}$ into two subsets $\mathbf{n}_1$ and $\mathbf{n}_2$, where

$$\begin{aligned}\mathbf{n}_1 &= \{j \in \mathbf{n} : |1-\gamma_0\lambda_{j_0}| > |1-\gamma_0\lambda_j|\}, \\ \mathbf{n}_2 &= \{j \in \mathbf{n} : |1-\gamma_0\lambda_{j_0}| = |1-\gamma_0\lambda_j|\}.\end{aligned}$$

Clearly, for all $j \in \mathbf{n}_1$,

$$\frac{|1-\gamma_k\lambda_j|^{2k}}{|1-\gamma_k\lambda_{j_0}|^{2k}} \to 0,\ k \in \mathbf{P}_2\ .$$

The set $\mathbf{n}_2$ contains a subset $\mathbf{n}_2'$ on which yet another subsequence $\{\gamma_k : k \in \mathbf{P}_3 \subset \mathbf{P}_2\}$ can be extracted so that

$$\frac{|1-\gamma_k\lambda_j|^{2k}}{|1-\gamma_k\lambda_{j_0}|^{2k}} \to c_j > 0,\ j \in \mathbf{n}_2',\ k \in \mathbf{P}_3 \tag{2.8}$$

and

$$\frac{|1-\gamma_k\lambda_j|^{2k}}{|1-\gamma_k\lambda_{j_0}|^{2k}} \to 0,\ j \in \mathbf{n}_2 - \mathbf{n}_2',\ k \in \mathbf{P}_3.$$

Dividing the top and bottom of (2.7) by $|1-\gamma_k\lambda_{j_0}|^{2k}$, $k \in \mathbf{P}_3$ and passing to the limit in $\mathbf{P}_3$, we obtain

$$\gamma_0 = \frac{\sum_{j\in\mathbf{n}_2'} c_j \mathrm{Re}(\lambda_j)}{\sum_{j\in\mathbf{n}_2'} c_j |\lambda_j|^2}. \tag{2.9}$$

Note that in the above formula we may assume that $\sum_{j\in\mathbf{n}_2'} c_j = 1$. Indeed, if this were not the case, then we would replace each c_i by $c_i/\sum_{j\in\mathbf{n}_2'} c_j$. The above equation gives the following sequence of equations:

$$\begin{aligned}\sum_{j\in\mathbf{n}_2'} c_j\gamma_0 |\lambda_j|^2 &= \sum_{j\in\mathbf{n}_2'} c_j \mathrm{Re}(\lambda_j) \\ -\sum_{j\in\mathbf{n}_2'} c_j\gamma_0 \mathrm{Re}(\lambda_j) + \sum_{j\in\mathbf{n}_2'} c_j\gamma_0^2 |\lambda_j|^2 &= 0 \\ \sum_{j\in\mathbf{n}_2'} c_j \left(1 - 2\gamma_0 \mathrm{Re}(\lambda_j) + |\lambda_j|^2\right) &= \sum_{j\in\mathbf{n}_2'} c_j (1-\gamma_0 \mathrm{Re}(\lambda_j)) \\ \sum_{j\in\mathbf{n}_2'} c_j |1-\gamma_0\lambda_j|^2 &= \sum_{j\in\mathbf{n}_2'} c_j (1-\gamma_0 \mathrm{Re}(\lambda_j)).\end{aligned}$$

Now noting that for $j \in \mathbf{n}_2$, $|1-\gamma_0\lambda_j| = |1-\gamma_0\lambda_{j_0}|$, the last equation gives

$$|1-\gamma_0\lambda_{j_0}|^2 = \sum_{j\in\mathbf{n}_2'} c_j\left(1-\gamma_0\mathrm{Re}\left(\lambda_j\right)\right).$$

On the other hand, since $\gamma_0\mathrm{Re}(\lambda_j) > 0$ for all j, then

$$\sum_{j\in\mathbf{n}_2'} c_j\left(1-\gamma_0\mathrm{Re}\left(\lambda_j\right)\right) < \sum_{j\in\mathbf{n}_2'} c_j = 1.$$

This shows that $|1-\gamma_0\lambda_j| < 1$. In particular, $|1-\gamma_0\lambda_{j_0}| < 1$. Finally, we have

$$\begin{aligned}\|x(\gamma_0)\|_\infty &= \lim_{p\in\mathbf{P}_3}\|x(\gamma_p)\|_\infty \\ &= \lim_{p\in\mathbf{P}_3}|1-\gamma_p\lambda_{j_0}| \\ &= |1-\gamma_0\lambda_{j_0}| < 1.\end{aligned}$$

■

Observe that, although the minimizing value γ_0 need not be unique, the value of $\|x(\gamma_0)\|_\infty$ is the same by virtue of Lemma 2.2. Observe also that in formula (2.9), the coefficients c_j are dependent on γ_0 and as such this formula does not explicitly define γ_0. The explicit definition of γ_0 will be given in the following theorem.

Theorem 2.6 *Let $x(\gamma)$ be given by (2.2). One of the following two statements holds.*

1. *There exists a $i_0 \in \mathbf{n}$ such that the minimizing value γ_0 is given by*

$$\gamma_0 = \frac{\mathrm{Re}\left(\lambda_{i_0}\right)}{|\lambda_{i_0}|^2}. \tag{2.10}$$

2. *There exist $i_0, j_0 \in \mathbf{n}$ such that the minimizing value γ_0 is given by*

$$\gamma_0 = \frac{2\,\mathrm{Re}\left(\lambda_{i_0}-\lambda_{j_0}\right)}{|\lambda_{i_0}|^2-|\lambda_{j_0}|^2}, \tag{2.11}$$

where $|\lambda_{i_0}|^2-|\lambda_{j_0}|^2 \neq 0$.

Proof. As usual we consider the function

$$F(\gamma) = \sum_{i=1}^{n}|1-\gamma\lambda_i|^{2p}.$$

$F'(\gamma) = 0$ gives

$$\sum_{i=1}^{n} |1-\gamma\lambda_i|^{2p-2}\left(\gamma|\lambda_i|^2 - \operatorname{Re}(\lambda_i)\right) = 0. \tag{2.12}$$

Suppose $\gamma_p \to \gamma_0$, $p \in \mathbf{P}_1 \subset \mathbf{P}$, then equation (2.12) is valid for γ_0. If there is a $i_0 \in \mathbf{n}$ such that $\gamma_0 |\lambda_{i_0}|^2 - \operatorname{Re}(\lambda_{i_0}) = 0$, then this determines the value of γ_0 given in (2.10). If no such i_0 exists, then partition the set $\mathbf{n}$ into two subsets $\mathbf{n}_1, \mathbf{n}_2$ such that

$$\begin{aligned}\gamma_0 |\lambda_i|^2 - \operatorname{Re}(\lambda_i) &> 0 \text{ for } i \in \mathbf{n}_1,\\ \gamma_0 |\lambda_i|^2 - \operatorname{Re}(\lambda_i) &< 0 \text{ for } i \in \mathbf{n}_2.\end{aligned}$$

Rewrite (2.12) as

$$\begin{aligned}&\sum_{i\in\mathbf{n}_1} |1-\gamma_p\lambda_i|^{2p-2}\left(\gamma_p|\lambda_i|^2 - \operatorname{Re}(\lambda_i)\right)\\ =\ &\sum_{i\in\mathbf{n}_2} |1-\gamma_p\lambda_i|^{2p-2}\left(-\gamma_p|\lambda_i|^2 + \operatorname{Re}(\lambda_i)\right).\end{aligned}$$

By passing to a subsequence $\mathbf{P}_2 \subset \mathbf{P}_1$, if necessary, we may assume that there exists a $i_0 \in \mathbf{n}_1$ and a $j_0 \in \mathbf{n}_2$ such that, for sufficiently large p, we have

$$\begin{aligned}\left|1-\gamma_p\lambda_{i_0}\right| &\geq \left|1-\gamma_p\lambda_i\right|,\ i \in \mathbf{n}_1, p \in \mathbf{P}_2\\ \left|1-\gamma_p\lambda_{j_0}\right| &\geq \left|1-\gamma_p\lambda_i\right|,\ i \in \mathbf{n}_2, p \in \mathbf{P}_2.\end{aligned}$$

The last equation may then be rewritten as

$$\frac{\left|1-\gamma_p\lambda_{i_0}\right|}{\left|1-\gamma_p\lambda_{j_0}\right|} = \frac{\left(\sum_{i\in\mathbf{n}_2}\left|\frac{1-\gamma_p\lambda_i}{1-\gamma_p\lambda_{j_0}}\right|^{2p-2}\left(-\gamma_p|\lambda_i|^2 + \operatorname{Re}(\lambda_i)\right)\right)^{\frac{1}{2p-2}}}{\left(\sum_{i\in\mathbf{n}_1}\left|\frac{1-\gamma_p\lambda_i}{1-\gamma_p\lambda_{j_0}}\right|^{2p-2}\left(\gamma_p|\lambda_i|^2 - \operatorname{Re}(\lambda_i)\right)\right)^{\frac{1}{2p-2}}}.$$

Passing to the limit over $\mathbf{P}_2$ and using Lemma 2.3, we obtain

$$\frac{|1-\gamma_0\lambda_{i_0}|}{|1-\gamma_0\lambda_{j_0}|} = 1,$$

from which we get

$$-2\operatorname{Re}(\lambda_{i_0}) + \gamma_0|\lambda_{i_0}|^2 = -2\operatorname{Re}(\lambda_{j_0}) + \gamma_0|\lambda_{j_0}|^2.$$

We claim that $|\lambda_{i_0}| \neq |\lambda_{j_0}|$; otherwise, the last equation gives $(\lambda_{i_0}) = \mathrm{Re}(\lambda_{j_0})$. This, together with the fact that $\mathrm{Im}(\lambda_{i_0}) \geq 0, \mathrm{Re}(\lambda_{j_0}) \geq 0$, implies that $\lambda_{i_0} = \lambda_{j_0}$. But, by the choice of i_0, j_0, we must have $\lambda_{i_0} \neq \lambda_{j_0}$, a contradiction. Hence, we may solve for γ_0 to obtain (2.11). ■

The value of γ_0 given by equation (2.10) occurs if, for example, the matrix T has one complex repeated eigenvalue. Whether there are other situations where the optimizing value γ_0 is given by (2.10) is not clear to us. Also, it is not clear from Theorem 2.6 how the values i_0, j_0 may be chosen. However, we may form the two lists

$$\left\{\mathrm{Re}(\lambda_i) / |\lambda_i|^2\right\}_{i \in \mathbf{n}}$$

and

$$\left\{2(\mathrm{Re}(\lambda_i) - \mathrm{Re}(\lambda_j)) / \left(|\lambda_i|^2 - |\lambda_j|^2\right)\right\}_{i \in \mathbf{n}},$$

compute $\|x(\gamma)\|_\infty$ for each element in these lists, and take the value at which the minimum norm is achieved. Note also that Theorem 2.6 reduces to Theorem 2.4 if the λs are real, in which case, of course, $i_0 = 1, \; j_0 = n$.

3 Examples

In this section we give two examples to illustrate the foregoing theory. In the first example the eigenvalues (or the entries of $x(\gamma)$) are taken to be real, while in the second example they are taken to be complex.

Example 3.1 Suppose $x(\gamma) = 1 - \gamma[1, 2, 8, 12, 20]^t$. In this case the optimal value for γ_0 is $2/21 = 0.09523$. Figure 1 gives the plot of $\|x(\gamma)\|_\infty$ versus γ. It verifies that $\|x(\gamma)\|_\infty$ is minimized at $\gamma_0 = 0.09523$.

Example 3.2 Suppose $x(\gamma) = 1 - \gamma[0.5 + 0.87i, 1 + i, 2 + 3i, 3 + 4i, 8]^t$. A plot of $\|x(\gamma)\|_\infty$ versus γ is given in Figure 2.

The list of values $2(\mathrm{Re}(\lambda_i) - \mathrm{Re}(\lambda_j)) / \left(|\lambda_i|^2 - |\lambda_j|^2\right)$ to three decimal places is

$$1.007, 0.250, 0.208, 0.238, 0.182, 0.174, 0.226, 0.167, 0.235, 0.256,$$

the third value of which is the optimizing value.

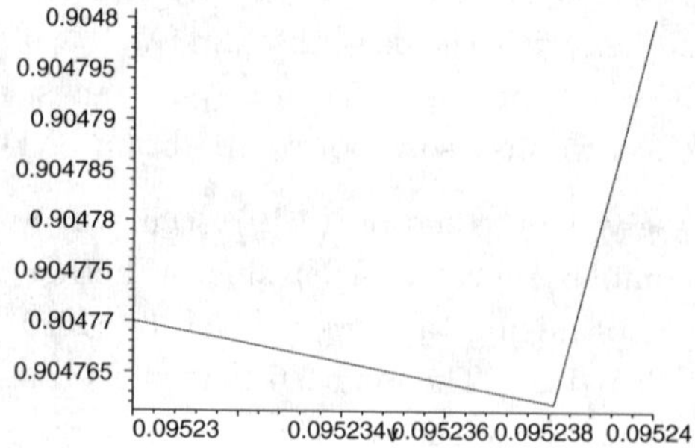

Figure 1: $||x(\gamma)||_\infty$ versus γ for a system with real eigenvalues.

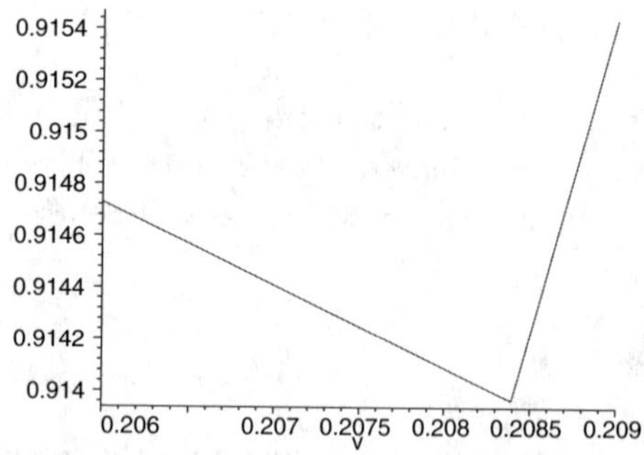

Figure 2: $||x(\gamma)||_\infty$ versus γ for a system with comples eigenvalues.

Acknowledgment. The first author would like to acknowledge the excellent research facilities provided by King Fahd University of Petroleum & Minerals.

References

[1] M. El-Gebeily, W. Elleithy, and H. Al-Gahtani, Convergence of the domain decomposition finite element-boundary element coupling methods, *Computer Methods in Applied Mechanics and Engineering*, **191** (2002), 4851-4867.

[2] N. Kamiya and H. Iwase, BEM amd FEM combination parallel analysis using conjugate gradient and condensation, *Engineering Analysis with Boundary Elements*, **20** (1997), 319–326.

[3] C. Lin, E. Lawton, J. Caliendo, and L. Anderson, An iterative finite-boundary element algorithm, *Computers and Structures*, **39**(5) (1996), 319–326.

[4] J. Stoer and R. Bulirsch, *Introduction to Numerical Analysis*, Springer-Verlag, Heidelberg, 1980.

[5] R. Varga, *Matrix Iterative Analysis, Series in Automatic Computation*, Prentice Hall, Englewood Cliffs, NJ, 1962.

[6] O. Zeinkowicz, D. Kelly, and P. Bettes, The coupling of the finite element method and boundary solution procedures, *International Journal of Numerical Methods in Engineering*, **11** (1997), 355–375.

Chapter 19

ON A SPECIAL CLASS OF SWEEPING PROCESS

M. Brokate

Technical University of München

P. Manchanda

Gurunanak Dev University

Abstract

The goal of this chapter is to discuss two variants of a special class of Moreau's sweeping process along with a certain current developments on this theme.

1 Introduction

In 1973, Moreau introduced the sweeping process [10, 11, 12]. It describes the movement $\xi = \xi(t)$ of a point in a Hilbert space H introduced by a time-dependent, closed convex set $C = C(t)$ according to

$$-\dot{\xi}(t) \in N_{C(t)}(\xi(t)),\ \xi(0) = \xi_0, \tag{1.1}$$

where $N_K(x)$ denotes the normal cone to a convex set K at a point x. The evolution variational inequality

$$\langle \dot{v}(t),\, w - v(t)\rangle \geq \langle f(t),\, w - v(t)\rangle\ \forall w \in \Gamma,\ \nu(t) \in \Gamma,\ v(0) = v_0, \tag{1.2}$$

with $\Gamma \subset H$ closed and convex, constitutes a special case of (1.1), if we set

$$\xi(t) = v(t) - \int_0^t f(s)ds,\ \xi_0 = v_0,\ C(t) = \Gamma - \int_0^t f(s)ds. \tag{1.3}$$

In the same manner, the evolution quasi-variational inequality

$$\langle \dot{v}(t),\ w - v(t)\rangle \geq \langle f(t),\ w - v(t)\rangle\ \forall w \in \Gamma(v(t)),\ v(t) \in \Gamma/v(t)),\ v(0) = v_0 \tag{1.4}$$

becomes a special case of the **implicit** or **state-dependent sweeping process**

$$-\dot{\xi}(t) \in N_{C(t,\xi(t))}(\xi(t)),\ \xi(0) = \xi_0, \tag{1.5}$$

if we set

$$\xi(t) = v(t) - \int_0^t f(s)ds,\ \xi_0 = v_0$$

$$C(t,\xi) = \Gamma(\xi + \int_0^t f(s)ds) - \int_0^t f(s)ds. \tag{1.6}$$

While the sweeping (1.1) has been studied extensively, see [7, 15], much less is known about the implicit process (1.5), see [2, 8, 15]. Kunze and Monteiro Marques [8] have proved the existence of a solution for (1.5), if C satisfies a Lipschitz condition with respect to the Hausdorff distance,

$$D_H(C(t,\xi),\ C(s,\eta)) \leq L_1||t - s|| + L_2||\xi - \eta|| \tag{1.7}$$

if $L_2 < 1$, and give examples for non-existence if $L_2 > 1$. However, no matter how small L_2 is chosen, uniqueness may fail to hold. It has been proven by Ballard, see complete references in [2], in the context of quasi-static friction problems. The sweeping process plays an important role in elastoplasticity and dynamics for unilateral problems, see [7, 15] for detailed references.

A variant of Moreau's sweeping process (1.1) has been studied, see [15]. A similar variant of sweeping process (1.5) has been also discussed in the same paper.

In 2003, Brokate, Krejči and Schnabel [2] considered two problems, namely, a variational inequality where the constraint Z depends on an additional given function, and a quasi-variational inequality in a similar situation. These are respectively special cases of (1.1) and (1.5) and a more general form of evolution variational inequality (1.2) and evolution quasi-variational inequality (1.4).

Lions [9] studied parallel algorithms for evolution variational inequalities and indicated the possibility of such algorithms for quasi-variational inequalities. Furati and Siddiqi [5, 6] have studied parallel algorithms for evolution variational and quasi-variational inequalities arising in superconductivity. Siddiqi, Manchanda, and Brokate [15] have mentioned the study of parallel algorithms for Moreau's sweeping process as an open problem.

The main goal of this chapter is to discuss the possibility of developing parallel algorithms of evolution variational and quasi-variational inequalities studied by Brokate, Krejĉi, and Schnabel [2]. Variants of these inequalities will also be discussed.

2 A Special Class of Moreau's Sweeping Process

Let H be a separable Hilbert space H endowed with a scalar product $\langle .,. \rangle$, a norm $||x|| = \sqrt{\langle x, x \rangle}$ for $x \in H$, and a family of bounded convex sets $Z(\rho) \subset H$ with a smooth boundary parametrized by $\rho \in K \subset X$, where X is a reflexive Banach space endowed with a norm $|| \cdot ||_X$ and K is a convex closed set with non-empty interior K^0. We assume throughout this chapter that there exists $0 < c \leq C$ such that

$$B_c(0) \subset Z(\rho) \subset B_C(0) \quad \forall \rho \in K. \tag{2.1}$$

Problem P: For given functions $u \in W^{1,1}(0, T; H)$, $r \in W^{1,1}(0, T; K)$ and an initial condition $x_0 \in Z(r(0))$, we look for a function $\xi \in W^{1,1}(0, T : H)$ such that

(i) $u(t) - \xi(t) \in Z(r(t)) \qquad \forall t \in [0, T]$,

(ii) $u(0) - \xi(0) = x_0$,

(iii) $\langle \dot{\xi}(t), u(t) - \xi(t) - y \rangle \geq 0 \qquad \forall y \in Z(r(t))$ for a.e.

The problem (P) represents a constitutive stress-strain law where the yield function depends on the temperature.

Problem (I): Let $g : [0, T] \times H \times H \longrightarrow K$ be a function satisfying the following hypothesis.
(G): $g : [0, T] \times H \times H \longrightarrow X$ is continuous, and $g(t, u, \xi) \in K$ for each $(t, u, \xi) \in [0, T] \times H \times H$. Its partial derivatives $\partial_t g, \partial_u g, \partial_\xi g$ exist and satisfy the inequalities

$$||\partial_\xi g(t, u, \xi)||_{\mathcal{L}}(X, Y) \leq r, \tag{a}$$

$$||\partial_u g(t,u,\xi)||_{\mathcal{L}}(X,Y) \le w, \qquad (b)$$

$$||\partial_t g(t,u,\xi)||_Y \le a(t), \qquad (c)$$

$$||\partial_\xi g(t,u,\xi) - \partial_\xi g(t,v,\eta)||_{\mathcal{L}}(X,Y) \le C_g(||u-v|| + ||\xi-\eta||), \qquad (d)$$

$$||\partial_t g(t,u,\xi) - \partial_t g(t,v,n)||_Y \le b(t)(||u-v|| + ||\xi - n||)$$

for every $u, v, \xi, \eta \in H$, and a.e. $t \in]0,T[$ with given functions $a(t), b(t) \in L_1(0,T)$ and given constants $r, w, C_g, C_u > 0$ such that

$$\delta = CL_0\, r < 1,$$

where C, L_0 are as in (2.1) and L_2 is given below.

For a given function $u \in W^{1,1}(0,T;H)$ and an initial condition $x_0 \in Z(g(0,u(0),u(0)-x_0))$, we look for a solution $\xi \in W^{1,1}(0,T;H)$ of the implicit problem denoted by Problem (I)

(i) $u(t) - \xi(t) \in Z(g(t,u(t),\, \xi(t))\ \forall t \in [0,T]$

(ii) $u(0) - \xi(0) = x_0$

(iii) $\langle \dot{\xi}(t), \quad u(t) - \xi(t) - y\rangle \ge 0 \quad \forall y \in Z(g(t,u(t),\xi(t)))$ for a.e. $t \in]0,T[$

It may be observed that the implicit sweeping process (1.5) becomes a special case of Problem (I) if we set $u = 0$, $g(t,u,\xi) = (t,\xi)$, $x_0 = -\xi_0$ and $C = -Z$. On the other hand, (I) becomes a special case of (1.5) if we set $C(t,\xi(= u(t) - Z(g(t,u(t),\xi))$ and $\xi_0 = u(0) - x_0$. The quasi-variational inequality (1.4) in subsumed under (I) by setting

$$\xi(t) = v(t) + u(t), \quad u(t) = \int_0^t f(s)ds, \quad x_0 = -v_0,$$

$$Z = -\Gamma, \qquad g(t,u,\xi) = \xi - u.$$

In application (I) is more appropriate than (1.5) due to the fact that it is easier to handle the dependence of ξ and u with respect to standard function spaces rather than dealing with metric properties of the set valued mapping $t \to C(t,\xi)$ as in the case of (1.5).

The mapping $M_Z : H \longrightarrow R_+$, where Z is a closed convex subset of H, formula

$$M_Z(x) = inf\left\{s > 0; \frac{x}{s} \in Z\right\} \qquad (2.3)$$

is called Minkowski functional associated with Z.

The polar set Z^* to Z is defined by the formula

$$Z^* = \{x \in H; < x, y > \leq 1 \quad \forall y \in Z\}. \tag{2.4}$$

$(\mathbf{L_1})$: Let $M(\delta, x)$ be Minkowski functional associated with $Z(\rho)$, then following conditions hold.

The partial derivatives $\partial_\rho M(\delta, u) \in X', \partial x\, M(\delta, u) \in H$ exist for every $x \in H \backslash \{0\}$ and $\delta \in K^0$, and the mappings

$$J(\rho, u) = M(\rho, u)\partial_x M(\rho, u) = K^0 \times H\backslash\{0\} \to H \tag{2.5}$$

$$L(\rho, u) = M(\rho, u)\partial_\rho\, M(\rho, u) : K^0 \times\; H\backslash\{0\} \to H \tag{2.6}$$

admit continuous extensions to $x = 0$ and $\rho \in K$.
$(\mathbf{L_2})$: For every $x, x' \in B_C(0)$ and $\rho, \rho' \in K$ we have

$$|L(\rho, x)| \leq L_0 \tag{2.7}$$

$$|J(\rho, x) - J(\rho', x')| \leq C_J(|\rho - \rho'|_Y + |x - x'|) \tag{2.8}$$

$$|L(\rho, x) - L(\rho', x')|_{Y'} \leq C_K(|\rho - \rho'|_Y + |x - x'|) \tag{2.9}$$

with some fixed constants $L_o, C_J, C_K > 0$.
(E): There exists a function $\mu : K \times H \longrightarrow R_+$ and a constant $k > 0$ such that

$$|\mu(\rho, x) - \mu(\sigma, x)| \leq k|\rho - \sigma|_Y \qquad \forall \delta, \sigma \in K, \quad x \in H \tag{2.10}$$

$$\frac{||x||}{M(\delta, x)} = \mu(\delta, x) \quad \forall \delta \in K \quad \forall x \in H\backslash\{0\}. \tag{2.11}$$

A typical example is $Z(\rho) = \rho\; Z$ for $\rho \geq 0$, where $Y = R$, $K = [c_1, C_1]$ for some $C_1 \geq c_1 > 0$, $Z \subset H$ is a bounded convex closed set with $B_{c_2}(0) \subset Z \subset B_{C_2}(0)$ for some $0 < c_2 \leq c_2$. Then (2.10) and (2.11) are trivially satisfied, as μ is linear in δ.

The following theorems have been proven by Brokate, Krejĉi, and Schnabel [2].

Theorem 2.1 *Let Hypotheses (2.1) and (E) hold. Then problem* (ρ) *admits a unique solution* $\xi \in W^{1,1}(0, T; H)$ *for every given functions* $\xi \in W^{1,1}(0, T; H), r \in W^{1,1}(0, T; K)$ *and every initial condition* $x_0 \in Z(r(0))$.

Theorem 2.2 *Let Hypotheses* $(L_1), (L_2)$, *and G hold. Then for every* $u \in W^{1,1}(0,T;H)$ *and every* $x_0 \in Z(g(0,u(0),u(0)-x_0))$ *there exists a unique solution* $\xi \in W^{1,1}(0,T;H)$ *to Problem (I) in the set*

$$\Omega = \left\{ \eta \in W^{1,1}(0,T;H) : ||\dot{\eta}(t)|| \le \frac{1}{1-\rho}((1+CK_0w)||\dot{u}(t)|| \right.$$

$$\left. +CK_0a(t) + \rho||\dot{\eta}(t)||) \right\}$$

$$\dot{\eta}(0) = u(0) - x_0 \ a.e.$$

Theorem 2.3 *Let* $(L_1), (L_2)$ *hold, let* $(r,u),(s,v) \in W^{1,1}(0,T;K) \times W^{1,1}(0,T;H)$ *be given, let* $\xi, \eta \in W^{1,1}(0,T;H)$ *be the respective solutions to Problem (P), and set* $x = u - \xi$, $y = v - \eta$. *Then there exist positive constants* C_0, C_1 *such that for every* $\lambda > 0$, *every* $(r,u),(s,v) \in W^{1,1}(0,T;K) \times W^{1,1}(0,T;H)$ *with* $\int_0^T (||\dot{u}|| + ||\dot{r}||_Y)dt \le \lambda$,

$$\int_0^T (||\dot{v}|| + ||\dot{s}||_Y)dt \le \lambda, \text{ and every } x_0 \in Z(v(0)),$$

$y_0 \in Z(s(0))$, *the respective solutions* $\xi, \eta \in W^{1,1}(0,T;M)$ *of Problem (P) satisfying the inequality*

$$\int_0^T ||\dot{\xi}-\dot{\eta}||dt \le C_0 e^{C_0}(||x_0-y_0||+||r(0)-s(0)||_Y + \int_0^t (||\dot{u}-\dot{v}||+||\dot{r}-\dot{s}||_Y \, dt). \tag{2.12}$$

Theorem 2.4 *Let the assumption of Theorem 2.3 be fulfilled. Then there exist positive constants* C_2, C_3 *such that for* $\lambda > 0$, *every* $u \cdot \in W^{1,1}(0,T;H)$ *with* $\int_0^T ||\dot{u}||dt \le \lambda$, $\int_0^T ||\dot{v}||dt \le \lambda$, *and every* $x_0 \in Z(g(0),u(0),u(0) - x_0), y_0 \in Z(g(0,v(0),v(0)-y_0)$, *the respective solutions* $\xi, \eta \in W^{1,1}(0,T;H)$ *of Problem (I) satisfy the inequality*

$$\int_0^T ||\dot{\xi}-\dot{\eta}||dt \le C_3 e^{C_{2\lambda}}(||x_0-y_0||+||u(0)-v(0)||+\int_0^T ||\dot{u}-\dot{v}|| \, dt). \tag{2.13}$$

3 A Variant of a Special Class of Moreau's Sweeping Process

Siddiqi, Manchanda, and Brokate [15] have studied the existence and uniqueness of solution of the following variant of the Moreau's sweeping process.

Find $u : [0,T] \to H$, where H is a separable Hilbert space such that

$$-u(t) \in N_{C(t)}(u'(t)), \text{ a.e. in } [0,T], u(0) = u_0. \tag{3.1}$$

Manchanda and Siddiqi, see, for example, [15], have also studied the following variant of the state-dependent sweeping process:

$$-u(t) \in N_{C(t,u(t))}(u'(t)) \text{ a.e. in } [0,T], u'(0) = u_0 \in C(0,u_0). \tag{3.2}$$

Under appropriate conditions (3.2) has a solution, but it is not unique.

Variants of Problem (P) and Problem (I) similar to (3.1) and (3.2) can be introduced, and a natural problem is to establish theorems similar to Theorems 2.1 to 2.4. The formulation of such theorems is under investigation and will be formally announced with proofs separately.

Parallel algorithms for evolution variational and quasi-variational inequalities have been investigated by Lions [9], Furati and Siddiqi [5, 6], and Siddiqi [14]. Siddiqi, Manchanda, and Brokate [15] announced the study of a parallel algorithm for Moreau's sweeping process as an open problem. The general case is still open. However, parallel algorithms for Problem (P) and Problem (I) are an active consideration and will be reported soon by us.

References

[1] A. Bensoussan and J.L. Lions, *Impulse Control and Quasi-Variational Inequalities*, Gauthiers-Villars, Paris, 1984.

[2] M. Brokate, P. Krejĉi, and H. Schnabel, On Uniqueness in Evolution Quasi-variational Inequalities, Preprint Zentrum Mathematik, TU Munchen, D-80290 Munchen, Germany.

[3] M. Brokate and A.H. Siddiqi, Sensivity in the rigid punch problem, *Adv. Math. Sci. Appl., Gakkotosho, Tokyo*, **2** (1993), 445–456.

[4] G. Duvaut and J.L. Lions, *Inequalities in Mechanics and Physics*, Springer-Verlag, Berlin, 1972.

[5] K.M. Furati and A.H. Siddiqi, Fast algorithms for the bean critical state model for superconductivity, To appear in *Numer. Funct. Anal. Optim.*

[6] K.M. Furati and A.H. Siddiqi, Parallel algorithms of quasi-variational inequality arising from type-II superconductivity model, To appear in *Numer. Funct. Anal. Optim.*

[7] M. Kunze and M. Monteiro Marques, An introduction to Moreau sweeping process, In *Impacts in Mechanical Systems Analysis and Modelling*, Edited by B. Brogliato, LN Physics 551, Springer-Verlag, Berlin, pp. 1–60, 2000.

[8] M. Kunze and M. Monteiro Marques, On parabolic quasi-variational inequalities and state-dependent sweeping process, *Top. Methods Nonlinear Anal.*, **12** (1998), 179–191.

[9] J.L. Lions, Parallel algorithms for the solution of variational inequalities, *Interfaces and Free Boundaries*, (1999), 3–16.

[10] J.J. Moreau, Probléme d évolution associé á au convex mobile d́un espace Hilbertien, *C.R. Acad. Sci. Paris, Series A-B*, **273** (1973), A791–A794.

[11] J.J. Moreau, Evolution problem associated with a moving convex set in a Hilbert space, *J. Diff. Eq.*, **26** (1977), 347–374.

[12] J.J. Moreau, Numerical aspects of the sweeping process, *Computer Meth. Appl. Mechanics Eng.*, **177** (1999), 329–349.

[13] U. Mosco, Some introductory remarks on implicit variational problems, in *Recent Developments in Applicable Mathematics*, Edited by A.H. Siddiqi, Macmillan India Limited, Delhi-Bangalore-Madras, pp. 1–46, 1994.

[14] A.H. Siddiqi, On current developments in evolution variational inequalities, *International J. Math. Sci.*, **2** (2003), 347–365.

[15] A.H. Siddiqi, P. Manchanda, and M. Brokate, On some recent developments concerning Moreau's sweeping process, in *Trends in Industrial and Applied Mathematics* Edited by A.H. Siddiqi and M. Kocvara, Kluwer Academic Publishers, Dordrecht, pp.339–354, 2000.

Index